Contents

Human Resource Management in Construction Projects

Books

Although construction is one of the most labour-intensive industries, human resource management (HRM) issues are given inadequate attention. Furthermore, the focus of attention with regards to HRM has been as a centralised head-office function – yet most problems and operational issues arise on projects. To help redress these problems this book examines both the strategic and operational aspects of managing human resources within the construction sector. The book is aimed at project managers and students of project management, who, until now, have been handed the responsibility for HRM without adequate knowledge or training.

The issues addressed in this book are internationally relevant, and are of fundamental concern to both students and practitioners involved in the management of construction projects. The text draws on the authors' experience of working with a range of large construction companies in improving their HRM activities at both strategic and operational levels and is well illustrated with case studies of projects and organisations.

Dr Martin Loosemore is Professor of Construction Management and Associate Dean (Post Graduate Studies), Faculty of the Built Environment, University of New South Wales, Sydney, Australia. He has published numerous internationally refereed articles and books, has advised the Australian government on HR legislation and reform in the construction industry and is a director of a management consultancy which specialises in risk management.

Dr Andrew Dainty is a Lecturer in Construction Management, Department of Civil and Building Engineering at Loughborough University, UK. He has researched widely in the field of construction HRM on projects funded by the EPSRC, ESRC and European Union. He works closely with industry, and has undertaken a range of research, training and consultancy projects with construction companies of all sizes.

Dr Helen Lingard is a Lecturer in Construction Management at the University of Melbourne, Victoria, Australia. On completion of her PhD she worked for a British contracting firm, advising on the occupational health and safety issues arising on large civil engineering projects in Hong Kong. Dr Lingard moved to Australia in 1996. Since then she has undertaken extensive research into the impact of job demands in the construction industry on individual employees' well-being.

Human Resource Management in Construction Projects

Strategic and operational approaches

Martin Loosemore, Andrew Dainty and Helen Lingard

Spon Press
Taylor & Francis Group

LONDON AND NEW YORK

First published 2003
by Spon Press
11 New Fetter Lane, London EC4P 4EE

Simultaneously published in the USA and Canada
by Spon Press
29 West 35th Street, New York, NY 10001

Spon Press is an imprint of the Taylor & Francis Group

Typeset in Sabon by Taylor & Francis Books Ltd
Printed and bound in Great Britain by St Edmundsbury Press,
Bury St Edmunds

British Library Cataloguing in Publication Data
A catalogue record for this book is available from the British Library

Library of Congress Cataloging in Publication Data
Loosemore, Martin, 1962–
    Human resource management in construction projects : strategic
    and operational approaches / Martin Loosemore, Andrew Dainty
    & Helen Lingard.
    1. Construction industry–Personnel management. I. Dainty,
    Andrew. II. Lingard, Helen. III. Title.
HD9715.A2 .L656 2002
624 '.068 '3–dc21   2002010918

ISBN 0–415–26163–5 (hbk)
ISBN 0–415–26164–3 (pbk)

Illustrations

Preface

The construction industry is one of the largest global employment sectors, providing work for a significant proportion of the labour market and accounting for a significant share of the world gross domestic product (GDP). The industry also represents one of the most risky, complex and dynamic industrial environments. A construction project relies on skilled manual labour supported by a management framework, which has to coordinate many professional, construction and supplier organisations whose sporadic involvement will change through the course of the project. The fragmentation and dynamism of this process and the need to integrate a wide range of occupational cultures renders construction one of the most complex project-based industries in which to apply good human resource management (HRM) practices.

Organisational behaviour and management texts abound which explain the central importance of the HRM function to organisational performance. However, few have considered the specific context and challenges that project-based industries present. Moreover, very few texts have sought to capture effective approaches to the HRM function within the construction sector, or to examine how these activities could be adapted and implemented in a way which improves the performance and job satisfaction of the industry's workforce. In this book we aim to address this shortfall by providing an informative exploration into how modern construction organisations manage the various aspects of HRM. As well as holding up a mirror to current practice, we also cast a critical eye over current approaches to managing people in the industry. We have attempted to identify new and alternative strategies for managing people that enable managers and organisations to survive the increasingly competitive and volatile construction industry labour market.

We would like to think that this book is different to existing texts on HRM in the construction industry. First, this book focuses on the project level and is designed to assist construction project managers to manage the HRM function effectively. In construction companies responsibility for HRM is almost entirely devolved to project managers, few of whom have any formal training in this area. Consequently most see it as a nuisance and as a secondary maintenance function which does not contribute directly to the attainment of project objectives. It is not surprising, therefore, that the construction industry has a poor – and sometimes appalling – record and public image in many areas of HRM. This leads us to the second unique feature of this book, which is the link it draws between HRM and organisational performance. We see this as the key to positioning HRM at the centre of the construction debate and thereby to releasing the significant untapped productive potential of the industry's workforce. If we can persuade project managers of the strong link that exists between HRM performance and project performance we will have succeeded in our aim. We believe that HRM is the forgotten reform in the construction industry and that recognition of its importance is long overdue. Another feature of this book is the array of subjects tackled. Many HRM books predictably follow the standard HRM cycle of recruitment, selection, development, orientation, induction, training, appraisal, compensation, promotion, demotion and termination. While we do not neglect these important issues, we restrict them to one chapter and thereby release the remaining eleven chapters to consider, in detail, issues of contemporary interest such as equality and diversity, empowerment and employee participation, occupational health and safety, work–life balance and industrial relations, etc. These topics are enlivened by numerous case studies which provide practical examples of best practice around the world – particularly in Asia, Australia and the UK. As the construction industry becomes truly global, the need to understand and contrast practices, performance and standards in different countries will become increasingly important to managers.

Finally, in writing this book we have also attempted to maintain a balance between theory and practice. We have done this by drawing upon mainstream organisational behaviour and HRM theory and exploring its application within the unique context that the industry presents. As the case studies and discussion contained within the text emphasise, we do not believe that there is a single definitive model for managing people in the construction industry. There is no magic formula for resolving the significant HRM challenges which the construction industry poses, but we hope this book will help managers develop and explore approaches which reflect the particular needs of their organisation, project and employees.

Martin Loosemore, Andy Dainty and Helen Lingard

Acknowledgements

The authors would like to thank the following people and organisations for their advice and assistance with this book. Their kindness and help with information and comments were invaluable in creating what we hope is a worthwhile contribution to the literature:

- Mr Mike Baines – Shepherd Construction Ltd
- Mr Andrew Ferguson – Construction, Forestry, Mining and Energy Union, Australia
- Mr David Higgon – Multiplex Construction Ltd
- Mr Alastair Keith – Birse Construction Ltd
- Mrs Heather Loosemore – Loosemore Risk Consultants
- Mr Jon McCormick – Multiplex Asset Management
- Mr Sean Quinn – MJ Gleeson Group PLC
- Ms Ani Raiden – Loughborough University
- Mr Charlie Reilly – Multiplex Asset Management
- Mr Chris Reynolds – Baulderstone Hornibrook Pty Ltd
- Ms Ayu Suartika – University of New South Wales, Australia
- Ms Kay Wragg – MJ Gleeson Group PLC
- Mr Tony Welch – Galliford-Try PLC

1 Introduction

The challenges of managing people in construction

Introduction

The ability to attract, retain and develop talented employees is a key feature of successful businesses. People are an organisation's most valuable asset and this is especially true in relatively low-tech, labour-intensive industries such as construction. However, people also represent the most difficult resource for organisations to manage. Unlike physical assets, people have their own individual needs which must be met and idiosyncrasies which must be managed if they are to contribute to organisational growth and development. People are individuals who bring their own perspectives, values and attributes to organisational life, and, when managed effectively, these human traits can bring considerable benefits to organisations (Mullins 1999). However, when managed poorly they have the potential to severely limit organisational growth and threaten the viability of a business. There are countless examples of corporate and project crises in the construction sector which have arisen as the result of people's behaviour, and it would seem that human resource management (HRM) has the potential to eliminate more construction risks than any other management approach (Loosemore 2000). More importantly, HRM has the potential to release a significant amount of productive potential in the construction industry, which has remained untapped because of widespread ignorance of good practice in this area.

In this chapter we aim to outline the aspects of the construction industry's structure and culture that render it one of the most problematic industries in which to manage people effectively. We set the context for the examination of HRM issues in construction by discussing how the industry militates against the utilisation of many proven techniques and approaches. Given the propensity for construction companies to devolve much of the responsibility for HRM to project-based managers, we believe it is essential that all managers understand the challenges they face, the HRM tools at their disposal and the limitations of their application.

Background

The construction industry accounts for a sizeable proportion of world-wide economic activity. For example, in Europe it accounts for some 10 per cent of gross domestic product (GDP) and in Australia it employs about 8 per cent of the nation's workforce (Proverbs *et al.* 1999). Productivity and profitability increases within construction would therefore have substantive benefits to the broader global economy. For example, in Australia it has been reported that if the construction industry increased its effectiveness by 10 per cent this could lead to an increase of up to 2.5 per cent in GDP (Stoeckel and Quirke 1990).

Despite recent advances in technology and production management techniques, construction remains one of the most people-reliant industrial sectors. Human resources represent the large majority of costs on most projects, and the industry employs an extremely diverse range of people from a wide range of occupational cultures and backgrounds, including people in unskilled, craft, managerial, professional and administrative positions. These diverse groups of employees operate as an itinerant labour force, working in teams to complete short-term project objectives in a variety of workplace settings. Hence the industry's project-based structure is made up of many disparate organisations which come together in pursuit of both shared project objectives and individual organisational objectives. These objectives are not necessarily compatible and they might not align with people's personal objectives, which can lead to competing demands on those working within project-based environments. These features make construction one of the most challenging environments in which to manage people effectively, to ensure that they contribute to organisational success.

In recent years there has been a widespread realisation that construction must improve its HRM performance before it can improve its overall efficiency, productivity and cost effectiveness. In the UK, for example, successive government-initiated reports have recommended action on improving the management of people as the cornerstone of strengthening its business and management practices (Latham 1994; Egan 1998). However, despite widespread calls for improvement and warnings that to avoid action will threaten the future competitiveness of the industry, the perception remains that HRM is a peripheral function which has a tenuous relationship with business success.

The challenges of managing people in construction

Texts abound on managing people at both strategic/organisational levels and in the workplace. Both the organisational behaviour and the HRM literature are well-established academic disciplines, in which many approaches and techniques have been developed for ensuring that people are managed and developed in ways which align with organisational goals and strategies. So why is there a need for a book which specifically examines HRM practices in the construction industry? The answer to this question is that the industry presents particular challenges that have the potential to undermine the applicability and effectiveness of the HRM function. In particular, the complexity and dynamism of the industry's project-based culture threaten to undermine the applicability of many central tenets of the HRM strategy that have been applied successfully in more stable sectors. Thus, before we critically examine the industry's practices it is important to understand the difficult context in which managers operate.

The nature of the industry's products and services

Construction activity is extremely diverse, ranging from simple housing developments to highly complex infrastructure projects. However, all types of construction project, regardless of size, have some common characteristics, which include the following:

- *Their unique, one-off nature*: unlike other sectors, where prototypes can be tested before real production gets underway, construction projects tend to be one-off, unique organisations that are designed and constructed to meet a particular client's product and service needs. This can lead to significant risks for people working on a project, which largely arise from learning-curve problems associated with new work activities and ever-changing workplace relationships.
- *Their tendency to be awarded at short notice*: many construction projects are awarded following a period of competitive tendering, where possibilities for thorough planning are often limited. Having been awarded a contract, a design consultancy or contractor has to mobilise a project team comprising an appropriate blend of skills and abilities to meet the project demands quickly. The resourcing function may need to respond to sudden changes in workload, as there can be no guarantee of how much work will be being undertaken at any particular time (Hillebrandt and Canon 1990).
- *Their reliance on a transient workforce*: construction projects are, for the most part, constructed *in situ*. Even with the increased use of off-site fabrication and the wider use of prefabricated components, the final product is normally assembled and completed in the required site location. This necessitates the employment of a transient workforce which can move from one project location to the next. This transience poses many problems for workers, such as longer working days, more expense in travelling to work and managing work–life balance issues, since their families may not be as mobile. Transience also arises *within* projects, since the composition of teams normally changes during different project stages, involving people from many organisations, backgrounds and locations.

- *Increasingly demanding clients*: in recent years there has been a steady increase in the quality of service and product expected by clients procuring construction work. For example, in Australia it has been estimated that construction projects are being delivered in about half the time they were ten years ago. Inevitably, this requires a considerable commitment from those working in the industry, which tends to manifest itself in unsafe working practices, long working hours and increased levels of stress (*Respect for People* 2000).
- *A male-dominated culture*: construction is one of the most male-dominated industries in virtually every developed society. Men dominate both craft trades and professional and managerial positions within the sector. This reliance on male employment leads to many challenges, such as skills shortages caused by recruiting from only a portion of the population, difficulties in the management of equal opportunities and workforce diversity, and considerable challenges in terms of creating an accommodating atmosphere in which individuals' diverse skills and competencies are fully utilised (Dainty *et al.* 2000a, 2000b).

These challenges require construction companies to balance project requirements with competing organisational and individual employee expectations, priorities and needs. It is the industry's inability to manage these competing demands effectively which has caused many of the enduring problems which plague the industry today. As we will see in later chapters, focusing on project and organisational requirements at the expense of human needs will result in employee dissatisfaction, reduced commitment, industrial conflict, increased turnover, more accidents, de-professionalisation, recruiting problems and a continued poor public image.

The project-based nature of construction activity and the devolution of the HRM function to the project manager

We have pointed out that construction is a project-based industry and that this involves bringing together different combinations of clients, designers, constructors and suppliers for relatively short periods of time. In construction the multidisciplinary characteristics of its project teams present particular challenges for managers attempting to secure appropriate staff for projects at different stages and based in geographically dispersed locations. In effect, construction projects form autonomous business units with their own multifunctional teams and objectives, and this inevitably means that line managers must take on responsibility for aspects of the HRM function. This devolution of HRM responsibility, often without any proper training or central support, is a special characteristic of the industry. For example, construction projects demand attention to a variety of human, technical and financial variables. However, the training and education of line managers, and therefore project management strategies, have traditionally focused on the issues of structuring and planning operations, with relatively little attention being paid to the human resource factor in defining a project's success (Belout 1998). Mullins (1999) points out that line managers are typically trained as specialists in their own discipline but do not receive the people-management skills needed to manage their function effectively. He argues that production-oriented management skills are seldom enough to meet the psychological needs which define an effective employment relationship. Whilst some managers with a natural ability to manage people are able to engender an effective team spirit and collective responsibility towards the production function, less effective managers risk the breakdown of team relationships and, ultimately, deterioration in organisational commitment.

A recurring theme in this book is that the construction industry relies on the abilities and skills of line managers to a greater extent than probably any other sector. Construction companies require managers of projects to make HRM decisions that align with the overall strategic philosophy of the organisation and which meet the operational staffing requirements of their individual team. One of the problems with this approach is the inevitable tension which develops between the short-term objectives of the project and the longer-term strategic needs of the wider organisation. For example, project managers may be aware of the need to provide training for their staffs' personal development, but the time and resource pressures of a project and the day-to-day problems that arise may prevent them from doing so. Inevitably, since project success is nearly always measured in monetary terms, people-related issues become a second priority to the core procurement challenges of meeting time, cost and quality targets.

The variable demand for construction products and services

The level of economic activity within the construction industry is highly sensitive to wider economic activity, and construction has always suffered from being one of the first industries to be affected by an economic downturn and one of the last to recover from it. This stems from the propensity for businesses to curtail construction plans if they face the threat of an economic downturn due to the large amounts of capital expenditure involved in such projects. Conversely, in the event of an upturn in business, construction is likely to be one of the last investments committed to, because of the need to assure a relatively certain period of economic stability and growth ahead. This cycle of peaks and troughs in construction demand makes it very difficult for companies to retain directly employed workforces and make long-term investments in its core professional staff. Consequently, most construction companies adopt a flexible model of HRM, in which they employ the bulk of their workforce on temporary contracts or as subcontracted labour (see Chapter 4). This has led to an industry structure in which small businesses vastly outnumber large firms. For example, according to the Australian Bureau of Statistics (ABS) Business Register data, only 12 businesses classified as 'general construction' (Australia New Zealand Standard Industrial Classification [ANZSIC] code 41) employ 500 or more employees (ABS 1998a). The majority of Australian construction firms are small businesses, with 97 per cent of general construction businesses employing less than 20 employees and 85 per cent employing less than five employees (ABS 1998a). Similarly, in the UK only 1.6 per cent of construction firms employ over 25 people (Druker and White 1996a). Smaller firms are unlikely to have a specialist HRM department, which requires that owners and operational managers must perform the HRM function without specialist input. Unfortunately, this industry structure is extremely difficult to monitor and control, and this is one of the reasons why the construction industry is renowned for poor HRM practices in areas such as training, safety, exploitation of illegal migrant workers, and avoidance of tax payments, workers' compensation payments and other legal rights. The dangerous result of not being able to control such practices in such a competitive industry is that other companies are forced to lower their performance to the lowest common denominator in order to survive. The large construction firms who employ these small companies, construction clients and government agencies have an extremely important responsibility to ensure that these practices do not occur.

The shrinking labour market and the image of the construction sector

Like any industry, construction has to compete for its workforce from the limited pool of people who are able and willing to work. Historically this has not been a problem for the construction industry. However, a sustained downturn in population growth in many developed countries and changes in gender demographics affecting traditional recruitment sectors have made this market more competitive and raised the real possibility of the industry being affected by skills shortages in the near future. For example, in Australia, men's overall labour force participation fell from 84 per cent in 1966 to 73 per cent in 1998, while women's rose from 36 per cent to 54 per cent in the same period (ABS 1998b). Increasing difficulties in recruitment are also being reflected in increasing workloads for existing industry employees. For example, recent figures from the UK have indicated that fewer managers and professionals work in the construction industry now than in the early 1990s (see CITB 2002). It is essential that such concerns be elevated to the top of the industry's strategic agenda if the industry's future labour demands are to be met.

Labour market demographics present a concern for all industrial sectors, as falling birth rates lead to competition between sectors for an increasingly limited pool of job candidates. It is inevitable that in this environment less attractive industries will be unable to recruit their share of high-quality school and university leavers and will eventually suffer from skills deficiencies. Clearly, high-achieving individuals are likely to gravitate towards industries and sectors which are seen as offering good wages, good working conditions and good career opportunities, and as being the most glamorous and attractive to work in. Thus, the unattractiveness of construction as a career choice has become a topic of concern and debate amongst the industry's various bodies and training organisations. For example, in the 1999 edition of the American *Jobs Rated Almanac*, civil engineering fell from 18th to 70th position in expressed job preference and 14 construction trades were rated in the bottom ranks. The reasons put forward for the construction industry's poor public image have been numerous, and they include:

- the site-based and hence itinerant work patterns, which result in job insecurity or require many construction workers continually to relocate in pursuit of new project opportunities;
- the poor on-site working conditions, health and safety record and employee welfare provision within the industry;
- the industry's association with manual, blue-collar occupations rather than more highly regarded white-collar positions;
- the male-dominated and discriminatory 'macho' culture that is commonly portrayed as the way the industry operates.

Given the shrinking labour market and image problems of the industry, it is clear that further economic growth is likely to lead to severe shortages in both traditional and new skills areas (Agapiou *et al.* 1995b). Whilst it is incumbent on those working within the construction sector to work to improve the image and attractiveness of construction as a career option, the reality is that the industry has a long way to go in improving its stereotypical image. Indeed, research has shown that it is currently regarded as having an occupational status similar to that of a cottage industry (Gale 1994).

Employee turnover and retention

In the recruitment climate outlined above it is becoming increasingly important for construction organisations to retain their professional employees in order to remain competitive. Employee turnover, or 'wastage', is an extremely important issue for construction companies' strategic HR planning, yet a culture of mobility has emerged in the industry which has led to a workforce of corporate mercenaries that coldly drift from job to job with little sense of loyalty to their employers. This should be a worrying development for any company that takes training seriously, although admittedly such companies are likely to experience less labour turnover. This relates to the cost of training staff to the point where they are sufficiently productive to generate income. For example, American estimates suggest that after 10 years of service an employer will have invested a minimum of US$600,000, in terms of salary, benefits, recruitment and training costs, for an employed engineer (Maskell-Pretz 1997). Within the UK there are concerns that staff turnover may increase even further as staff shortages intensify and competition between different employers increases. For example, in a recent UK survey 42 per cent of construction professionals said that they were actively looking for new positions (J. Ford 1997).

Thus the need for companies to retain their staff seems set to become a major HRM issue in the construction industry of the future. Without an increase in labour resources, only companies offering competitive salary packages, good working conditions and exciting career opportunities will be able to satisfy their labour requirements. Indeed, recent reports have suggested that skills shortages are already leading to increased salary levels (Cargill 1996). However, increased remunerative costs lead to competitive labour markets, which has inflationary effects on the cost of construction work (Agapiou *et al.* 1995b). For this reason, Briscoe (1990) predicted that UK construction companies may start to lose projects to other countries if wage levels increase as a result of skills shortages, which may threaten the future growth of the UK industry. Thus it is essential that the industry addresses its image problems and begins to recruit from non-traditional sectors such as women and, in some countries such as the UK, ethnic minority groups. These issues are addressed in more detail in Chapter 8.

Subcontracting and self-employment in construction

The construction industry relies on subcontracting for the majority of its production effort. Hence the construction industry comprises a large number of small and medium-sized enterprises which operate in a subordinate productive role to larger 'main' contractors. This is more of a construction characteristic in some countries than others. For example, in the UK self-employment is higher than in any other European country at around 45 per cent, compared to 10 per cent in Germany and 18 per cent in France. The rise of self-employment amongst construction workers in the UK can be traced back to the 1980s and early 1990s, when a political agenda of de-unionisation, a philosophy that small is beautiful and favourable tax reforms made self-employment a lucrative option for many skilled workers. This more flexible structure was perfectly suited to the fluctuating workloads of the construction industry and led to major structural changes where the majority of the workforce became self-employed. However, this also produced problems of reduced control, which lie at the heart of many of the industry's inefficiencies today. For example, tax evasion has become a major problem in the construction industry and research indicates that small companies are less likely than larger organisations to have well-developed operational HRM policies. Research also suggests that small businesses do not manage occupational health and safety risk as effectively as larger businesses and may be unaware of their responsibilities under occupational health and safety law (Lingard 2002). These factors present difficulties for the prevention of occupational injuries and disease, which likely contribute to the higher incidence of occupational injury in small construction firms (McVittie *et al.* 1997). In essence, many of the changes brought about during the late 1970s in the UK were arbitrary, politically motivated and cosmetic, and many industry employees, far from being small entrepreneurs, are in effect disguised waged labour (Rainbird 1991). Although the late 1990s saw authorities begin a clampdown on bogus self-employment, this is likely to take a long time to manifest itself in the form of significantly greater levels of direct employment in the industry.

Training, employee development and knowledge creation

Training, personal development and knowledge creation lie at the very heart of achieving a motivated workforce and an efficient, effective, creative and innovative industry which has a positive public image. Training is the most effective way to maintain, update and enhance the intellectual capital of the industry's workforce and to ensure that its activities contribute positively to the well-being of society as whole. However, investment in these areas remains at a relatively low level compared to other industries. For example, in Singapore the skills level of the workforce is 23 per cent, in Western Australia and America it is 40 per cent, in Hong Kong it is 30 per cent and in Japan it is 60 per cent (MMS 1999). This stems in part from the low-tech and macho culture of the industry, which has always placed more value on brawn than brains. It also arises from the high degree of self-employment described above, where so many companies are too highly geared to make expensive investments in training, are too separated from clients to benefit from any innovations and have too short-term a perspective to make long-term investments in developing their employees. However, construction professionals are now more educated and sophisticated than at any time in the past and have higher expectations of their employers to provide for their personal career development (Druker and White 1996a). Therefore, those companies which fail to meet the psychological expectations of employees stand to lose their most able and ambitious personnel to their competitors in an increasingly competitive marketplace for good people.

Schein (1992) warns that modern organisations have changing needs which will not necessarily remain compatible with the needs of the individual. Consequently, organisations which operate in dynamic environments (such as construction) will need to be proactive in managing their employees' career development and in ensuring that the needs of the individual are aligned with the needs of the organisation. Organisations can effectively control the personal development of their employees in line with organisational development by establishing a career management process. This ensures a satisfactory career for the employee, optimum use of human resources and employees who match the needs of the business (R. Thompson and Maybe 1994).

Communication

Communication is probably the most important enabler of effective HRM practices, but, when poor, it also has the potential to severely limit their effectiveness. HRM communications must have an internal and external dimension. The internal dimension must focus on ensuring effective communications between managers and workers in different parts of an organisation, particularly project staff and central HRM departments. In contrast, the external dimension should focus on communications with external interest groups such as governments, pressure groups, local communities and unions. These represent important stakeholders in all organisations, and ignoring them can lead to numerous problems. Loosemore (2000) found that in construction projects defensiveness to outsiders can be a problem. On some projects the pressures, cohesion, loyalties, focus and momentum that can develop become so intense that the workers effectively seal themselves off from the outside world, considering outsiders as an unnecessary distraction and even covering up problems that may expose internal weaknesses to them. However, ironically, the occasional involvement of outsiders who are unfamiliar with a project is often the most effective means of detecting potential problems. Their exclusion only increases a project's crisis-proneness.

Employee relations

In its broadest sense, employee relations concerns the process of establishing and negotiating the terms and expectations of the employment relationship. This process is especially important for employees in an industry that is renowned for its unsafe and unfair practices. Traditionally, this negotiation was undertaken by trade unions on behalf of members in the form of collective agreements, although recent legislative changes have provided greater flexibility in allowing for non-union collective bargaining, individual contract-based employment and project-specific labour agreements. Provisions for collective agreements still exist in many countries and tend to be negotiated on a voluntary basis. They involve employers' associations, which represent construction firms, and trade unions, which represent the interests of workers who are their members. These are high-level negotiations on general working rules and rates of pay that are then recommended as national frameworks of pay and conditions. For example, in Australia, collective agreements are established in the Australian Industrial Relations Commission (AIRC). This is a court of law which conciliates and, if necessary, arbitrates between union, government and employer representatives to set basic rates of pay and working conditions for an industry.

Such industry-wide agreements contrast markedly with the flexible approach adopted on most individual projects, in which terms and conditions of employment may be agreed around particular project requirements and needs. This contrast requires that project managers are skilled in managing employee relations and in negotiating appropriate working-rule agreements, as well as being cognisant of national agreements and their impact on the employment relationship. Unfortunately, all too often this process is neglected and many vulnerable employees find themselves being exploited by unscrupulous employers in the name of making a quick profit. In most instances these people are unaware of their rights and are too intimated to make a formal complaint. This is becoming a particular problem, with ever-larger numbers of illegal immigrants targeting the construction industry as an easy place to find work without too many questions being asked.

Having said this, it could reasonably be expected that in an industry relying on a high degree of self-employment and subcontracting, the importance of actively managing employee relations in a formal sense would be fairly limited. However, the construction industry has a long tradition of collective bargaining structures and industry-wide agreements that still affects the nature of the employment relationship. Furthermore, in countries like Australia the construction union is still powerful and highly active, and needs to be consulted in many HRM decisions. Thus, as with so many aspects of the HRM function, employee relations is also an especially complex field in construction.

Equal opportunities and diversity

Providing equality of opportunity for all of those working in a particular industry or organisation should form a cornerstone of good employment practices. Discriminating against people on the grounds of their gender, race/ethnicity, age or disability leads to an under-utilisation of people's skills and talents and to a stifling of workforce diversity, which could promote innovation and improved working. Despite this, the vast majority of those working in construction in developed countries are male. In the UK, for example, the industry employs around 1.75 million people, of whom under 10 per cent are women (Court and Moralee 1995). This makes it the most male dominated of all major industrial sectors. There is also a demonstrable under-representation of employees from ethnically diverse backgrounds (Cavill 2000). Recently, however, the UK construction industry has begun to recognise the limitations associated with recruiting from such a limited labour pool. Whilst the motives for this shift may be somewhat opportunistic, being driven by the need to cope with demographic changes in the labour market rather than a genuine concern for equity, there is also a growing realisation that diverse workforces present other potential benefits. The experiences of other sectors have shown that improving opportunities for women leads to a workforce that is better informed, and to organisations which are more adaptable, closer to their customers and more responsive to market changes (Coussey and Jackson 1991). In the UK, improving the industry's performance on diversity formed key recommendations of recent government-initiated reports and subsequent action taken to respond to their recommendations (Latham 1994; Respect for People 2000).

In Australia, which has the largest foreign-born workforce outside Israel, the problems of equal opportunities occur in a completely different context. Here, the construction industry has absorbed a relatively high proportion of immigrants into its workforce. Consequently its managers are affected more than most by Australia's multiculturalism and its attendant problems and opportunities. However, there is disturbing evidence that the construction industry has not addressed this challenge effectively. For example, Loosemore and Chau (2002) found that, amongst other things, ethnic minorities in the construction industry had been subjected to racist comments in work and felt that their employment opportunities were less than those of their white colleagues. In other countries there is also evidence that few construction companies have formal equal opportunities policies, that ethnic minorities and women are under-represented at senior managerial levels, that discrimination, sexism and racism are rife, and that the construction industry suffers from a poor public image in the eyes of ethnic minority groups and women (Bagilhole *et al.* 1995; Lim and Alum 1995).

Despite increasing acknowledgement of the need to improve equal opportunities in construction, achieving improvement is unlikely to be easy for project managers. Ingrained and institutionalised age and gender prejudices, excessive mobility requirements, informal selection procedures, stereotyped assumptions, unspecified job criteria and the 'old boy' network have all been shown to have an impact on minority groups' under-achievement in paid work (Hansard Society 1990).

Health, safety and welfare

Few areas of HRM can be as important as managing people's health, safety and welfare at work. However, despite advances in occupational health and safety legislation, research and management techniques, construction remains one of the most dangerous sectors to work in. Moreover, it remains one of the few sectors where occupational health and safety performance is not improving year on year. Whilst every jurisdiction will have its own occupational health and safety legislation and enforcement mechanisms, safety management on site has to be tailored to respond to the individual hazards that each project presents. Furthermore, management strategies must be tailored to the unique combination of individual employees working on the project.

The role of HRM in contributing to safe working cannot be overstated, but safe working is far from easy to achieve. In particular, health and safety must be made a top priority if it is to be taken seriously and incorporated into all management-system procedures to ensure a consistent approach. However, the fragmented delivery mode and high levels of self-employment in construction inevitably mean that employees must bear some degree of responsibility for their own health and safety. Furthermore, the commercially oriented, male-dominated, macho culture of construction is unlikely to promote a safety-conscious attitude amongst employees. The acceptance of risk-taking as 'part of the job' and the belief that accidents 'happen to others' have been identified as sources of unsafe behaviour on many construction sites (Lingard 2002). These attitudes will only be changed if employers demonstrate that risk-taking is unacceptable and that safe working is a non-negotiable condition of employment. This requires high safety standards, safety issues driven by senior managers, effective systems, continuous training and education programmes to alter attitudes and behaviours, and effective induction and communication strategies to ensure an awareness of occupational health and safety issues and their importance to the project management team.

Conclusions

In this chapter we have outlined the complex and problematic context of HRM within the construction industry. Exploring the management of people in the construction industry without acknowledging these wider issues would oversimplify the challenges facing construction organisations in effectively managing the HRM function. In particular, project-level HRM practices should not be viewed in isolation from the wider organisational and sectoral contexts in which they are embedded. These will define and shape the ways in which operational policies are put into practice and will determine how effective they are. With this in mind, this book is designed to identify appropriate methods of managing people at the project level which accord with wider organisational and social objectives. The aim is to showcase elements of good practice, but also to cast a critical eye on existing HRM practices in order to identify how these could be managed more effectively. We explain the interaction of the various HRM functions, and how project and business performance can be improved through the effective management of people. Many of our ideas are drawn from mainstream management theory to ensure a sound theoretical foundation to the book, and our aim has been to translate them into practical tools to operate within the sector.

Discussion and review questions

1   What are the key HRM challenges facing construction companies post-2000? How are these likely to differ from those faced at the end of the last century?
2   Define elements of the industry's structure, culture and operation that could militate against the effective management and development of people within the industry.
3   What additional challenges does the construction project environment present that other project-based sectors do not have to cope with?

2 The development of modern organisational and management theory

This chapter provides a brief overview of the development of management and organisational theory and its influence on contemporary HRM theory. The aim of this chapter is to provide a theoretical framework for the exploration of HRM issues in later chapters. It charts the development of management thought and organisation theory and explains how these have influenced the ways in which modern organisations approach the HRM function.

Introduction

The development of modern HRM has been punctuated by different schools of thought, which have explained the existence, purposes and functioning of organisations in different ways. This range of theories is often difficult to reconcile and modern management thought remains a constantly changing cocktail of different ideas, many of which provide a unique view of organisations and their problems. A key feature of the study of management and organisational behaviour is that it is difficult to identify a single solution to a particular problem (Mullins 1999). Many seemingly contradictory theories exist, some of which may seem on first sight to have little relevance or application to the modern organisation. This can render the study of management frustrating and confusing. However, gaining an understanding of management theory is important because many of the early ideas on management underpin modern approaches to the HRM function.

It would be counterproductive for managers to attach themselves to one organisational theory in the belief that it can solve all of their problems. It is more appropriate to develop an understanding of them all, within the social, economic and political context in which they developed. This enables managers to assess their value in a highly dynamic world. As should have become apparent from Chapter 1, few industries are as subject to change and uncertainty as construction, and so it is incumbent on managers to be sensitive to their environment and adopt different strategies and approaches which are appropriate to their particular circumstances.

The seeds of contemporary management thought

The growth of management as a self-conscious discipline is a nineteenth-century construct, although Adam Smith's *Wealth of Nations* (1776) indicates some attention to organisational productivity and efficiency at the very beginnings of the industrial revolution in Britain. For instance, it was Smith who first documented the benefits associated with the division of labour, noting that:

> [A] workman not educated to do this business, could scarce, perhaps with the utmost industry, make one pin in a day. But in the way in which this business is now carried on, not only is the whole work a peculiar trade, but it is divided into a number of branches, of which the greater part likewise are peculiar trades, ... and the important business of making a pin is divided into about eighteen distinct operations. ... I have seen a small factory, where ten men only were employed [who] could when they exerted themselves make among them about twelve pounds of pins in a day. There are in a pound upwards of four thousand pins.
>
> (Smith 1776/1986: 109)

Smith's ideas laid the seeds for large-scale factory production in Britain, which eventually led to the demise of traditional cottage-based industries during the industrial revolution. However, with the development of such techniques came unsafe and inhumane working environments characterised by long hours and low pay, for men, women and children as young as 5 years old. Furthermore, for the first time society was introduced to the phenomenon of unemployment as traditional skills became increasingly marginalised by new management techniques and technologies such as the steam engine. In response there developed problems of social unrest, which manifested itself in organised campaigns of violence towards factory owners and machine breaking, a phenomenon which eventually became know as Luddism, after a mythical leader called Ned Ludd. While the working conditions which led to this social unrest may seem a world away from modern industry, we shall see later in this book that in terms of managerial and technological innovation there are many similarities between today's construction industry and that which existed at the dawn of the industrial revolution. It is clear that modern-day managers should remember the lessons of the industrial revolution and take heed of the social implications of failing to meet employees' expectations.

We shall return to this theme later, but in the meantime it is worth noting how early managers responded to these problems of industrial conflict. In particular, Robert Owen, a self-made cotton manufacturer, began to think seriously about the human implications of managerial decisions. He developed the view that people were the product of their surroundings and were therefore improvable. In his factory of 2000 people (including 500 pauper children) he experimented with paternalistic management by providing his workforce with housing, free education and a subsidised village store. He also increased the age at which children could work from 5 to 10 years old and reduced the working day from 14 to 12 hours. On the harder side, he operated a curfew, rigorously enforced cleanliness standards and punished drunkenness with fines. While Owen was primarily driven by the search for productivity improvements, his innovations were of immense importance because they proved, for the first time, that workforce welfare was linked to profits and productivity. Indeed, this idea that the interests of workers and managers were not necessarily at odds was eventually taken further by Charles Babbage (1832), who was also concerned with the social costs of increasing mechanisation and urbanisation. He was the first to develop practical mechanisms such as profit-sharing schemes to involve all interests in the prosperity of an enterprise and to link workforce welfare with company profits.

Thankfully, today's working conditions in developed countries bear no resemblance to the misery of those experienced by workers in the early industrial revolution. However, later in this book we shall see that the modern-day construction industry still has a long way to go in improving working conditions for its manual workers. Furthermore, we will show how poor working conditions are no longer the preserve of the working class, in that professionals are increasingly subjected to a working environment which places their health at risk through workplace stress and difficulties in maintaining an effective work–life balance.

The influence of mass production on the early construction industry

Although the early building industry did not directly experience the benefits of factory production, it was profoundly affected by these early developments in management thought. Until the industrial revolution the construction industry was essentially craft based and founded on simple tried and tested traditional technologies and production methods. Projects were procured in a fashion which is similar to today's design and build projects, where a master builder would be commissioned to design, construct, manage and maintain a building directly for a client. However, the industrial revolution presented many new challenges associated with increasing urbanisation and the development of new production technologies and materials. For example, the pressures of producing large numbers of houses to accommodate the growing populations of new towns and cities led to the development of off-site fabrication, standardisation and modularisation in building design, and to the development of new technology such as cement mixers. Moreover, innovations in the science of materials, such as steel and concrete, led to the development of structural engineering and specialist subcontracting, and to the growth of the construction professions such as architecture and quantity surveying. Collectively, these developments had a fundamental impact on the way construction projects were managed. In particular, this increased speciali-sation and fragmentation resulted in a greater need for management of the construction process in order to reintegrate these increasingly disparate components into a cohesive working team.

The process of industrialisation was also a turning-point for the civil engineering industry, which began to develop in response to the need for infrastructure to support increasing industrialisation and urbanisation. For example, between 1822 and 1900 there were 22,000 miles of rail track laid in the UK, which employed literally millions of navvies in subcon-tracted gangs. As Coleman points out:

> [T]he making of a railway was organised in this way: first the company ... appointed an engineer, say Robert Stephenson or Brunel – to devise a route, specify the works to be done, superintend their construction and to be responsible to the company for the whole venture. The company then invited tenders for part or whole of the work and appointed a principal contractor or contractors to carry out these works.
>
> (Coleman 1965: 51)

In this sense, the method of procurement was similar to modern-day construction management. Indeed, it also seems that little has changed in relation to the exploitation of subcontractors, since the legal contracts used were rather onerous and the methods of management very primitive. Essentially, each subcontractor was typically set a target of filling a certain number of wagons with spoil per day. Normally the wagons were towed manually by the labourers as they progressed and the digging occurred by hand. On average, this meant that each man had to dig about 20 tonnes per day, regardless of conditions!

Contemporary management theory

During this period of history, change and uncertainty were constant features of life, much as they are today. Indeed, it is generally agreed that we are currently experiencing a second industrial revolution which is similar in scale to the first, differing only in that its origins are in elec-tronics rather than mechanics. This has resulted in a second and ongoing period of fundamental change in the construction industry in terms of its professional boundaries, procurement and contractual arrangements, tech-nology, design, specialisation, fragmentation and standardisation. Once again, this change is driving the need for managerial skills in order that this be managed effectively and that the industry and its organisations benefit from the transition.

However, while there are undoubted similarities between the original and modern industrial revolutions, modern managers face very different problems to those faced in the early eighteenth century. For example, the power balances between operatives and managers within organisations were very different in the eighteenth century than they are today. Furthermore, in the eighteenth century managers had little other than their own experience to draw upon in making decisions and were largely uneducated in the discipline of management. In the eighteenth century concerns were primarily with production capacity, and technologies were enthusiastically embraced with little concern for their human implications. In essence, modern managers and their workforce are more educated than their predecessors, the power balance between them is more equal, management is based on different values to those of pure engineering, and the world is far more competitive and egalitarian. In response to this changing industrial environment management has developed considerably since the pioneering work of Smith, Owen and Babbage. Essentially, three broad themes have emerged, the earliest emphasising *production efficiency*, the second emphasising *human behaviour* and the third emphasising organisations as *systems*. Each school of thought is discussed below.

Production efficiency: the 'classical' approach

This school of thought which consists of *scientific management, administrative management* and *bureaucratic management* has its origins in America and represents an extension of engineering principles to the management of organisations. Sometimes known as the 'classical' approach, these theorists thought of organisations in terms of their purpose and structure in order to understand how their methods of working could be improved (Mullins 1999). America experienced its industrial revolution after Britain and on a much larger scale. For example, one of Ford's earliest factories in Detroit employed over 70,000 people, posing very different managerial problems to anything experienced before. This led to the rapid professionalisation of management in America, particularly under the auspices of the American Society of Mechanical Engineers (ASME), which sought to redress the neglect of management in other important institutions such as the American Society of Civil Engineers (ASCE) and the American Institute of Mining (AIM).

Scientific management

The initial momentum for the ASME's concern with managerial skills came from Frederick Taylor, who became its president in 1906. His early experiences of work as a machinist shaped his low opinion of other people. In essence, Taylor believed in the mediocrity of the masses and that the average human being had an inherent dislike of work and responsibility. He also believed that most people preferred to be directed, had little ambition, were primarily motivated by monetary and materialistic needs, wanted security above all, and would only work under external coercion and control. Being an engineer and having experienced science's success in the manipulation of physical materials, he strove to apply the same principles to the control of people. In essence, Taylor believed that uncertainty was a major cause of organisational problems and that through science it could be eradicated.

In metaphorical terms, Taylor's ideal organisation was a goal-seeking, machine-like entity with parts working in harmony towards a commonly recognised set of objectives. Problems were caused by an inability to deal with uncertainty, which produced internal conflict between managers and workers. In Taylor's opinion managers were responsible for ensuring that this undesirable phenomenon did not occur and while studying for a master's degree he developed the 'task system' to achieve these aims. The principles underpinning the task system have become known as Taylorism or scientific management. At the centre of this approach (which is described in F. W. Taylor 1911) was 'time study'. This involved breaking jobs down into simple tasks and constructing standard times for them, based on observations of people at work. Efficiency could be increased by setting targets based on these standards, closely supervising people to ensure their attainment and paying people according to this.

Later, time study was developed further into 'motion study' by Frank Gilbreth, a bricklayer by trade and also a prominent member of ASME at the same time as Taylor. He was intimately involved in the development of the first mechanised concrete mixers, conveyors and reinforcing bars, and he extended Taylor's work into the detailed study of people's movements (motion study). The end result of a motion study was a set of prescriptive flowcharts describing the correct way to undertake every aspect of a particular task. Gilbreth argued that by slavishly following such flowcharts, wastage and bad practices could be eradicated from the construction industry. In his own words:

> [T]o be pre-eminently successful, a) a mechanic must know his trade; b) he must be quick motioned and c) he must use the fewest possible motions to accomplish the desired result. ... Tremendous savings are possible in the work of everybody. ... But the possibilities of benefits from motion study in the trades are particularly striking because all trades, even at their best, are bungled.
>
> (Gilbreth 1911; quoted in Springel and Myers 1953: 55)

Motivated by similar values, Henry Gantt (1919), the inventor of the slide rule, was also a disciple of Taylorism. Gantt developed the Gantt chart, which has become hugely influential in construction planning.

However, Gantt moderated Taylor's ruthless emphasis on efficiency with greater attention to training and method as opposed to accurate measurement. Taylorism is often criticised by today's modern management writers. His view was to treat people as machines, so that one method of performing a task should apply to every worker. This overly simplistic view of people does not take into account any variation in levels of skill, motivation or social interaction between different workers. However, despite this simplistic and outdated view, many aspects of scientific management prevail today in the construction industry. Some of these have been carried through into modern HRM thinking and will be discussed in later chapters.

Administrative management

In parallel to the above developments, another engineer called Henri Fayol was seeking to apply scientific methods to management in France. However, Fayol's 'administrative management' differed from scientific management in that it focused on managerial efficiency rather than workers' efficiency. According to Fayol, managers were the supreme co-ordinating authority with ultimate responsibility for preventing problems or, in their undesirable eventuality, steering the organisation to recovery. According to Fayol, there were two levels of managerial activities, namely *day-to-day* and *governance*.

Day-to-day management involved six activities, namely technical (production, manufacture, adaptation), commercial (buying, selling, exchange), financial (optimum use of capital), security (protection of property and persons), accounting (stocktaking, balance sheets, cost, statistics), managerial (planning, organising, command, coordination, control). Governance referred to the overall coordination of these activities and represented the original conception of what we now call strategic management. Importantly, Fayol was also concerned with the lack of management theory and education, arguing that:

> [W]hilst the greatest effort is being made, and profitably so, to spread and perfect technical knowledge ... management does not even figure in syllabuses of our colleges ... why? Is it that the importance of managerial ability is misunderstood? No ... the real reason for the absence of management teaching in our vocational schools is the absence of theory; without theory no teaching is possible.
>
> (Fayol 1949: 14)

Some would argue that this is still the fundamental problem holding back advances in construction management today (Betts and Lansley 1993; Runeson 1997). It is certainly true to say that mainstream management theory has taken a long time to become accepted in construction education and practice.

Bureaucracy

Max Weber (1947), the father of bureaucratic management, was a political economist and the first to recognise the importance of organisational power and authority. He conceived the ideal organisation as one in which people would do exactly as specified out of respect for authority. Such an organisation would be characterised by:

> highly specialised, routine operating tasks, very formalised procedures ... a proliferation of rules, regulations and formalised communication throughout .... large-sized units at operating level; reliance on the functional basis for grouping tasks; relatively centralised power for decision making; and an elaborate administrative structure with a sharp distinction between line and staff.
>
> (Weber 1947; quoted in Mintzberg 1983: 164)

Essentially, Weber claimed that for an organisation to run smoothly people should specialise, there should be a clear chain of command, employees should be hired and fired on the basis of scientific performance appraisals and tests, managers should specialise in management, and there should be formal rules and procedures to follow at all organisational levels. Weber was concerned with creating a well-run, predictable organisation in which consistent decisions were made on established facts and principles, and in which people were rewarded according to their performance and expertise. The allocation of privileges within this system should be in accordance with the rules and procedures established for its operation.

Unfortunately, the business environment in which Weber developed his theory was very different to today's, and we now tend to associate bureaucracy with red tape, inefficiency and slowness. This is because the rules and procedures that must underpin such a system can become more important in their own right than as a means to an end. This can have the effect of stifling innovation and creativity within an organisation, and lead to a lack of concern for and responsiveness to employee needs. In HRM terms, overly bureaucratic systems can lead to restrictions in the psychological growth of the individual employee, causing feelings of failure, frustration and conflict. Modern organisations have largely tried to disassociate themselves from bureaucratic models of management, although elements of Weber's approach can still be found in the operating structures of some construction companies (see Chapter 4). Indeed, bureaucracy should not be dismissed outright. In some important areas of human resource management, such as occupational health and safety (see Chapter 9), there is a need for companies to adopt strict procedural and administrative controls.

Human behaviour

Over time, the ruthlessness and sterility of scientific management induced a great deal of resentment in managers and operatives alike. Not only was the autonomy and professionalism of managers reduced by their new role of merely imposing prescribed rules and procedures, but their insulation from the workforce prevented them from building relationships with the people they sought to manage. Furthermore, as frontline representatives of organisations they increasingly came into conflict with workers, whose animosity grew because of de-skilling and relentless demands for productivity improvements. Out of these tensions grew a new school of management thought which incorporated a psychological element into management theory. This was the birth of human resource management, and in its earliest form it was referred to as the *collectivist movement*, the aim being to base management on a better understanding of the forces which shape human behaviour in organisations.

Collectivism

The collectivist movement was led by Mary Parker Follet (1924) and Elton Mayo (1933). Their work is still very influential in modern management thinking and emphasised the role of groups within organisations, but in particular the psychological factors which caused people to consent to group membership and conform to group norms. They believed that if dedication to the group could be achieved, then the collective group energy could be greater than the sum of its constituent members. 'Synergy' is the popular term which has become associated with this phenomenon in modern management texts.

In essence, collectivists saw organisations as collections of individuals with a free will who were inextricably linked in a complex and dynamic social network. This meant that there was little sense in scientific management's habit of breaking down organisational functions into separate constituent parts, analysing them in isolation and then artificially reconstructing them. Furthermore, in contrast to the perception in scientific management that industrial conflict was caused by economic motives and was a sign of system failure, collectivism considered its causes to be sociological and psychological and to be an inevitable aspect of organisational life to be accommodated rather than suppressed. Indeed, if properly managed, conflict could perform positive organisational functions, a view which is gaining in popularity in construction project management. However, Loosemore *et al.* (2000) found that the necessary managerial attitudes and skills do not yet exist in the construction industry to harness the potential positives of conflict.

An important contribution to the collectivist movement was also made by Chester Bernard (1938), who concerned himself with the functions of the executive. In this way he closely mirrored the work of Fayol. However, Bernard made an important attempt to bridge the divide between the requirements of formal organisation and the needs of the socio-human system. In doing this he placed more emphasis on informal systems within organisations, arguing that they needed as much management as formal systems. In Bernard's view the ignorance of the informal aspects of organisations had been the downfall of scientific management as the process of industrialisation made the business environment more complex, uncertain and volatile. Paradoxically, since scientific management was the prime driver of industrialisation, it has been responsible for its own demise.

Bernard also widened the concept of organisation to include investors, suppliers, customers and clients – or, in modern management terminology, the supply-chain *stakeholders*. Finally, he also realised that individuals had a certain degree of autonomy and that an organisation was made up of varying *interest groups*, with *goals* that would not always coincide with the organisation's. Groups were important in management because members of a group became constrained to accept its values and goals and would not always obey a manager's requests. In this sense, Bernard's portrait of an effective manager was one who was sensitive to varying interests within an organisation and who could arbitrate between them to avoid conflict. Furthermore, it was important to develop a human resource policy which attracted suitable individuals and motivated them by providing a shared set of positive values, so that people voluntarily wished to make the organisation a success. This was one of the earliest references to the concept of organisational culture, which is attracting considerable attention in contemporary management thinking, particularly in construction, with the increasing attention to the concept of trust in many countries (Latham 1994; COA 1998). However, it is important to note that the concept of organisational culture assumes a homogeneity of workforce which does not reflect the complexity of the construction industry environment. Rather than a mechanism for attaining top-down control, it should be managed to harness positive input from distinct groups of individuals in the workforce.

The influence of psychology: neo-human relations

As interest in human behaviour in organisations grew, so did the influence of psychology. This was because the link between satisfaction and worker productivity was not always a clear and positive correlation. In particular, Abraham Maslow (1943) and Frederick Herzberg (1959) made attempts to understand the forces which motivated people at work and the way in which individual adjustment, group relations and leadership styles impacted on worker motivation. For example, Maslow argued that people were driven by needs which were organised into a hierarchy, each appearing on the satisfaction of a lower-level need. People's most basic needs were physiological (food, warmth); then came safety, love, self-esteem and self-actualisation needs. Although it is now recognised that there may be limited transferability of this hierarchy across different cultures, the managerial implications of Maslow's hierarchy are still relevant. Arguably, the most significant implication was that managers with a scientific stance could not satisfy a person's needs because the imperatives of control reduced individuals' autonomy and the potential for self-esteem. Indeed, it was arguable whether any manager could provide for every aspect of someone's needs, meaning that there would come a point where a manager would lose control over motivation. Recognising non-work sources of motivation is therefore important in modern HRM.

Herzberg's contribution was the *motivation-hygiene theory*, which was based on the discovery of motivators and demotivators. Removing the latter merely acts to reduce dissatisfaction but is incapable of increasing satisfaction. For example, bureaucratic procedures are a demotivator and their reduction merely reduces demotivation. No amount of reduction will provide satisfaction. Herzberg's main criticism of scientific management was that it focused almost entirely on demotivators and was therefore incapable of motivating a workforce, only reducing its levels of demotivation. According to Herzberg, people have a natural desire to achieve and scientific management prevented them from doing so. A manager's job should be to correct the mismatch between job content and personal aspirations through better job design. D. McGregor (1960) argued that management style is a function of the manager's attitude towards the nature of work behaviour. He proposed two contrasting models of managerial approach, based on assumptions about work and people – Theory X and Theory Y. Managers who advocate Theory X are job centred and work on the assumption that the average human being is stuck in early adolescence, has an inherent dislike of work and responsibility, cannot be trusted, prefers to be directed, has little ambition, wants security above all, and will only work under external coercion and control. These managers, he was at pains to show, accurately represented current management practice and were not merely an intellectual construct. In McGregor's view, so long as the assumptions of Theory X continued to underpin managers' actions they would fail to discover, let alone utilise, the huge potential of the average human being. This is because Theory X management deprives people of opportunities to satisfy the higher-level social needs of self-esteem and self-actualisation which had previously been identified by Maslow. He went on to argue that the problem with Theory X assumptions was that they attributed uncooperative and antagonistic behaviour in organisations to human nature, whereas it was in fact a natural human response to the deprivation of social needs, which its advocates induced through their managerial style.

These criticisms led McGregor to develop an alternative set of assumptions about people, which he termed Theory Y. He argued that such assumptions would lead to a more employee-centred style of management capable of fully exploiting the creative and productive potential of people. These assumptions were in stark contrast to those of Theory X and were that the expenditure of mental and physical effort in work is as natural as play and rest; that the average human being was not necessarily averse to work but would see it as a source of reward or punishment depending upon controllable conditions; that control and threats of punishment are not the only way to bring about dedication to organisational goals – indeed, people will exercise self-direction and control under proper conditions and can learn not only to accept but to seek responsibility; and that the capacity to exercise imagination, ingenuity and creativity is relatively widely, not narrowly, distributed in the population and is typically only partially utilised. Theory Y offers sharply different implications for managers, suggesting a strong possibility for human growth and development within organisations and a considerable degree of potential creativity, which could be tapped by thoughtful and sensitive managerial strategies that aligned individual and organisational goals. As Culp and Smith (1992) argue, if you treat people as if they were what they ought to be you will help them to become what they are capable of being. In this sense, the employee-centred style places the responsibility for efficiency squarely in the manager's lap and any blame for an uncooperative workforce on management's methods of organisation and control. This is a bitter pill for managers to swallow and takes courage to accept.

Contemporary management theory

The importance of the human relations movement was not realised until the 1970s, when a consensus was reached that attention to the psychological well-being of workers is also in the interests of good business. This led to a period of employee-centred management which emphasised open communication and teamwork via the introduction into organisations of job-enlargement and participatory decision-making schemes. These are known within the HRM literature today as *job enrichment* and *empowerment*, both of which will be explored later in this book. The aim was to help workers achieve their higher-level needs, and managerial effectiveness became measured by the degree to which a manager's responsibility was accepted by subordinates.

Continued research in this area has led to the recognition of informality within organisations, the acceptance of instability, conflict and change, the recognition of the need for adaptive qualities, and the realisation that the inflexible imposition of rules and procedures can be counterproductive, particularly during times of change and uncertainty. This has been a major advance in management thinking, which continues to shape contemporary theory in mainstream management, and in construction management to a lesser extent. Essentially, contemporary management theories conceive organisations as irregular, ambiguous, self-organising social systems comprising a multitude of interdependent people with varied and changing interests (Tsouskas 1995). These people attribute unique meanings and interpretations to their world and at best only partially respond to direction and managerial stimuli. In this sense, management is seen as a social activity of coordinating purposeful individuals who attribute unique ideas, meanings and interpretations to their worlds. This cultural variability must be taken into account, which means that management can no longer be seen as a mechanical activity designed to control some objective and static social scene. Managers should not treat people in isolation, and should move away from a reliance on predictive causal models which promise quick-fix solutions and grossly oversimplify the complexity and variability of people within organisations.

The organisation as a system

Whereas classical approaches emphasised the technical requirements of the organisation, excluding the people working within it, and the human relations approaches emphasised human psychological and social aspects of work, excluding the organisation, the systems approach attempts to reconcile both approaches by addressing the interrelationships of structure *and* behaviour (Mullins 1999). Ludwig Von Bertalanffy (1950) is often cited as the founder of this school. As a biologist, he used the approach to gain an insight into the functioning of the human body. From such specialised origins, the systems perspective has proved valuable in a wide variety of fields, including construction management. Through this perspective, an organisation is seen as a combination of interdependent parts or subsystems which collectively make up the whole. In metaphorical terms the organisation is seen as a living organism which has to adapt continually and naturally to changes in its environment in order to survive. The value of systems theory to the study of organisations is its ability to simplify complex situations by considering its subcomponents (subsystems). However, this is not done in isolation, and systems theory is particularly concerned with the relationships and interdependencies between these subsystems.

Closed and open systems

The *closed-systems* and *open-systems* perspectives have emerged as the two main strands of thought within systems theory, the latter emerging as the most popular, particularly within construction management. While the closed-system perspective conceives an organisation as independent of its environment, the open-systems perspective emphasises the interdependence between the organisation and its environment. An open system takes inputs/resources from its environment, transforms them in some way and then sends outputs back to that environment. A key aspect of open-systems theory is that an open system depends upon its environment for sustenance and must be responsive to it in order to survive. In construction it is a perspective which has proved very useful in explaining organisational effectiveness and for understanding why organisations fail.

The systems approach involves the study of the relationship between interdependent technical and social variables; changes in the technical system will impact on the social system and vice versa. From this the idea of the socio-technical system was developed and it can be seen to have clear relevance to the construction sector, where technological advances could bring about changes in the integration of groups and the sociological properties of working methods. While systems theory provided new insights into these issues, Mills (1967) warned that its strength, which is its ability to simplify complex situations, is also its primary weakness. He argues that the simplicity of such models can create a blinkered approach which underplays the way that human emotions, needs, aspirations and behaviours reverberate through the pattern of work. An example of the limitations of a systems approach is in controlling the risk inherent in complex systems. While a probabilistic, statistical approach can be used to determine the likelihood of failure of technical system components, the reliability of human beings who interact with the system is almost impossible to model. This criticism is important because the issue of management control is fundamentally a problem of human behaviour. Silverman (1970) also criticised systems theory from a metaphorical viewpoint, asserting that the perspective of organisations as independent living entities capable of their own actions is incorrect. Rather, '[o]rganisations do not react to their environment, their members do' (1970: 37). The emphasis must therefore be upon people not systems, a view reflected by Tsoukas (1995), who argues that the systems approach has the damaging effect of making managers seem independent of the people in their organisation.

Contingency theory

Contingency theory evolved from the open-systems perspective and refined it. Pioneers of this approach delved more deeply into the relationship between organisations and their environments in the belief that there was no one best way to organise for all situations. The contingency approach does not define any optimum state for an organisation, as it sees structure and success as being *contingent* upon the stability of an organisation's environment, the nature of its workers, the technology used in production and the routines of the tasks performed. In general, harder/scientific approaches are more effective when workers do not want autonomy and undertake routine tasks, using high-tech production-line-type technologies in a stable environment. Softer, more humanistic approaches work in the opposite conditions. This revelation has been a major advance in management theory and has guided much contemporary construction management research. This is particularly so in the area of procurement, where there have been many attempts to identify the determinants of the most suitable procurement system for differing types of project requirements.

In contingency terms, managers should seek to maximise the degree of *fit* between an organisation's internal structure and the demands of its environment. Problems occur when a major misfit develops between an organisation and its environment and thereby introduces the need for radical change. Logically, it follows that the causes of an imbalance can be either organisational (internal) or environmental (external) and that it can only be addressed by changing one or both of these components. The concept of fit led to considerable efforts being made to describe the organisational characteristics which suited different environmental conditions. For example, Burns and Stalker (1961) distinguished between organic and mechanistic forms of organisation. Organic structures are characterised by informal authority structures, free vertical and horizontal flows of information, flexible attitudes and a high commitment to task. In contrast, mechanistic structures emphasised vertical information flow, rule-bound behaviour, formal authority and little commitment to task. Burns and Stalker found that as the uncertainty of the environment increased, so did the appropriateness of the organic form. Until recently these ideas have had little influence on the construction industry, which slavishly followed a highly mechanistic model of management. The psychological and cultural origins of this management style are explained below, although it is important to note that there is still considerable resistance to move to an organic mode.

Other researchers, such as Lawrence and Lorsch (1967), considered the concepts of differentiation (divisive forces) and integration (cohesive forces). They found that to ensure efficiency, the level of differentiation should be a direct reflection of environmental complexity but also should be accompanied by an equivalent level of integrative effort. In these terms, it is evident that, while the construction industry has differentiated in response to increasing complexity (as seen in the growth of subcontracting and, more recently, professional diversity), it has struggled to provide the necessary levels of integration to counter its adverse effects (NEDO 1988; Latham 1994; Egan 1998). In theoretical terms, this is the challenge which faces the construction industry today.

Whilst contingency theory has reinforced and extended the idea that appropriate management style is a function of situational and organisational factors, it is not without criticism. For example, Stopford and Wells

(1972) highlighted the problem of structural change lagging behind contingency factors, arguing that an environment needs to become highly dynamic before a response is made. In other words, minor problems will be ignored until they reach crisis proportions and force a change in strategy. This phenomenon is often vividly illustrated in the realm of construction safety, where accident investigations often reveal that the early warning signs of accidents were ignored until it was too late (Loosemore 2000). A further problem with contingency theory is the confusion surrounding the exact relationships between structural and environmental factors. Indeed, Mintzberg (1979) questioned the reliability of some of the more abstract organisational concepts developed by the contingency school to explain this. For example, he argued that concepts such as decentralisation and participation are difficult to measure. This implies that the contingency perspective provides managers with little control over their destiny when faced with a crisis. More specifically, Booth (1996) is critical of its failure to consider adequately the way in which organisations can influence their environments. From a problem-solving point of view this is a major issue, in that environmental control and manipulation can substantially reduce or even reverse a crisis.

Finally, there has also been criticism of contingency theory from behaviouralists, who claim that the contingency approach is a step backwards towards the prescription of scientific management. This is because it is based on highlighting some rationality and causal order between organisational structure and success. Furthermore, it has the effect of making managers seem independent of their organisation, and thereby clouds the importance of the relationship between the manager and the organisation he or she manages. It is important to note that, as well as achieving a good fit between organisational structure and the business environment, it is also vital to achieve a fit between people and their organisation.

Conclusions

This chapter has provided a basic introduction to the development of mainstream management and HRM thought in the context in which it occurred. The various schools of thought and the controversies surrounding them have been discussed to provide a more informed perspective from which to assess their relative value. While management was once a science, it is now understood to be more of a craft or art form which rests heavily on a manager's understanding of human behaviour. This is reflected in the development of the broad classification of management theory through classical, human relations, systems and contingency approaches. This is a useful classification for the manager to understand in order to be able to gain an understanding of the nature of different approaches used within modern organisations. However, arguably the most important advance has been the realisation that *there is no one best way to manage*. In this sense, managers must assess the various approaches in terms of their appropriateness to the problems faced and the context in which they arise. This heuristic can also be applied to HRM approaches, as is discussed in the next chapter.

Discussion and review questions

1   Evaluate the role that management theory has to play in informing the practice of management in today's construction industry.
2   Explain the limitations of classical approaches to management theory in considering the HRM function of the modern construction organisation.
3   Critically assess the relevance of the contingency approach to the management of a large construction organisation. Identify the environmental influences that need to be taken into account in determining a suitable management approach.

---

3 Human resource management theory

Strategic concepts and operational implications

Organisational success is dependent upon the effective management of people. At the heart of this process is an organisation's HRM strategy, which should seek to align HRM practices with the wider organisational objectives. In this chapter we introduce strategic HRM (SHRM), examining the thinking underlying it and the impact it has on key operational issues concerned with implementing effective people-management practices.

Introduction

The application of theories of organisational behaviour to the context of people management has, in the last three decades, led to the development of various schools of thought, such as personnel management, industrial relations (IR), HRM and, more recently, SHRM. The broad function of personnel management has been in existence since the industrial revolution, where it fulfilled the welfare-officer role, supporting underprivileged factory workers through the provision of various benefits (L. Hall and Torrington 1998). The personnel-management role continued to expand throughout the mid- to late twentieth century, to embrace the industrial relations function necessitated through the expansion of the trade union movement. However, it was not until the latter part of the twentieth century that a formal and specialist personnel function became necessary in organisations to embrace manpower planning and management-development needs. The personnel-management model dominated during the late 1970s and early 1980s, focusing on procedures and control, administration of employment contracts and job grades, and collective bargaining. However, despite a growing recognition that employee welfare was important to organisational effectiveness, the function had little strategic involvement until the late 1980s. During the 1980s organisations came to realise that people were a resource that needed proper management at a strategic level, and it was out of this realisation that the term 'human resource management', or HRM, emerged (Redman and Wilkinson 2001).

Today the role of the human resource manager is broader and more complex than that of the personnel manager. Notably, the function of human resource management is considered a core managerial function rather than a specialist support function, and it is considered good practice for anyone with managerial responsibility to be involved with employee-management issues (Cornelius 2001). This transition from specialist support function to mainstream management activity can be seen to have occurred in many modern organisations and has paralleled the wider acceptance of HRM as a way of securing competitive advantage.

Defining HRM

The 1990s has witnessed a gradual redefinition of the HRM function from a personnel welfare advisory role to a performance-driven core management activity. However, a precise definition of HRM is difficult since the issue has been subject to considerable debate within management literature and no one has yet provided a single, authoritative definition of what the concept embodies. This may be due in part to the uncertainty of whether the change in title from personnel management to HRM has necessarily coincided with an actual change in the way that organisations manage people. Consequently, there has been a great deal of debate amongst leading researchers and practitioners as to what the distinction between the terms is and what this actually means in practice.

In essence, proponents of HRM contend that, whereas personnel management is workforce centred, and therefore directed at employees' needs, HRM is resource centred, and therefore aimed at meeting management's human resource needs. In this context, Legge (1989) suggests that HRM emphasises the development of the management team (with line management having responsibility for employee-related issues) and has a more strategic emphasis (i.e. is considered as a senior management activity). Storey (1993) takes this further by identifying four key aspects that categorise the differences between HRM and personnel management. These concern *beliefs and assumptions, strategic aspects, line management* and *key levers*, and are subdivided under 27 separate headings in Table 3.1.

Storey's analysis draws a distinction between the traditional and contemporary views of the people-management function, and distinguishes HRM as a process which aligns the needs of the organisation and those of the employee. By placing people-management activities at the centre of organisational strategy, it raises the profile and importance of people and human resource managers as a key competitive resource for an organisation. Thus, an appropriate definition of HRM could be:

A managerial perspective, with theoretical and prescriptive dimensions, which argues for the need to establish an integrated series of personnel policies consistent with organization strategy, thus ensuring quality of working life, high commitment and performance from employees, and organizational effectiveness and competitive advantage.

(Huczynski and Buchanan 2001: 673)

Table 3.1   Key differences between HRM and personnel management

No.	Dimension	Personnel Management and IR	HRM
(A) Beliefs and assumptions			
1	Contract	Careful delineation of written contracts	Aim to go 'beyond contract'
2	Rules	Importance of devising clear rules/mutuality	'Can-so' outlook; impatience with 'rule'
3	Guide to management action	Procedures	'Business-need'
4	Behaviour referent	Norms/custom and practice	Values/mission
5	Managerial task vis-à-vis labour	Monitoring	Nurturing
6	Nature of relations	Pluralist (groups)	Unitarist (individuals)
7	Conflict	Institutionalised	De-emphasised
(B) Strategic aspects			
8	Key relations	Labour–management	Customer
9	Initiatives	Piecemeal	Integrated
10	Corporate plan	Marginal to	Central to
11	Speed of decision	Slow	Fast
(C) Line management			
12	Management role	Transactional	Transformational
13	Key managers	Personnel/IR specialists	General/business/line managers
14	Communication	Indirect	Direct
15	Standardisation	High	Low
16	Prized management skills	Negotiation	Facilitation
(D) Key levers			
17	Selection	Separate, marginal task	Integrated, key task
18	Pay	Job evaluation (fixed grades)	Performance related
19	Conditions	Separately negotiated	Harmonisation
20	Labour–management	Collective bargaining contracts	Towards individual contracts
21	Thrust of relations with stewards	Regularised through facilities and training	Marginalised (with the exception of bargaining for change models)
22	Job categories and grades	Many	Few
23	Communication	Restricted flow	Increased flow
24	Job design	Division of labour	Team work

Table 3.1 continued

25	Conflict handling	Reach temporary truces	Manage climate and culture
26	Training and development	Controlled access to courses	Learning companies
27	Focuses of attention for interventions	Personnel procedures	Wide-ranging cultural, structural and personnel strategies

Source: Storey (1993) 'The take-up of human resource management by mainstream companies: key lessons from research', *The International Journal of Human Resource Management*, Vol. 4(3), 529–53. Published by Taylor & Francis Ltd (http://www.tandf.co.uk/journals).

This definition acknowledges the need to bring together organisational and employee needs through the HRM function. It suggests that any failure to align these needs, which are often conflicting, will lead to a breakdown in employer–employee relations and hence will threaten the future growth of an organisation. This contemporary view of HRM as a core enabler of the achievement of corporate goals has raised its profile to that of a strategic imperative, which has implications for all business processes within an organisation. This may explain the increasingly high profile of HRM issues within modern organisations.

A critical examination of the HRM concept

Despite numerous studies into the nature of HRM and what it represents, it still remains a widely criticised and ambiguous concept. Most importantly, its contribution to organisational performance remains unclear and is not well understood. Indeed, many critics of HRM see nothing new in the concept, seeing it as a cynical new labelling exercise to refresh the flagging fortunes of personnel managers, unnecessarily elevating their status to boardroom level. The criticisms of the HRM concept are broadly that:

- it is an ambiguous term more symptomatic of portraying an image than in delivering real benefits to organisational performance;
- it treats conflict within organisations in a simplistic manner, since the concept relies on a unitarist conformity and commitment to organisational goals and values which may not exist in reality;
- the ordered HRM models are incompatible with the complex, emergent realities of organisational strategy.

Critics of the HRM concept suggest that, rather than adding value to the business through its strategic integration with managerial objectives, the reality is that HRM can remain a disappointingly mechanistic function. They suggest that the theory of HRM represents a false and unobtainable image for personnel managers to aspire to, because aligning so many competing needs within a single approach is bound to be problematic. Furthermore, they suggest that HRM is merely a 'dressing up' of the personnel function that masks the realities of what remains a market-driven, organisation-centred activity.

The argument surrounding personnel v. HRM has been polarised by the debate surrounding the 'hard' and 'soft' views of HRM. The hard view of HRM focuses on the resource (or cost) dimension, whilst the soft view emphasises the human-input dimension. However, Sisson (1994) argues that the language of soft HRM can be used to mask what is in reality a hard HRM approach. Below Sisson's rather cynical analysis of the rhetoric embodied by the soft HRM concept is shown against the hard HRM realities that reflect the approaches of organisations managing the function today.

Table 3.2   Soft HRM rhetoric hiding hard HRM reality in the construction industry

Soft HRM rhetoric	*Hard HRM reality*
Employees first	Market pressures
Efficient production	Lean production
Flexibility	Re-engineering, scientific management
Core and periphery	Outsourcing, reducing commitments
Devolution	Delayering/reducing middle management
Right-sizing	Redundancy/downsizing
New working patterns	Part-time instead of full-time jobs
Empowerment	Devolving risk and responsibility
Training and development	Multiskilling, doing more with less
Employability	No employment security
Recognising individual contributions	Undermining trade union bargaining
Teamworking	Reducing the individual's discretion

Source: Adapted from Sisson (1994)

Although this could be seen as a somewhat disparaging view of HRM, it probably reflects the approach taken towards people management by many modern businesses in the construction industry (Druker *et al.* 1996). For example, increasing salary levels has traditionally been used as a key retention strategy, despite its ineffectiveness in securing the long-term commitment of employees (see Cargill 1996; Knutt 1997c). If the hard HRM approach reflects reality, then the construction industry has a long way to go in achieving what proponents of contemporary HRM want to achieve in the management of employee/management relationships.

As with all debates in management, there is no wrong or right answer and individuals must make up their own mind as to the value of the HRM approach in their own specific context. However, in developing and implementing such strategies it is important to remain cognisant of the fact that many people within organisations, both managers and their employees, remain sceptical of the new HRM function and its ability to contribute to business objectives. As will be discussed later in this book, for many managers in construction the HRM function is not seen as a strategic-level activity and integral to the direction and growth of the business, but as a necessary liability that must be tolerated. This presents a considerable cultural challenge to those trying to implement innovative HRM approaches in the construction industry.

Strategic HRM

One of the aspects that defines the transition from 'personnel' management to HRM is the need to integrate HR planning within the strategy of the organisation. Considering HRM as a strategic function rests on the belief that an organisation's human assets offer it a sustainable source of competitive advantage. Indeed, some take the radical view that SHRM offers organisations the main source of competitive advantage in the long term. So what precisely do we mean by SHRM? Armstrong suggests that SHRM:

> is concerned with the development and implementation of people strategies which are integrated with corporate strategies and ensure that the culture, values and structure of the organisation and the quality, motivation and commitment of its members contribute fully to the achievement of its goals.
>
> (Armstrong 1991: 81)

SHRM comprises a set of practices designed to maximise organisational integration, employee commitment, flexibility and quality of work (Guest 1987). It differs from traditional HRM in its emphasis on relationships between people, structures, strategy and the environment external to the organisation (Fombrun *et al.* 1984). Adopting SHRM is a clear acknowledgement that corporate objectives and human resource objectives are inexorably linked, rather than in conflict. A prerequisite of success is widespread acceptance throughout an organisation of the importance of people and their contribution to corporate goals. Given the relative importance of people in the construction industry when compared to many other high-tech industries, the SHRM philosophy is advocated in this book. The enormity of the cultural shift which is needed to achieve this will become evident as this book develops, and the issue of culture and how to change it will be a recurring theme. Nevertheless, from this point onwards the aim of this book is to provide a comprehensive framework and a set of techniques for construction companies to align their people-management function with corporate objectives. Ironically, for an industry driven by economic rationalism, efficiency and productivity there is more to gain than most in managing human resources effectively. However, before progressing we need to understand more about current SHRM thinking and this is discussed in more detail below.

Models of SHRM

In recent years many competing models of SHRM have been developed, which have all tried to capture the ways in which organisations can align HRM practices with their wider strategic objectives. Just as it is difficult to define a generally accepted definition for HRM and its role within organisations, there is no universally agreed model of how the SHRM function should operate. However, there are several models which gained widespread acceptance throughout the 1990s. These are the Michigan, Harvard and Warwick models of SHRM, which have become known by the locations of the researchers who developed them. Each of these models is explained below and their practical implications discussed.

The Michigan Model

The Michigan Model of SHRM is shown in Figure 3.1. It approaches the SHRM function in a rather dispassionate manner, treating people like any other resource that should be managed in such a way as to maximise utility whilst minimising cost. The Michigan Model emphasises the interaction of functional aspects of the SHRM role, such as selection, appraisal, rewards and development. These must be linked to (or aligned with) the strategy of the organisation in order that SHRM practices support the strategic direction of the organisation.

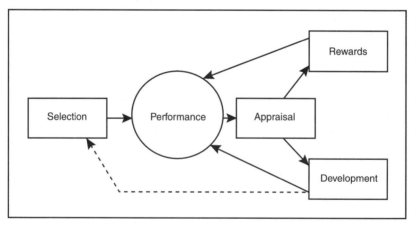

Figure 3.1   The Michigan Model of HRM

Source: Devanna, M.A., Fombrun, C.J. and Tichy, N.M. in C.J. Fombrun, N.M Tichy and M.A. Devanna (eds) copyright © 1994, John Wiley, New York. This material is used by permission of John Wiley & Sons, Inc.

Recently the applicability and utility of this model have been criticised as being limited in the context of modern organisations because it treats the SHRM function as a closed system, ignoring the hostile business environments in which many companies operate. Thus, just as the contemporary view of organisations described in Chapter 2 sees their operation as being contingent upon external (or environmental) influences, SHRM strategies should be capable of taking into account these external factors in the management of people. In practical terms, this means that for organisations operating in highly dynamic environments there would be little point in developing SHRM policies which emphasise adherence to strict protocols and procedures. In order to be aligned with the need for flexibility, any strategy must allow for change and some degree of control by line management (this is discussed in detail in Chapter 4).

The Harvard Model

The Harvard School has made a major contribution to the development of SHRM by providing a useful open-systems model of how SHRM policy influences other organisational functions and is constrained by stakeholder interests and situational factors (see Beer *et al.* 1984). Stakeholders are people who are influenced by or can influence the operations or outputs of an organisation (Anthony *et al.* 1996). They could include senior managers, employees, shareholders, external pressure groups, customers and suppliers. The Harvard School Model is illustrated in Figure 3.2 and provides an important link between SHRM decisions, the business environment and an organisation's performance (Huczynski and Buchanan 2001: 678).

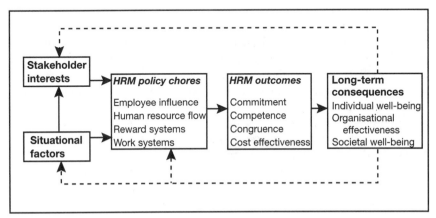

Figure 3.2 The Harvard Model of HRM

Source: Beer *et al.* 1984; reprinted with the permission of the Free Press, an imprint of Simon & Schuster Adult Publishing Group, from *Managing Human Assets*, M. Beer, B. Spector, P.R. Lawrence, R.Q. Mills and R.E. Walton, © 1984 by The Free Press.

The ability to take account of situational and stakeholder interests is particularly important in construction projects, which employ a wide range of interest groups and often have a major impact upon the general public. Here the mutual interdependence of those involved, those affected and those who can influence the project's outcome must be taken into account through appropriate SHRM policy decisions. For example, every project will require decisions to be taken as to whether to employ project team members directly or to hire in agency staff for the duration of the work. This decision will be contingent upon both the wider situational factors (such as the state of the internal and external labour market) and the wider interests of the organisational stakeholders (the employees already employed and those both internal and external to the organisation who stand to be affected by the project's success). These factors should inform and influence SHRM policy decisions, which in turn will define the success or otherwise of the approach adopted.

Although the Harvard Model has had an enduring influence on SHRM developments, it also has shortcomings. For example, although it acknowledges environmental and stakeholder influences, the nature of the causal chain suggested by the model is unclear. In other words, it does not explain how the four policy areas of employee influence, human resource flow, reward systems and work systems are influenced by the environmental/ stakeholder factors and how they affect SHRM outcomes. Thus it has limitations in terms of explaining how HRM should be considered as a strategic function.

The Warwick Model

A problem for researchers considering the strategic role of SHRM is that the majority of the accepted models were developed within an American context. Approaches outside of the USA demand a perspective on the SHRM function that reflects the particular cultural context that exists in different countries. Thus a third important SHRM model is the Warwick Model (see Hendry and Pettigrew 1990). This model, shown in Figure 3.3, differs from the others discussed above in reflecting European traditions and management styles. The model comprises five interrelated elements, which allow an analysis to be made of how external factors impact upon the internal operations of an organisation. The model recognises the wider context in which SHRM operates, and emphasises the full range of tasks and skills that define HRM as a strategic function.

The main contribution of the Warwick Model is that it incorporates culture and business outputs into the SHRM framework. Each box within the model reflects a particular context within which the organisation operates and shows how strategic change impacts on the SHRM function. The Warwick Model encompasses a wide range of interrelated, complex tasks, each taking place within the wider internal and external contexts of the organisation. This implies that SHRM should be viewed as a strategic func-

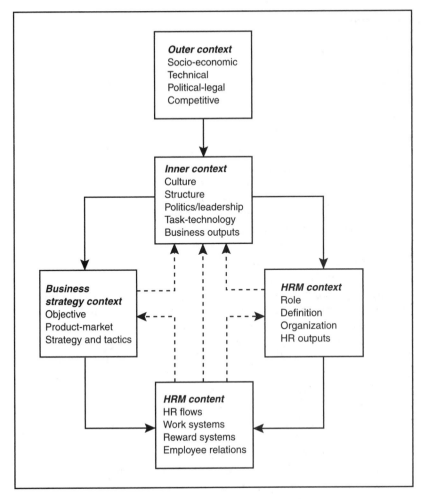

Figure 3.3 The Warwick Model of strategic change and HRM

Source: Hendry and Pettigrew (1990), 'Human resource management: an agenda for the 1990s', *The International Journal of Human Resource Management*, Vol. 1 (1), 17–43, Taylor & Francis (http://www.tandf.co.uk/journals).

tion and suggests the need for effective SHRM approaches to engage with the 'soft' elements of SHRM if business strategy objectives are to be met. These will form recurring themes throughout the remainder of this book.

Core components of SHRM

Regardless of which SHRM model best reflects the realities of managing the function within the context of modern organisations, it is clear that SHRM strategy must align with and support business strategy if it is to be seen as successful. This recognition of the wider context in which the

SHRM function must operate, together with the acknowledgement of the interdependence of business strategy with SHRM, is what differentiates contemporary perspectives of the function from the more mechanistic personnel role. According to Anthony *et al.* (1996), SHRM has six key characteristics:

- *It recognises the outside environment*: this comprises a set of opportunities and threats to the organisation that must be recognised and taken account of by the strategic decision-making process. They can include social, demographic and labour-market changes, legislation, economic conditions, technology, political forces, etc. All of these factors can impact on an organisation's ability to recruit, develop and retain people who will take the organisation forward.
- *It recognises competition and labour-market dynamics*: these affect wage/benefits levels, unemployment rates and working conditions, and define the necessary conditions that an organisation must provide to remain competitive in the labour market.
- *It has a long-range focus*: a strategic focus implies that consideration is given to the long-range direction and objectives of the organisation. This will depend on the management philosophy of the organisation regarding where it wants to position itself.
- *It has a decision-making focus*: this means that the organisation consciously chooses to direct and commit its human resources in a particular direction.
- *It considers all stakeholders*: a strategic approach demands that the organisation take account of the views and interests of all stakeholders, internal and external.
- *It is integrated with corporate strategy*: perhaps the most important characteristic is that HR strategy should be integrated with the firm's overall corporate strategy. For example, if a company sets out to grow rapidly and to dominate a particular market, then the strategy should be rapidly to acquire new human resources with the necessary skills in order to achieve that goal.

By satisfying these criteria organisations can ensure that HRM policy makes a strategic contribution to the achievement of its business goals. The first step in developing an SHRM plan should be to clarify an organisation's short-term, medium-term and long-term direction. Developing policies to achieve these goals then involves analysing an organisation's human resource strengths, weaknesses, opportunities and threats (a SWOT analysis) and identifying where changes need to be made. Armstrong (1991) defines these areas as:

- *Culture change*: developing an appropriate organisational culture which is closely aligned with corporate goals;

- *Organisation design*: developing the organisational roles and relationships to take account of new developments;
- *Organisational effectiveness*: developing organisational capabilities in terms of teamwork, communication, productivity and customer service, and improving the ability to manage change;
- *Resourcing*: recruiting, training and developing the people required to achieve the organisation's strategic objectives;
- *Performance management*: using performance appraisal to monitor the performance of employees;
- *Reward management*: developing compensation mechanisms which convey to employees the organisation's values and performance expectations by rewarding appropriate levels of performance in these areas;
- *Motivation*: developing an intrinsically and extrinsically motivated workforce through financial and non-financial rewards;
- *Commitment*: developing a feeling of 'mutuality' within the organisation where the needs of individuals are integrated with those of the organisation;
- *Employee relations*: developing strategies for reducing conflict between employees and management, and increasing cooperation between different groups within the organisation
- *Flexibility*: developing a structure, systems and techniques which allow the organisation to respond flexibly to change.

The operation and implementation of effective SHRM practices in construction

If strategic involvement is taken as the factor differentiating personnel management and SHRM, then the evidence indicates that few construction companies take a truly long-term view and adopt SHRM policies (Druker *et al.* 1996). Rather, they adopt fairly reactive approaches to staffing and managing people. Before exploring the ways in which construction companies should approach the SHRM function, it is necessary first to establish the various activities that are encompassed within it. These activities need not be within the domain of the human resource manager – the reality for many construction companies is that they will not employ such a person, and even when they do the vast majority of people-management tasks will still be devolved to line managers working in the project environment. The problem when there is not a separate and formal SHRM role is that these disparate activities will remain so. We will discuss this further in Chapters 4 and 5. Nevertheless, in this section we briefly explore the generic aspects of the HRM function which need to be considered by those given power and responsibility for the management of people within organisations. This will act as a precursor to the more specific examination of how the strategic and operational aspects are managed in construction in Chapters 4 and 5.

The achievement of SHRM priorities is dependent on an organisation creating the right structural and cultural conditions for them to take place. According to Mullins (1999), this involves:

- designing an effective organisation structure;
- staffing the structure with suitable people;
- managing the employment relationship effectively.

Collectively, these should ensure the creation of positive and productive human relationships and high levels of motivation, commitment and morale. The challenge in each of these areas is to maintain a balance between organisational goals and the well-being of employees, as is discussed briefly below.

Designing an effective organisation structure

An effective SHRM policy can only take place within an effectively designed organisational structure. This can be viewed at two levels:

- the overall operation of the organisation in terms of how it manages and distributes work amongst various employee groups and functions in pursuance of the strategic goals;
- the design of the organisation in terms of the hierarchies, roles and relationships.

The organisation of work has been the subject of constant debates since the early work of Frederick Taylor (see Chapter 2). However, since the 1980s debate has centred around the need for organisations to retain a degree of flexibility in terms of both the numbers of people that they employ and the functional skills that employees provide. This principle was developed by Atkinson (1984a) through his 'flexible firm' model (described in detail in Chapter 4 in relation to how large construction companies manage their employees in a way which allows them to take account of dynamic and changing markets). The process of defining work roles and relationships can be summarised under the concept of job design, which seeks to specify the contents, methods and relationships of particular roles in order to satisfy both organisational requirements and the needs of the individual job-holder. This means that the function of designing and specifying job roles must ensure that fulfilling and motivating positions are developed which accord with the strategic objectives of the organisation.

Armstrong (1991) identifies four factors which are important to effective job design: the process of intrinsic motivation; the characteristics of task structure; motivating characteristics of jobs; and the implications of group activities. In order to reconcile organisational and individual needs in relation to these factors, techniques such as job rotation, job enlargement, enrichment and empowerment can be used. Job design can also be used as an important tool in engendering organisational change (Torrington and Hall 1995). However, this is contingent on the quality of planning and preparation and the level of integration with other organisation structures, as will be explored in Chapter 4.

The physical shape of the organisation, or its design, will be contingent upon many factors, both internal and external. The perspective of contingency theory (explained in Chapter 2) suggests that there is not a single ideal structural design that will apply to an organisation, but that there will be a range of choices that will be contingent upon the situation being analysed. This means that managers need to understand their business environment and what structure best fits it at any point in time. Since different parts of an organisation manage different elements of an organisation's business environment, it is natural to find different organisational forms within the same organisation. The challenge is in their smooth integration. Contemporary construction companies tend to adopt fairly fluid structural forms which allow them to cope with competing functional and project demands. The organisation structures typically used by construction companies are discussed in depth in Chapter 4.

Staffing the structure with suitable people

This staffing function is also known as 'employee resourcing' and forms one of the most challenging aspects of the SHRM function. The major components of employee resourcing are recruitment and selection, deployment and team formation, performance management, retention and training, career development, dismissal and redundancy (Taylor 1998).

Employee resourcing activities aim to ensure that the right numbers of employees with the right skills and competencies are in the right place at the right time. This inevitably results in a balancing act, in which managers have to consider longer-term strategic considerations while providing immediate solutions for the shorter-term operational issues (Beardwell and Holden 1997). Thus, the resourcing function embodies much of what many people would traditionally consider as forming the core aspects of SHRM.

Managing the employment relationship

It is clear from the preceding discussion that there is a need for the organisation to take great care in balancing organisational and individual employee needs if it is to successfully fulfil its strategic objectives. However, a problem here is the trend amongst many modern businesses to devolve responsibility for operational aspects of SHRM to line managers. This is particularly the case in geographically dispersed project-based industries such as construction, where line managers have the responsibility for many of the day-to-day aspects of the SHRM function. This inherently leads to a loss of direct control of the SHRM function and a reliance on the adherence of autonomous line managers to desired policies and practices. In this situation line managers have a particularly important responsibility to balance their staff needs and those of the organisation. The maintenance of this relationship forms a cornerstone of many of the approaches advocated within this book for managing people within the construction project environment.

Psychological contracts

Whilst formal employment contracts can define many aspects of the employee–employer relationship, they cannot delineate every aspect, and socially constructed expectations and obligations fill the gaps that are left. These less formal expectations are known as *psychological contracts*, which describe the beliefs of each party as to their mutual obligations within the employment relationship (Herriot 1998). The implication of psychological contracts is that, in addition to the 'hard' areas of the employment contract that have to be met, a 'soft' set of expectations held by the individual also have to be organised and managed.

Acknowledging the existence of psychological contracts is important in terms of understanding employee relations, since it allows employment contracts to be seen as a two-way exchange process rather than as being imposed by employers (see Herriot and Pemberton 1997). Issues covered in a psychological contract include individual differences, interpersonal interaction, motivation, leadership and management style, group/team dynamics, change and empowerment (Makin *et al.* 1996). From a functional perspective, psychological contracts accomplish two tasks: they help to predict the kinds of outputs which employers will get from employees, and what kind of rewards the employee will get from investing time and effort in the organisation. Psychological contracts are extremely powerful if used effectively but extremely dangerous if abused. A breach, break or violation of the psychological contract is a serious issue from which it is difficult to recover, because it will result in reduced trust and goodwill – both of which are essential to the effective functioning of organisations. A *breach* of a psychological contract is where the employee sees their organisation as failing to meet one or more of their obligations. Consequently, understanding the psychological contract, and the obligations that employees see as being owed by their employers, is fundamental in understanding the causal factors behind employee turnover.

Within the relationship defined by the psychological contract both the employer and the employee inform, negotiate, monitor and then renegotiate or exit the employment relationship. As such, psychological contracts represent a reciprocal and dynamic deal, which evolves as the employer's commitment changes (Sparrow and Marchington 1998). According to Rousseau (1995), the content of psychological contracts can be anywhere along a continuum bounded by two distinct types: relational contracts (long-term, open-ended relationships within unitary organisations which lead to the exchange of loyalty, trust and support); and transactional contracts (short-term relationships set in pluralistic organisational contexts and characterised by mutual self-interest). Regardless of where psychological contracts fit within this continuum, they should be seen as interactive and dynamic (Herriot and Pemberton 1997). Employers should not assume that they will remain static in the context of continually changing employment relationships.

Conclusions

This chapter has provided an introduction to the concept of SHRM. We have introduced some of the relevant aspects of SHRM, examining both the thinking underlying SHRM decision-making and the impact that this has on the key operational issues concerned with implementing effective people-management practices. These have been explored through the critical examination of popular models of the strategic HRM function. All the models are controversial and widely debated in terms of their representation of reality. Nevertheless, they do highlight the difficulties inherent in trying to take account of all of the relevant factors that influence effective SHRM practices.

SHRM embraces the issues of workforce management and control through the management of flexibility, workforce empowerment and diversity. This trend emphasises the individualisation of the employment contract, providing a focus on the individual employee's psychological contract instead of traditional IR issues. Although this has improved our understanding of the dynamics of how SHRM policies impact on the performance of those working for the organisation, many SHRM objectives become problematic in the dynamic context of project-based sectors. The particular difficulties inherent within the construction industry and the approaches that can be taken to mitigate these difficulties are discussed gradually in subsequent chapters.

Discussion and review questions

1 List and discuss the possible influences on the HRM strategy of large construction companies operating within the current volatile economic climate.

2 Using Storey's table of the key differences between HRM and personnel management concepts (pp. 34–5), evaluate the approach

of an organisation with which you are familiar and discuss to what extent it has embraced the new HRM orthodoxy.
3   Discuss the particular difficulties faced by construction firms trying to ensure that their employees' psychological contract needs are met by the organisation.

4 Strategic approaches to managing human resources in the construction industry

In this chapter we will explore how construction companies and their managers approach the HR function from a strategic perspective. Initially, we discuss the basis of SHRM thinking applied to construction and identify approaches that can be adopted when formulating an SHRM strategy. Next, we examine the ways in which construction organisations have developed and structured themselves in order to cope with the inherent difficulties of operating in such a dynamic and competitive environment. Finally, we explore the alternative strategies that can be adopted by individual managers in the management of people and explore the impact that these can have on operational efficiency.

Introduction

Organisations operate within dynamic environments and are affected by both external and internal forces which constantly impact on their competitiveness and ability to profit from their commercial environment (Anthony *et al.* 1996: 3). Managers must constantly monitor, review and adapt their business strategies to ensure that they can respond to these forces effectively. Any change in an organisation's strategy will demand parallel changes to its structure, processes, physical resources and human resources. Of these factors, the HRM function in particular is crucial for coping with change, as it transcends each of these areas. For this reason, construction firms of all sizes have to develop flexible and strategic approaches to HRM in order to manage their day-to-day activities effectively.

Developing an SHRM strategy

Thinking strategically about HRM demands that an organisation look beyond the here and now to consider the external and long-term factors likely to impinge upon its business over the next few years. Anthony *et al.* (1996) suggest six key characteristics of a strategic HRM approach which provide a framework of requirements for SHRM formulation, and we discuss them in relation to construction below.

Recognising and responding to the environment

Central to strategy formulation is a recognition that construction companies do not operate within a vacuum and that the outside environments present opportunities and threats to the future development of the business. The role of the HRM strategy is to capitalise on the opportunities and to mitigate the threats through its people-management policies. For example, a construction company may see an opportunity in government-sponsored infrastructure projects over the next five years. By exploring its human resource capabilities it may find that it has an overcapacity of skills in general building construction, but not enough in heavy civil engineering. By proactively retraining its managers and operatives and recruiting people with the requisite skills it can put itself in a position to exploit the new market opportunities.

Recognising and responding to labour-market dynamics

The labour market is as competitive as the commercial market, which means that attracting, rewarding, deploying and retaining people should be a primary focus within an SHRM strategy. In construction this is a particularly complex issue as the workforce is itinerant and the industry relies on a wide range of different skills, ranging from craft-operative to support services and production-management functions. To be effective, the organisation must develop a strategy that responds to the labour-market situation in which it operates. For example, a construction company may recognise a future national shortage of quantity surveyors and seek to address this potential shortage by reviewing its recruitment and training activities, perhaps by sponsoring students or actively recruiting surveyors through targeted campaigns. Alternatively, it could retrain some of its other staff in surveying skills to offset the shortage.

Considering all organisational personnel

Every individual working for the organisation must be considered as part of the overall strategy, regardless of their gender, race, physical ability or seniority. Workforce diversity is a strength which should be planned and managed effectively to harness its productive potential. However, this is not easy and the problems of ensuring equal opportunities for all are discussed in more detail in Chapter 8. Nevertheless, a construction company must make appropriate adjustments for the different tasks that people perform within the organisation. The key to avoiding any potential problems is to do this sensitively and fairly and with an awareness of the market for each unit of labour. Considering employees as a homogeneous group is not possible, as well as being unwise and unnecessary, and will mean that the organisation will operate dysfunctionally. By considering the labour market for specific professions and how this is changing over a five-year period, the competitiveness of the pay levels should be ensured. Similarly, if particular individuals are identified as necessary to the future development of the organisation, then it may be appropriate to consider remunerating them at a higher level than some of their colleagues in similar positions.

Taking a long-range view

Taking a three- to five-year view of the HRM situation is vital if the organisation is to cope with the external environment and labour market. The problem in construction is that the cyclical nature of the market makes this difficult and engenders a short-term view on almost all business functions. Furthermore, the traditional competitive tendering system creates so much workload uncertainty that it would be foolish to recruit until a project has been won. Finally, the devolution of HRM responsibilities to project level in many companies does nothing to engender a long-term attitude. To most project managers, the furthest they need to think, or are expected to think, is the end of the project. Nevertheless, despite these barriers to SHRM a company needs to take a long-term perspective which overarches individual projects if it is to staff itself proactively rather than reactively and thereby secure the most talented labour force available. For example, a construction company may identify a strategic priority as being to move into a new market over a five-year period, which would highlight the need to start recruiting and retraining staff with new skills now. This strategy does not dictate how the organisation is to achieve this target, but provides a long-term vision and overall direction that will influence future recruitment, training and development activities.

Focusing on choice and decision-making

Strategising involves looking into the future and planning the best way forward from a range of alternative options. When considering HRM strategy, a choice has to be made about the future direction of an organisation and how its employees will help it to achieve this goal. For example, an organisation may look at three distinct market opportunities – housing refurbishment, public new-build housing and private new-build housing. A strategic decision cannot be made on market opportunity alone, as it often is, but must also consider the available skills and resources to take advantage of it. Too often companies decide to enter a market without having fully considered the HRM implications of doing so. The result is workforce intransigence and extra pressure and stress for those who have to shoulder the burden of a lack of human resources. If the organisation is badly placed from an HR perspective to exploit the most lucrative market opportunity, then it must consider in advance whether it should refocus on an alternative market or reorient its workforce to exploit the opportunity. This emphasises that all strategic decisions involve a degree of choice and compromise, but it is the quality of these decisions that will ultimately dictate the success of the business.

Integrating SHRM with overall corporate strategy

The final requirement of SHRM formulation is that it aligns with the firm's corporate strategy. For example, if a construction company wants to build a reputation for quality rather than low cost, then it will need to begin changing the culture of its workplace. This may involve retraining staff, recruiting new staff who will champion the new philosophy, putting new reward systems in place, etc.

Formulating an SHRM strategy

Thinking strategically about the HRM function demands that SHRM activities are viewed in the context of where the organisation is and where it wishes to be in the future. Thus, as was briefly discussed in Chapter 3, any development of an SHRM approach must be preceded by an analysis of the alignment between an organisation's current position and its desired future direction. Some key questions to ask in such an analysis are:

- Is the corporate culture of the organisation compatible with the overall business direction and performance expectations of the modern industry context?
- Are the people currently employed by the organisation suitably skilled and flexible enough to cope with any forecasted change in the organisation's environmental position?
- Has the organisation taken account of external factors that could impinge on its ability to recruit and retain high-quality staff?

The answers to these questions should result in strategies to close any gap identified. Armstrong's (1991) HR strategy formulation matrix (Table 4.1) provides an analytical approach to help do this. This simple technique works by applying elements of SHRM (listed vertically on the left) to organisational objectives (listed horizontally along the top). Table 4.1 presents some of the issues facing organisations with the core strategic directions discussed in Chapter 3 as an example of how this technique can be used.

When completed, Table 4.1 will represent an HRM strategy to help close the gap between current and future positions. The next stage of the strategy formulation process is to devolve responsibility for each element of the HRM strategy to those most able to deliver it. Whilst some elements will be best managed centrally by senior managers or a dedicated HRM department, the majority are likely to be devolved to line managers, partic-

Table 4.1 HR strategy formulation matrix

	Market share expansion	*Increased emphasis on quality*	*Increased use of ICT*	*Etc.*
Organisation structure	Develop a fluid divisional structure and specialist sub-divisions for coping with niche markets	Product specialisation and defined roles for the quality manager	IT specialists within divisions to ensure the integration and alignment of organisational approaches	
Resource requirements	Specialist skills needs	Recruit quality managers	Retain and develop IT skills in existing staff	
Training and staff development	Customer care and business development	TQM training for all line management and project-based staff	Specialist IT training for operations staff	
Performance management	Rewards for repeat business through improved performance	Reward the achievement of quality targets and a zero-defects culture	Build in IT literacy to the performance management system as a core competency area	
Etc.				

Source: Adapted from Armstrong, 1991: 86

ularly where they have considerable autonomy for the management of employees, as they do in construction.

The final stage of strategy formulation is to monitor the impact that the implemented strategies have on delivering the organisation's objectives. This information must be fed back into the strategic planning process in order to shape, review and adjust strategies where necessary. In construction, the monitoring function is particularly important, as strategies are unlikely to remain static for long. It is likely that unpredictable workloads, workforces and projects will result in wildly fluctuating human resource requirements that necessitate continual action as part of the organisation's resourcing strategy. Needless to say, effective communication structures are also a prerequisite for effective strategy development, and they are explored later in this book.

Approaches to HRM in construction

In this section we will briefly explore some of the common strategic approaches to HRM used by construction companies to cope with the business environment in which they exist.

Construction companies as 'flexible firms'

Construction companies face a dilemma in having to maintain staffing levels that can comfortably deal with the cyclical demands of the industry's market whilst maintaining organisational growth and development. Given the high levels of competition and resulting low profit margins that characterise the industry's operation, avoiding workforce overcapacity has been a core priority for most construction companies over the last 30 years. It is inevitable, therefore, that most organisations have contracted outside labour to allow them to maintain flexibility within their workforce. In markets characterised by fluctuating workloads there are clear financial advantages in this approach. Savings for employers using contracted-out labour include National Insurance contributions, the administrative costs of making tax deductions from the employee, payments for sickness or holidays, etc. (Druker and White 1996a). Contractors who rely on labour-only subcontracting and the hiring of self-employed operatives also avoid the costs of training their direct employees. Whether this is an advantage is debatable, since well-trained labour who are aligned with an organisation's culture and strategy are likely to be more motivated and productive. However, perhaps the principal advantage to construction employers is the flexibility which contracted-out labour forces provide. In times of economic downturn or recession the organisation can quickly offload their indirect workforce and then rehire them when required. This prevents them employing non-productive staff between contracts, thereby ensuring the maximum possible output from their workforce.

A theoretical organisational framework for explaining how companies cope with employee resourcing in dynamic industrial environments is provided by Atkinson's (1984a) 'Flexible Firm' model (see Figure 4.1). This influential and widely debated illustration of flexibility within organisations represents an approach to workforce management that encompasses three types of flexibility: functional, numerical and financial. *Numerical flexibility* is the ability of an organisation rapidly to expand or contract to cope with fluctuating workload demands through the use of short-term contracts, subcontracted and outsourced labour. *Functional flexibility* refers to multiskilling and the ability of employees to switch between different tasks. *Financial flexibility* refers to flexible pay systems based on local conditions as opposed to nationally negotiated rates. Most construction firms are functionally and numerically flexible, although in some countries financial flexibility is constrained by collective bargaining, where nationally negotiated pay agreements define the remuneration for particular jobs.

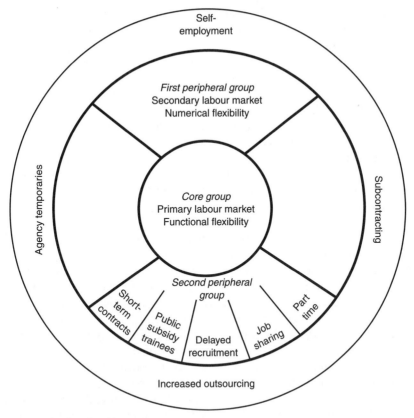

Figure 4.1 The Flexible Firm

Source: Developed from Atkinson (1984a). Reprinted with the permission of Institute for Employment Studies.

Atkinson's model divides workers according to whether they are in the 'core' or the 'periphery'. Core workers are drawn from the primary labour market, have permanent contracts and offer functional flexibility for the organisation. Peripheral workers do not have permanent contracts and provide more numerical and financial flexibility. Implicit within Atkinson's model is that different flexibilities are required in order for an organisation to cope with dynamic staffing environments. For example, a reduction in numerical flexibility places simultaneous demands on the core group to compensate through their functional flexibility. Hence, in times of labour shortages core staff may find themselves having to take on unfamiliar roles or supporting colleagues in areas that they are not used to dealing with. A similar scenario might be brought about by a reduction in financial flexibility. In other words, if wages rise to a level where taking on more staff is not possible, then existing workers will have to do more work. This provides a powerful way of understanding and defining people's roles and importance to an organisation.

Atkinson's is not the only model of how organisations can ensure flexibility in their approach towards employment and workforce maintenance. For example, Handy (1988) put forward the 'Shamrock' organisation. As with Atkinson's model, Handy's includes a professional core of key employees on permanent contracts who hold the information required for the key jobs. A second group comprise workers on shorter-term service contracts or subcontractors who support the core, followed by the third group, the part-time workers, who merely provide temporary support to cope with peaks and troughs in the demand cycle. Whilst there are advantages to employing large numbers within the latter two groups, there are also disadvantages in terms of the lack of commitment by staff towards the employer and the inability of part-timers to undertake significant coordinating responsibilities. Clearly, strategies have to be properly balanced to ensure that there are sufficient core workers to develop and manage the organisation.

As modern organisations now commonly adopt multiple and parallel forms of flexibility in response to pressures such as changing demographics, expectations, technologies and an increasingly uncertain business environment, Atkinson's model could be criticised for its inflexibility! The traditional forms of flexibility incorporated within the model (i.e. multiskilling, financial flexibility and time flexibility) have been supplemented by geographical flexibility, organisational flexibility and cognitive flexibility (Sparrow and Marchington 1998). All of these are necessary in the context of running a national, multidivisional construction business where staff are regularly redeployed in order to resource an ever-changing portfolio of projects and supporting functions. Furthermore, Atkinson's model is based on the assumption that professional and managerial employees form the core labour group who are retained as full-time employees. However, with the increasing acceptance of gender equity in society, professional jobs with flexibility are becoming more important. For example, in Australia 59 per cent of two-parent families have both parents in paid employment, and Australian statistics reveal that in dual-income couples 70 per cent of all mothers and 56 per cent of all fathers reported that they always/often felt rushed or pressed for time and could benefit from more flexible work arrangements (ABS 1999a). Not only is workforce participation changing, but so too are parents' expectations of their relationship with their children. In particular, fathers now expect to have a greater role in parenting. This means that even in the male-dominated construction industry work–life issues will become increasingly important in the future. For example, Russell and Bowman (2000) showed that fathers now spend more time with and are closer to their children than they were 15 years ago. However, 68 per cent of fathers said they did not spend enough time with their children, and 53 per cent felt that job and family interfered with each other. Interestingly, 57 per cent of fathers identified work-related barriers, such as expectations of longer hours and inflexibility, as being the critical factor preventing them from being the kind of father they would like to be. As the pressure from working parents mounts, companies will be forced to respond by introducing more flexible work practices or will ignore employees' requirements at their peril. It is increasingly likely that organisations will need to address these changing expectations in order to attract good people in their thirties. With reducing numbers of graduates in many construction professions, there is a strong argument for construction firms to provide flexible working arrangements as a means to retain valued 'core' managerial and professional employees. We will discuss work–life balance in more detail in Chapter 8.

While there are undoubtedly benefits to be gained from increased workforce flexibility, it is important to end this section on a note of caution. For example, the increased need for flexibility has contributed to a growing sense of job insecurity as organisations have undergone radical restructuring, delayering and downsizing (Sparrow and Marchington 1998). Job insecurity is known to be a source of job stress and burnout, which are, in turn, associated with diminished individual and organisational effectiveness. For example, Winch (1994) cites UK contractors' casual approach to workforce employment as a cause of the industry's poor safety record, whilst Debrah and Ofori (1997) see it as a primary cause of a lack of inter-trade cohesiveness amongst the workforce and of a lack of investment in developing trade skills within the industry. Thus, whilst flexibility may be an effective method of coping with fluctuating markets and dynamic staffing requirements, there remain many advantages to maintaining a large primary labour market within construction organisations. Indeed, many construction organisations are now recognising these dangers and increasing, once again, their numbers of core employees.

Organisational structures in construction

An organisation's structure defines the way in which roles, responsibilities, power, authority and control are allocated amongst its workforce. Without a structure the organisation would have no basis on which to control and organise its work and to manage the way in which it interfaces with other organisations in its supply chain. As we discussed in Chapter 2, contemporary organisation theorists have put forward the contingency approach to organisation design. This means that the design should take account of the organisation's size, technology, social and legal environments, markets, past history and culture. However, the main challenge of organisation design is to link together the various functions that constitute an organisation in such a way as to optimise the activities of the organisation whilst promoting its growth and development. Mintzberg (1979) divides these functions into five groups, namely: the *operating core* (those workers engaged in the main production activities of the organisation); the *strategic apex* (where the overall strategic management of the organisation occurs); the *middle line* (the line managers who effectively form the interface between the strategic apex and the operating core; the *technostructure* (specialists who support the main production function); and *support staff* (those who provide general and administrative support to those in production functions).

The most basic form of organisational structure which characterises the early life of most construction organisations is the craft structure (Naoum 2001: 89). These organisations are usually small and rely on an owner having centralised power and direct control over most functions. Organisational members are expected to have a multitude of skills, and operations are flexible in order to cope with a lack of resources and the inevitable change that smaller companies have to endure in response to market pressures and the potential for growing the business. HRM is highly personalised around relationships with the central figure. As a result of limited resources and a preoccupation with growth and survival, HRM strategies tend to be at best informal and at worst non-existent.

Greiner proposed a model (see Figure 4.2) of organisational evolution which shows how structures change as companies grow in size and which implies different approaches to managing human resources. He suggests that, initially, growth occurs through the creativity of the owner, who exercises entrepreneurial skill and ability. However, as companies grow they need more robust organisational hierarchies to cope with the increasing need for decentralisation and specialisation in operating practices. As Greiner's model suggests, growth results in a *crisis of leadership* in which

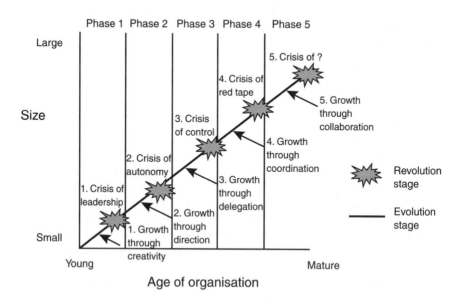

Figure 4.2   Greiner's model of organisational growth through evolution and revolution

Source: Greiner 1998: 58

the owner is unable to maintain involvement in and control of every activity undertaken by the firm. Greiner suggests that most owners cope with this by restructuring the firm, arranging it into departments and functions with section heads, the leader providing overall direction. During this period growth continues through the direction of the leader and HRM needs become more complicated, with more staff and the need to maintain and develop specialist skills. However, HRM responsibilities tend to remain centralised. As the firm continues to grow, a *crisis of autonomy* arises, as the owner cannot maintain autocratic control over the operation of these departments and functions. The organisation is now complicated and the departments and functions are highly specialised. The leader is therefore unable to retain an overview of complex and rapidly changing situations, and there may be a need to create a separate HRM function with responsibility for these matters. Through this delegation the organisation is able to continue to grow. However, as operations become increasingly complex and power is devolved, control issues emerge and at the centre, senior management experiences difficulty controlling the organisation's performance: a *crisis of control* occurs. In response to this the organisation implements measures to coordinate its activities, such as controlling budgets, formulating plans and procedures, and reviewing performance regularly. It is then that HRM policies might become formalised. However, further growth may lead to the development of an overly bureaucratic organisation which is unable to be innovative or take the initiative. If this occurs, what Greiner terms a *crisis of red tape* can occur. The only way for the organisation to continue to grow in such a circumstance is to adopt a more collaborative management style, in which shared values, teamwork, cooperation and creativity are fostered, thus preventing the stagnation of systems driven by red tape. In this environment HRM becomes hugely important as a way of motivating employees and breaking down barriers, stereotypes and divisions which have developed at a personal level as the company has grown into competing divisions. In many ways this is a return to the philosophy that drove the craft structure.

Clearly, different organisations are at different stages of maturity and development. However, the structure of most large construction organisations resembles Figure 4.3, being characterised by fairly rigid functionally based hierarchies. These effectively form a vertical chain of command, where managers towards the top of the hierarchy have line authority and power over those subordinates further down. Those in staff positions are employees in supporting roles who enable the line to carry out necessary functions (Langford *et al.* 1995: 65).

Figure 4.3 shows how under the line and staff approach the functional departments are grouped within separate, autonomous departmental units. Whilst this has clear advantages in terms of division of labour, it has disadvantages in that the organisation will find it hard to adapt and change in response to market demands. Furthermore, competition for resources can

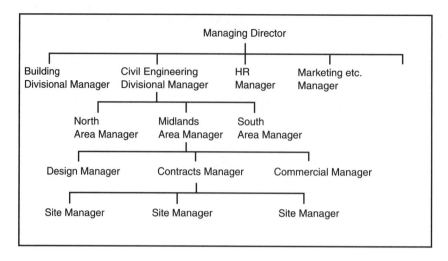

Figure 4.3   Line and staff organisational structure

arise between different departments, and the many functions can create coordination problems which slow down the speed and responsiveness of the organisation in dealing with change.

For organisations operating in a project-based industry such as construction, the matrix structure has become the most common organisational form. The matrix structure effectively adds a lateral dimension to the line and staff structure, allowing the integration of previously disparate functions for the purposes of specific projects (see Figure 4.4). This integration is facilitated by a project manager. If a project needs to be resourced, then a project manager is selected and draws from functional specialists in the various departments to make up the project team. The individuals can be chosen to reflect the particular needs of the project, allowing the manager to build a work group which accords with the particular needs of the client and the technical requirements of the project. Individuals deployed to teams maintain their vertical line-management relationship with their departmental superior, but will also have a shared responsibility to the project manager which flows laterally across the project structure. This structure is ideal for organisations operating in project-based industries, where multidisciplinary teams have to be formed quickly and then rapidly disbanded and redeployed elsewhere within the organisation.

The matrix structure provides many advantages for the construction organisation. Project teams are far more likely to comprise the necessary skills as people can be drawn from across the organisation and not only from a single operating division. Furthermore, it is flexible, as it facilitates individuals coming in and out of teams as required during the project's development. It is also good for employee motivation, as the individual is

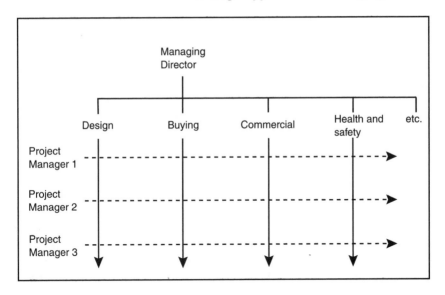

Figure 4.4   Example of a matrix organisational structure

clearly affiliated to a project whilst maintaining their professional identity within a functional department. However, there is also a key disadvantage, in that an employee can have divided and often confused responsibilities (and loyalties) towards their line and project managers. The structure also has a profound effect on career structures within organisations, and this must be taken into account by the HR manager. The project-based employee will have a choice of career routes through either a project or a functionally based route, both of which must be accommodated by the organisation if employees are to be retained. Furthermore, to prevent employees being labelled as project based and restricted in their career path there must be opportunities for people to move into the line structure. The problem is that line positions are relatively rare and senior when compared to project-based positions, and the opportunities for movement can be limited.

Responsibilities for the HRM function in construction companies

Whilst the work carried out within many organisations could be categorised as ongoing and/or repetitive, project-based industries have distinct characteristics, such as impermanence, uniqueness and uncertainty (Hamilton 1997: 64). This places particular pressures on the HRM function, insofar as it must respond to the particular project-specific HRM issues arising upon each contract. Inevitably, this means that companies operating within project-based sectors tend to devolve much of the HRM function to a project level. Thus, whilst the HRM department will probably develop the strategic priorities and direction of the HRM function, the way in which these are implemented at a project level is the responsibility of line managers in the production function. Indeed, regardless of the particular approach that a construction company adopts, the project-based structure of construction demands that managers responsible for activities at a functional level must take at least some responsibility for managing people-related functions at a site-based level. The fact that construction projects are often physically located well away from the head office means that construction line managers often act with even more autonomy than their counterparts in other industries and sectors.

Devolution of HRM responsibilities is a relatively straightforward and sensible strategy given the importance of aligning people management with the strategic objectives of the organisation at project level. However, despite the increase in the number of firms engaged in project-based working over recent decades, little research has been conducted into the most effective ways in which project managers can effectively manage the process. Problems are exacerbated by the inevitable short-term objectives, temporary nature of teams and role/authority ambiguity that the matrix organisational structure promotes. There is also a lack of training in HRM issues, a lack of time and resources, and severe pressure to deliver the project on time and within budget. Understandably, in this environment anything which is not seen as directly contributing to productivity is given lower priority. Hence, perhaps the greatest challenge for HRM in the construction industry is to highlight more forcefully the link between people's well-being and productivity, and to manage the division of responsibilities between HR and line managers in a way which ensures the optimal management of the HRM function.

It is not easy to define the precise role and responsibilities of the construction company's HRM departments and line managers. This will be determined by the size and structure of the organisation and how closely aligned HRM is with the strategy and philosophy of the organisation. Within the HRM department the allocation of HRM activities may be split amongst various specialists before being devolved to line managers within the organisation. Line managers make up the links in the line of command from strategic management to the operational production function at project level. They report upwards to senior managers for strategic direction and objectives, and have subordinates whom they direct in meeting these organisational objectives. This two-way communication process, both up- and downstream within the organisation, sometimes leads these important managers to be described as *middle managers*. In construction organisations, project managers represent key middle managers in the production function by heading up the principal operational production units of the organisation (projects). The way that the responsibilities and relationships are managed within most construction companies is that the HRM department acts in an advisory capacity to the line management – that is, line managers retain control of the staff within their work group (i.e. project or department) but draw on specialist advice when required. Mullins (1999) summarises this relationship in the model shown in Figure 4.5.

Mullins' approach emphasises that the personnel/HRM function should represent a *shared* responsibility between senior managers, line managers, staff and HRM specialists. Viewed in the context of a construction organisation, it is easy to see how particular aspects of the people-management function could be delineated amongst each of these groups:

- *Top management*: these senior managers and directors should identify the strategic objectives of the organisation, define project objectives and oversee the general direction and performance of the project teams. This will include specifying the extent to which line managers will take responsibility for HRM objectives.
- *Line management*: these project managers should handle day-to-day personnel-related matters such as the organisation of work responsibilities, employee performance, safety, work-based training and communication. Although they will have direct responsibility for

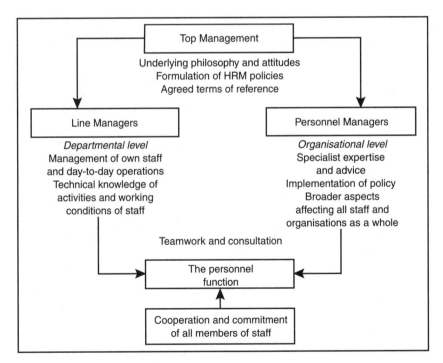

Figure 4.5   The personnel function

Source: Mullins 1999; *Management and Organisational Behaviour*, reprinted by permission of Pearson Education Limited

staffing and personnel decisions, they will be able to draw on the support of the specialist HRM managers.
- *Personnel management*: these specialist HRM professionals can provide services to the line/operational management at two levels. First, they provide specialist advice concerning people-management issues that line managers feel under-equipped to deal with. Good examples include disciplinary and grievance procedures, details of equal opportunities and employment legislation, recruitment advice, providing information on performance management and reward systems, etc. Their second role will be to manage directly elements of the personnel/HRM function that are best dealt with centrally rather than on a fragmented project-level basis. These include some aspects of recruitment, the development of formal HRM policies and procedures that demand a consistency of approach across the organisation. HR managers can ensure the effectiveness of management activities through their neutrality or slight detachment from core central-management activities. The objective view that this detachment affords them should allow an emphasis on people management to be maintained (Hall and Torrington 1998).

Within a large construction organisation the key to the success of this devolution of responsibilities is a partnership between HRM and line managers that remains *flexible* in order to allow it to cope with the inherent changes that beset all projects. For example, the construction labour-market climate is constantly shifting, with changing markets and advances in technology placing new demands on construction companies to recruit, retain, train and develop specialists in many areas. Similarly, training and development activities are influenced by changes in employment legislation, or particular industry and client expectations which individual project managers could not reasonably be expected to remain cognisant of. Other challenges concern the need to know and understand the individual project subculture, especially in terms of the industrial relations between operatives and management, and between the various members of the client, construction and procurement teams. The complexities of managing this change demand that a four-way dialogue is maintained between top managers, personnel managers, line managers and project staff. Merely devolving responsibility without ensuring a support structure for line managers is a recipe for disaster, and it is important that the value of a distinct HRM role within such organisations is not lost. However, no matter how well structured and managed the relationship between project management and HRM functions, the main factor in the ultimate success of HRM strategies at project level will be the attitude of the project manager. For this reason the following section looks in more detail at attitudes towards people in the construction industry.

Managers' attitudes towards people in construction

Managers can be categorised in terms of their attitudes towards people. For example, Likert (1961) developed a continuum of managerial styles ranging from 'job centred' to 'employee centred'. Job-centred managers focus on the task to be done and operate through formal rules and procedures, clearly defined hierarchies, specialisation, job separation and top-down information systems. They insist on loyalty, obedience and adherence to strict deadlines, and rely upon close supervision, monetary rewards and threats of punishment as a means of achievement. Job-centred managers believe that there is an underlying truth to the world, that organisations are stable static entities that respond in predictable ways to managerial stimuli, that there are clear-cut linkages between events and cast-iron mechanisms for controlling them. They relate closely to the principles of scientific management discussed in Chapter 2, and consequently base their management on the principles of measurement and control, seeking detachment from those they manage and having only a secondary concern with their well-being. In essence, job-centred managers focus on the task rather than the people doing the task. They treat people like cogs in a well-oiled machine, trying to eradicate any uncertainty that could arise from their idiosyncrasies, with the aim of creating a stable and predictable managerial environment.

In contrast, employee-centred managers focus on people rather than the job and recognise that their needs are not completely satisfied by monetary rewards. They believe that scientific management creates more uncertainty than it solves, by placing people in a restrictive environment where they behave dysfunctionally. They focus on the spiritual aspects of organisational life, working on the assumption that efficiency is achieved through the satisfaction of people's needs. The best mechanism for achieving this is a decentralised organisation characterised by flexibility, openness, collective responsibility, participative decision-making, trust and an absence of rules and procedures. Such managers consider that uncertainty is inevitable in organisational life and seek to accommodate it rather than suppress it. They achieve this by emphasising the value of lateral communication, groups and teamwork within organisations. Below we will explore the nature of these contrasting management approaches and how they can influence the efficacy of the production function within construction.

Employee-oriented construction project management

As discussed above, managerial styles can be broadly classified as being people oriented (employee centred) and task oriented (job centred). However, these styles of management represent the extreme ends of a continuum. In reality, most managers fall somewhere between these two extremes, combining elements of both styles in their own unique approach to management. While the employee-centred style immediately appears more attractive and ethical, if not more challenging, it would be wrong to classify one style as better than another. If the development of management theory has taught us nothing else, it is that *there is no one best way to manage*. Rather, it has been found that the most appropriate managerial approach depends upon four factors, namely: the *people* being managed, the *task*, the *technology* and the *environment*. Typically, the employee-centred (soft, non-scientific, organic) approach is most suited to people with a desire and capability for autonomy, who are undertaking a non-routine, complex task with low technology, unmechanised procedures in an uncertain, ever-changing environment. In contrast, routine tasks under-taken in a stable environment with highly mechanised technology by people who do not desire autonomy are more suited to a job-centred (scientific, mechanistic, hard) approach. Every construction project must therefore be treated differently.

However, while no two construction projects are exactly the same, there are common attributes which would imply that an employee-centred emphasis would be appropriate on most occasions. This is because, in comparison to other industrial production processes, the construction process is relatively 'small batch' in nature, being concerned with the production of one-off products of a relatively unique and prototype nature. Furthermore, despite considerable attempts at mechanisation, construction still remains a labour-intensive, low-tech production process which relies heavily on the creativity of professionals and craftsmen, who still value their autonomy and intellectual freedom in a relatively uncertain and unpredictable production environment. Indeed, a model developed by Perrow (1970) suggests that increased mechanisation in the construction industry must be pursued with care. His model sought to clarify the relationship between production technology and the nature of problems facing an organisation, which he categorised on two dimensions, namely *analysability* and *variety*. When organisations face a large number of unexpected problems variety is high. Conversely, it is low when the environment is predictable. Analysability is low when these problems demand experience, judgement and intuition in their solutions, and high if systems of well-developed procedures can be applied to their solution. Perrow also developed a typology of four tasks, namely *routine*, *craft*, *engineering* and *non-routine*. Routine tasks lend themselves to mechanised technologies, craft tasks to traditional technologies, engineering tasks to analytical technologies, and non-routine tasks lend themselves least to any type of technology – intuition and judgement playing a more prominent role. This is illustrated in Figure 4.6. The literature within construction suggests that the industry faces managers with an environment of high variety and low analysability, which means that the appropriate technology is non-routine. This suggests that an employee-centred strategy should be most effective for managing construction activities.

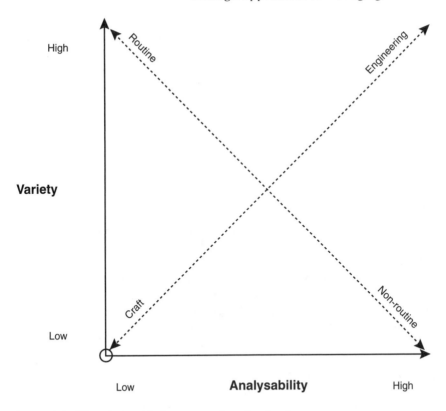

Figure 4.6 The relationship between task and technology

Justification for an employee-oriented style of construction project management is also provided by the business environment, which is becoming increasingly competitive and saturated with risk and uncertainty. There is no evidence to suggest that the construction industry is immune to such changes, which are producing a situation where the ability to innovate is increasingly crucial to success. As De Bono has pointed out, in an increasingly dynamic world it is 'far more useful to be skilled in thinking than to be stuffed with facts' (De Bono 1993: 25).

The necessity of intellectualism in modern management is in total contrast to the past, where success could be reliably achieved through the application of well-established and proven systems, processes and techniques which had been mechanically absorbed in a rote-learning environment. As Stacey argues, the application of this traditional approach to the current business environment is counterproductive, because in an unpredictable environment 'order leads to chaos where chaos would have led to order' (Stacey 1992: 63). The irony in Stacey's insight is his suggestion that traditional management approaches have been the cause of the uncertainty and instability which they have sought to eliminate or control – that is, their solution has become a part of the problem and this has led industries such as construction into a self-perpetuating spiral of scientification, which has become increasingly difficult to break. In scientific management, out-of-control situations are controlled by imposing more controls, and it is only now that we are beginning to accept that it was the controls which were creating the out-of-control situation in the first place. As Bolman and Deal (1995) warn, scientific styles of management merely provide the illusion of control and a sense of invincibility. They justify their position by citing new research surrounding the Polaris Missile Programme, where it has always been assumed that the use of detailed planning techniques such as the Programme Evaluation and Review Technique (PERT) played a part in its success. However, new research has highlighted a widespread ignorance of the plans and programme among the project personnel. Indeed, in complete contrast to previous studies, the project's success is now being attributed to the ingenuity of the project personnel in working around the plans while maintaining the external appearance that they were the main driving force behind their actions. This external facade was important because it fostered a myth that maintained the strong external support for the Polaris project and kept critics at bay. Therefore the main value of scientific management techniques was not in their contribution to efficiency. Indeed, in this sense they may actually be counterproductive. Rather, it is political, in that they are essential to maintaining the internal theatre and drama of organisational life, presenting a reassuring portrait of a committed and conscientious management which pacifies external constituencies:

> If management is making decisions, if plans are being formulated, if new units are being created in response to new problems, if sophisticated evaluation systems and control systems are in place, then the organisation must be well managed and worthy of support.
>
> (Bolman and Deal 1996: 94)

The scientific, job-centred culture of construction

The discussion above offers considerable justification for an employee-oriented style of management in construction, particularly for innovative and high-risk projects. However, invisible pressures within the industry continue to encourage managers to adopt a job-centred style, and it is to these pressures that we now turn. An explanation of why such inappropriate practices predominate must commence where most construction professionals begin their careers: in the education system. Following this, as they work in the industry they are further ingrained with scientific values through the industry's educational, procurement and contractual practices, as discussed below.

Education and knowledge

A brief scan of the content of dominant construction project management texts quickly highlights the job-oriented basis upon which much of construction project management thinking and practice has developed. Most construction texts dedicate page after page of predictable attention to the scientific techniques of planning such as PERT, network analysis, precedence diagrams, cost planning and control. While such techniques are valuable, there is relatively little attention to psychological, sociological and behavioural issues, which offer potential insight into the behaviour of people in the construction industry. The limited attention that is given to behavioural issues is more akin to 'social engineering', in that it is generally designed to discover some kind of rationality in people's behaviour to provide the basis for prescriptive managerial guidelines. Applications of behaviour modification, which are underpinned by Skinner's (1938) theory of behaviourism, are an example of such social engineering. Behaviour modification involves manipulating employees' receipt of contingent rewards so as to reinforce desired behaviours, such as safe behaviour or pro-environmental behaviour. However, research suggests that when applied in isolation behaviour-modification techniques are of limited effectiveness. This is because they fail to take account of other sources of employees' motivation, such as intrinsic beliefs and values. For example, a behaviour-modification experiment designed to improve solid-waste recycling behaviours was found to be only partially effective because employees' values were not aligned with the contractors' waste-management policies (Lingard *et al.* 2001). Indeed, even our leading academic journals, which seek to represent the cutting edge of construction project management thinking, cannot avoid the relentless pursuit of ever-more sophisticated scientific techniques, such as business process re-engineering, benchmarking, quality assurance and value engineering (Green 1999). The popularity of these trends merely reflects fashions in mainstream management which themselves are based on a traditional engineering culture of prescription and control (Richardson 1996).

Educational institutions also play an important part in reinforcing the industry's scientific values, because delivery is normally based within a formal classroom environment. In the majority of construction project management courses, contact hours for students are high, formal examinations still play the major role in assessment, and the learning experience is characterised by a relatively high degree of prescription compared with other disciplines such as the humanities. Within this system of delivery there is little time for thinking, and only a token gesture can be made to the creation of a flexible, inquisitive, questioning, student-centred learning environment which encourages students to debate contentious issues and develop their intellect.

Procurement practices

The construction industry's traditional procurement practices are also a strong source of its job-centred values. By procurement we mean the *who*, *what*, *when* and *how* of project organisation – that is, procurement systems essentially differ in who they employ, what those people do, when they are involved and how they are involved. For example, the parties employed in *design and build* and *traditionally* procured projects are essentially the same. What distinguishes the systems is merely what they do, when they do it and how they do it. In design and build the contractor is brought in earlier, on a different basis and has a broader range of responsibilities.

The traditional procurement system in the UK is underpinned by the Royal Institute of British Architects' (RIBA) Plan of Work, a sequential 13-stage framework describing the operations which comprise the traditional procurement process. This has been adopted in many other countries and continents – in Asia, Africa, New Zealand and Australia – and depicts a process which evolved in response to major historical events such as the industrial revolution, and in which the architect plays a central role (Bowley 1966). It starts with the client commissioning an architect to oversee the production process through design and construction. Once the client's requirements have been established and the feasibility of the scheme assured, a design team is assembled by the architect. The design process then moves through a number of phases during which the design becomes ever more detailed. The main contractor is not involved in the design process and is employed once the design has been substantially completed. This separation of design and construction as distinct phases of activity is one of the main defining characteristics of the traditional procurement process and alone is a significant stimulant to a job-centred style of management. This is largely because the lack of participation creates divisions, misunderstanding, suspicion, a lack of trust and the potential for conflict within the project team.

The main contractor is normally chosen by competitive tender after having priced a bill of quantities. This is a document which itemises all the elements of the contractor's work, which all tendering contractors are required to price to provide a common basis for their evaluation. While competition is beneficial, this employment procedure has become another inducement to job-centred values, largely because it has produced an overemphasis on price as the main employment criterion. While there are indications that this is being redressed by recent legislation, in the UK's public sector at least, the result has been the erosion of margins to unrealistically low levels. This in turn has led to a reduction of goodwill and flexibility in dealing with unexpected problems, an increase in claims consciousness in contracting organisations and a general deterioration of relationships within construction projects.

The bill of quantities, as a document, has also received considerable criticism for its structure and inflexibility. Not only has its finite detail been questioned in relation to the costs and time involved in producing it, but its accuracy has also been questioned, especially when it is produced from incomplete drawings, as has increasingly been the case. Furthermore, there is considerable rework involved for contractors in pricing it, and this can cause problems which may manifest themselves in disputes later in the project. In essence, the bill of quantities epitomises the obsession of scientific managers with detail that merely provides the illusion of control.

The result of the tendering process is the appointment of a successful contractor, with whom the client enters a direct contractual arrangement. The construction stage then commences and the bill of quantities becomes a contractually binding document against which the contractor is paid. The main contractor then takes responsibility for the production of the building and can subcontract portions of the work. This conveniently brings us on to the subject of construction contracts, which are perhaps the clearest clue to the industry's scientific value system.

Construction contracts

Contracts are essentially written documents which seek to ensure some element of predictability and control in people's actions during the course of a construction project through the legitimate power of the courts. While it is arguable that contracts were never designed to be managerial documents, there is little doubt that the legal implications of breach of contract do have a constraining effect on the actions of project participants. For example, most construction contracts stipulate periods within which certain information must change hands, periods within which parties must be paid, what should be included in those payments, detailed processes for changing the project's scope, determining it or claiming extras from other parties, etc. In most cases the complexity of contract documents in trying to cover every eventuality is extraordinary.

What this demonstrates is an assumption that uncertainty can be controlled, that people's interactions can be planned and standardised, that there is one best way to manage, that contract drafters know best and that project members cannot be trusted to interact in the correct manner. In other words, contracts have become a substitute for reflective management, and the underlying values are essentially scientific, which makes it difficult for a manager to employ an employee-centred style. Clearly, this assertion must be placed in perspective, since contracts do not attempt to control every aspect of project life and it would be wrong to argue that managers walk around with the contract in their hand. In reality projects are run by people, not by contracts. Nevertheless, contracts are an obsession of the construction industry and they do constrain managerial actions, particularly when relationships become strained, which is when an employee-oriented approach is most beneficial.

Perhaps a stronger inducement to job-oriented management is the detailed content of construction contracts, which is said to induce division, mistrust and conflict. For example, one of the most common criticisms of construction contracts revolves around their ambiguity, particularly in relation to risk distribution. This ambiguity is said to stem from their large volume, complex structure and legalistic language (Barnes 1991; Murdoch and Hughes 1992; Uff 1995). The argument has been that this creates misunderstanding and the under-resourcing of projects, which represent the prime causes of conflict in construction. Indeed, conflict has become so endemic within the construction industry that some have argued that it has become institutionalised, and such a confrontational environment is not conducive to the trust which forms the basis of an employee-centred approach to management.

Continuing the reference to trust and collective responsibility, there has also been criticism of the tendency of traditional contracts to separate risks rather than to share them amongst project participants. While the complete sharing of risks would be both unrealistic and irrational, it has been argued that a greater sharing of risks would reduce the conflicts of interest, insensitivity and selfishness that cause conflict in construction projects. Defenders of traditional contracts have argued that they have been developed by a representative body of all primary interests in the construction process, and that problems of interpretation arise from misadministration and a lack of expertise within organisations which sign up to these agreements. However, critics argue that it is unrealistic to expect such expertise to exist widely in an industry characterised by such a high number of small organisations. Such organisations will not have the resources or the inclination to employ lawyers at the tendering stage when there is only a limited chance of securing a project and thereby recouping any investment. Critics also argue that the collaborative process by which contracts are produced results in ineffective compromises, little choice, and in documents which mirror and therefore perpetuate the traditional divisions within the construction industry. In this sense, contract-drafting bodies actually perpetuate the problems they are trying to resolve through prescription. They also cause slow decision-making processes, which ensure that the contracts produced are invariably out of synchronisation with contemporary thinking, as well as economic, political and technological developments. There have been many other criticisms of traditional contracts which may encourage a job-oriented approach, such as their penal nature. Few contracts provide incentives to innovate and most focus instead on punishment for non-compliance, a classic reflection of a scientific managerial culture.

Changing the culture of construction: engendering employee-centred management strategies

The preceding discussion has suggested that procurement, contractual and educational practices within the construction industry have contributed to a scientific culture which encourages a job-centred style of management. This does not seem to suit the strategic needs of today's construction industry or the changing business world in which managers find themselves. Yet scientific management is still alive and flourishing in the construction industry and change will not come easily. In the confrontational environment of construction the party which unilaterally moves to a softer stance will invariably be punished. In this sense, any change must be negotiated and multilateral. Furthermore, construction managers will find it hard to make the transition to a more employee-oriented style because of a cultural low tolerance of ambiguity. The industry's practices have ensured that most managers rest easy in the belief that there is an underlying truth to the world, that organisations are stable static entities that respond in predictable ways to their actions, that there are clear-cut linkages between events and cast-iron mechanisms for controlling them. To imagine anything else is likely to engender a profound sense of insecurity by challenging everything they have been taught. The result is a continuing mismatch between the strategic needs of the industry and the way people are managed within it.

The only realistic solution is a long-term one, which will involve changing the way people think through new imaginative educational practices and course structures. However, educational change will also come hard because of government funding constraints – ultimately, *learning* costs more than *teaching*. Furthermore, there is strong support for the status quo from professional institutions, which have vested interests in the traditional procurement system and the modes of educational delivery that perpetuate it. While traditional construction courses are reassuringly familiar, predictable, easy to manage, easy to define, easy to deliver and easy to absorb for students, innovation in content and methods of delivery has a critical role to play in changing the culture of the construction industry. As De Bono argues, 'the purpose of education is to fill an empty mind with an open one, rather than a full one' (De Bono 1993: 63).

So the traditional and entrenched perspectives of the industry inhibit students' educational experience, even at university level. Evidence of this is provided by Lomas (1997), who notes that students are increasingly adopting a commercial view of the education process by basing their choice of subjects on how these will contribute to their future employment opportunities, rather than what is intellectually challenging and interesting. Furthermore, Denning (1992) identifies a general trend among university students to believe they will be more valuable to employers if they have less theory and more practical experience. It is not surprising that fewer and fewer students want to prolong their stay on campus. If this sentiment continues it is possible that the breadth, if not the depth, of education in the construction industry will suffer, as graduates continue to focus solely on subjects which they know will be attractive to future employers. Industry interests should not be permitted to dictate the educational experience of students at university and destroy the only opportunity they have to explore and develop their intellect in the widest possible sense.

The tension between academic research and theory and industry practitioners' views of what education should be is well documented. In the construction industry criticisms of academia as being out of touch with the 'real world' are well known and may be partly due to the industry's male-dominated, traditionally blue-collar culture. For example, Marks (2000) found the British working-class male culture to be anti-intellectual. According to Marks, this culture places significant emphasis on masculinity and physical labour, while education is seen to be effeminate and not 'real work'. The construction industry is vastly different from a university environment, in that many senior construction managers started their working lives as tradesmen and then progressed through the company ranks. This difference makes managing the university–industry interface particularly difficult in construction. However, it is important to manage it more effectively, since Gerber (2001) suggests that both practical knowledge and theory play an important part in learning how to perform effectively in a job.

While change may come hard, recent events may force it upon those who resist. For example, for some time there has been evidence that technological advances will force professional institutions to redefine their traditional roles and relationships towards a more managerially based, employee-oriented and less technique-ridden footing. Evidence of this trend was clear during the 1980s with the emergence of the project management role, which was fervently claimed as their own by a host of established institutions such as the Royal Institution of Chartered Surveyors (RICS), the Chartered Institute of Building (CIOB) and RIBA. Furthermore, changes in the structure of government funding arrangements which place more emphasis on the quality of teaching and research will force educational institutions to provide a more stimulating learning environment and to face up to the mistaken association of teaching with learning which has pervaded traditional approaches to construction education. In particular, the emphasis on research as a funding criterion will force traditional educationalists to realise that quality education does not mean high contact hours with students. Instead, the quality of education is directly linked to the quality of research and to a more student-centred learning environment.

Conclusions

An organisation's SHRM strategy should be determined by, and closely aligned with, its overall business strategy and direction. However, this chapter has suggested that a dominant paradigm of operation in construction companies makes this difficult. Structurally, the common approaches identified can be defended because they provide organisations with a flexible framework which enables them to cope with the dynamic, short-term resourcing priorities with which they are faced. However, the common approach to the management of people by line managers given responsibility for aspects of the HRM function is not as easily defensible. At this level of operation the industry's contractual, procurement and educational practices appear to have created a job-based tradition which undervalues people and thereby dehumanises the construction process. In this chapter we have argued that traditionally, there has been an over-reliance on the scientific principles of measurement and control, which neglect the spiritual aspects of organisational life and suppress the creative potential of people. Such an approach has failed fully to exploit the creative potential of the construction industry's workforce, has devalued people in the managerial equation and has grossly oversimplified the complexities and managerial challenges posed by an increasingly unstable world. A change to a more employee-oriented approach could release a huge untapped creative and productive potential in the construction industry. This is not a call for the abolition of construction contracts, the elimination of competition and a complete change in the way the industry delivers construction project management education. Rather, it is a call for a more thoughtful and flexible approach to the management of construction which gives greater attention to the needs of people and their inextricable interrelationships with project goals. In the construction industry this will demand a fundamental cultural shift, which will not be easy. In the long term the education system will play a crucial role in bringing this about. In the shorter term HR managers are in the front line and can also do much to assist the transition. They have at their disposal many tools with which they can manipulate culture, such as the recruitment process (by controlling the types of people that gain entry to the organisation), promotions (by controlling who reaches positions of influence within the organisation), induction and socialisation (through their influence over the social dynamics within the organisation), codes of practice/mission statements (by influencing the guiding influences on employees' work practices) and reward/appraisal systems (by rewarding traits such as intelligence, communication skills and capacity to cope with change, traits which value the personal strengths that employees bring to the organisation, rather than merely conforming to organisational rules and protocols) (see Brown 1995).

Discussion and review questions

1   List the external issues currently facing the industry that will have to be taken into account by construction organisations over the next five years if they are to remain competitive. Next, use the strategy formulation matrix approach to identify HR priorities for a medium-sized organisation looking to develop and expand its market share within your local construction market. You should identify, first, strategic imperatives for the organisation and, then, SHRM measures to address each factor through the firm's human resources.

2   Explore the organisational structure of a construction firm with which you are familiar. Identify the structural measures the organisation has taken in order to cope with the dynamic nature of the construction industry's labour market.

3   Discuss the techniques that HRM specialists can use to engender a more person-centred approach amongst managers given the devolved nature of the HRM function within most construction companies.

5   The mechanics of human resource management in construction

Resourcing, development and reward

Having explored the SHRM function and the generic approaches to managing it used by construction organisations, we now proceed to explore the day-to-day operational aspects of the function. These aspects are referred to as the *mechanics* of SHRM – the range of processes and operations that allow the organisation to manage and develop their workforce in a way which supports their SHRM goals. We will discuss these processes under three broad headings: employee resourcing, human resource development (HRD) and reward management. Under these headings we will explore how an organisation can manage these interrelated processes effectively and how information technology can be used to facilitate the integration of strategic and operational SHRM goals. Their importance cannot be overstated. Without effective management an organisation will be unable to attract, develop and retain the people that will enable it to achieve its strategic objectives.

Introduction

A successful organisation must harness the efforts of the human resources at its disposal. The greatest challenges for construction organisations are in finding good people and utilising them to their full potential in a way which contributes to the accomplishment of organisational objectives, or, in other words, managing the talent within the organisation. In order to achieve this they must train and develop, promote, retain and release their staff as appropriate. The processes involved can be linked within a continuous loop known as the SHRM cycle, which is influenced at every stage by an organisation's objectives and by external environmental forces.

Employee resourcing is concerned with planning for the needs of the organisation, defining the roles and responsibilities necessary to achieve these goals and then recruiting and selecting the people to fill these roles. *HRD* is concerned with managing the performance of employees in such a way that they contribute to organisational objectives. This includes training, management development, and the management of structures and career paths to ensure clear succession routes through the organisational hierarchy. *Reward management* is about defining the wage and benefit structures in a way which responds to the needs and expectations of employees and ensures their commitment to the organisation. Together these interrelated activities ensure the supply, development and motivation of employees, and their careful management is a prerequisite to achieving the human resource stability necessary for organisational growth and development. In most large organisations they are supported by information and communications technology (ICT) tools known as human resource information systems (HRISs). These range in sophistication from simple transaction processing and reporting tools to full decision-support systems that allow organisations to forecast and manage their resource requirements and future SHRM needs. For construction companies the management of the cycle depicted in Figure 5.1 can be problematical. The industry's susceptibility to economic cycles, the variability of construction projects, the autonomy of operational managers at project level and the need to cope with unexpected change put in place a problematic context for the management of people within the industry. In this context it is no surprise that the construction industry's SHRM performance record is poor in comparison to that of industries which use more sophisticated employment techniques.

To help construction organisations improve their SHRM performance the remainder of this chapter will discuss each stage of the SHRM cycle from the perspective of a construction organisation. We will also explore

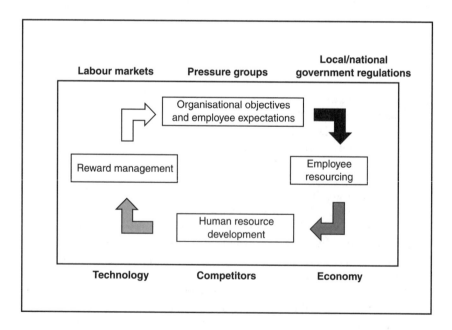

Figure 5.1   The SHRM cycle

the role of HRISs in facilitating the management of these stages, and how these can be linked to the strategic requirements of an organisation.

Expectations of the employment relationship

It is important to realise that people, unlike other resources, have their own needs and requirements, which may or may not be aligned with their employer's needs and requirements. Should they feel that their expectations are not being met, they can exercise their right to leave. Thus, before we explore the mechanisms used by organisations to manage the SHRM function it is important to gain an understanding of what employees require from the employment relationship. Failing to meet employees' psychological expectations will lead to a deterioration in commitment and trust, a decline in performance and, ultimately, high levels of staff turnover (see below).

Whilst fulfilling formal employment contract obligations is relatively straightforward, meeting psychological contract expectations is far more complex. It demands that an organisation gains a deep understanding of its employees' needs and how these can be reconciled with organisational objectives. Meeting employee needs in this way represents one of the fundamental challenges for the SHRM function, particularly in dynamic and fragmented project-based industries such as construction.

The formal employment contract

The formal contract of employment forms the cornerstone of the employment relationship. Such contracts define the terms and conditions of employment (i.e. what is expected from the employee) and what the employer will provide in return. The former usually comprises a job description (a summary of the employees' roles and responsibilities), the hours that they are contracted to work, and procedural aspects such as sickness procedures, required notice periods and employers' rights to change or amend their working conditions. The rewards offered to the employee in return for their commitment and outputs are stated in terms of salary levels, details of other benefits such as bonuses, company cars and pensions and holiday entitlements. As was discussed in Chapter 4, modes of working under such a contract are diversifying, with a marked increase in part-time and casual employment, temporary work, shift-working and teleworking (working from home). Furthermore, there has been a shift in recent years away from open-ended employment contracts to fixed-term arrangements where employment is terminated at the end of a stipulated period. In the construction industry firms tend to adopt flexible employment practices to cope with the uncertainty of their environment (see Chapter 4) and the duration of fixed-term contracts can be as short as a single project. However, it should be noted that employees on short-term contracts are unlikely to develop a sense of loyalty or commitment to an organisation without some kind of assurance that their efforts are likely to be rewarded with a permanent position in the future. Neither will such employees come to identify with the company's values to the extent that these values will exert an influence on their behaviour.

It is important to note that in most countries formal employment contracts contain more than the express terms identified above. The law of contract also recognises implied terms (those determined by custom or practice, or which are inferred by the express provisions of the contract) and those required by statute (those which general employment legislation demands that every employer and employee adhere to). For example, in Europe countries that are signed up to the EU Social Chapter require firms operating within these countries to adhere to maximum working-hour limits, safe working practices and minimum welfare standards and provision. Whilst these may not be explicitly referred to within the contract of employment, the employee can assume that they will apply in order that the employer meets their statutory requirements.

The psychological contract

The emphasis of contemporary approaches to SHRM is now on the *transaction* between the employee and the employer. This recognises the two-way relationship that exists within the employment marketplace and represents a shift of emphasis from seeing the relationship as a legal and formal one where an employee simply obtains work from an employer in return for a reward. It is increasingly being recognised that while formal employment contracts can define many formal aspects of the employee–employer relationship they cannot accommodate the expectations and obligations which each party has towards the other. This acknowledgement has led to the relatively recent development of the *psychological contract*, which attempts to define the *expectations* of the employee and employer.

Psychological contracts define the informal beliefs of each of the parties as to their mutual obligations within the employment relationship (Herriot 1998). These subjective aspects are important in understanding employee relations, because they allow employment contracts to be seen as a two-way exchange process, rather than one imposed by employers (see Herriot and Pemberton 1997). The incompatibility between employee and employer expectations has increasingly been recognised in recent years with the lack of long-term job security offered by employers in the current labour market. In construction the worldwide recession of the early 1990s showed that even middle-ranking white-collar workers could not be assured of the long-term stability of their positions and roles within the modern construction organisation.

Within the relationship defined by the psychological contract, both the employer and the employee inform, negotiate, monitor and then renegotiate or exit the employment relationship. In other words, psychological contracts represent a reciprocal and dynamic 'deal', which evolves as the employer's commitment changes. According to Rousseau (1995), the content of psychological contracts can be anywhere along a continuum from *relational contracts* (long-term, open-ended relationships within unitary organisations which lead to the exchange of loyalty, trust and support) to *transactional contracts* (short-term relationships set in plural-istic organisational contexts and characterised by mutual self-interest). Regardless of where psychological contracts fit along this continuum, they should be seen as interactive and dynamic (Herriot and Pemberton 1997). Employers cannot assume that they will remain static, as employment rela-tionships are contingent upon both internal and external pressures on the organisation.

Psychological contracts are important for construction organisations because, in addition to the 'hard' areas of the employment contract that have to be met, they acknowledge that a 'soft' set of expectations held by the individual also have to be organised and managed. This is particularly important in some sectors of the industry. For example, recent Australian research (Lingard and Sublet, in press) suggests that white-collar construc-tion employees place significantly greater value on their expectations of involvement, respect, recognition of achievement and a sense of personal growth than they do on 'hard' transactional rewards or benefits in an employment relationship. Any mismatch between the expectations of employees and the realities of what they receive can lead employees to feel cheated, and hence to a reduction in motivation and commitment. Maslach and Leiter (1997) identify six areas of mismatch between a person and his or her job that can lead to disengagement and reduced organisational effectiveness. These are as follows:

- *Work overload*: where downsizing and other strategies for increasing productivity intensify the work experience for employees, who work longer and harder with fewer resources. This results in a mismatch between employees' time and energy and the productivity expectations of the job.
- *Lack of control*: where employees do not have the capacity to set priorities for their day-to-day work or be involved in decisions about the use of resources. Without such control, employees, particularly professionals, cannot balance their interests with those of the organi-sation and lose interest in their work.
- *Insufficient reward*: where jobs bring less material rewards, prestige and job security than employees expect. Loss of intrinsic rewards asso-ciated with performing well, working with respected colleagues and enjoying one's work are increasingly important expectations. Losing the 'joy' of work impacts on employees' creative thinking and problem-solving abilities.

- *Breakdown of community*: lower job security leads to fragmented personal relationships at work and undermines effective teamwork. Conflict increases and a sense of isolation ensues. The utilitarian view implied in short-term contracts can threaten the social environment of the organisation. Such contracts draw on an individual's skills and energy without making a commitment to developing that person in the long term.

- *Absence of fairness*: employees need to perceive a workplace as fair, a place where trust, openness and respect prevail. A focus on short-term financial performance removes a sense of organisational community and can breed cynical mistrust of managers' actions.

- *Conflicting values*: where there is a major conflict between the values of the individual and those of the organisation. In the case of professionals there may also be conflict between their professional ethical codes and the commercial realities of corporate life. For example, as one engineer commented, 'I could join a building company but I'd have to leave my engineering ethics at the door.' Not doing what they believe to be right places well-intentioned and talented employees in a personally indefensible position from which they are likely to eventually withdraw.

Mismatches such as these are defined within the SHRM literature as psychological contract breaches and violations, which have serious implications for employee relations. However, an inherent problem with psychological contracts is that by their very nature they are hard to define until they have been breached or violated. Furthermore, research has revealed widely contrasting behavioural responses to psychological contract breaches, ranging from constructive to destructive and from active to passive (Rousseau 1995). Thus different employees may attempt to deal with similar psychological contract breaches in a variety of different ways. This variability in employee reactions means that organisations should attempt to create developmental approaches to the HR function which meet employees' personal career needs and expectations.

In the construction industry understanding psychological contract expectations could make a significant contribution to improving workplace morale, motivation, retention and, indirectly, areas such as safety performance. However, understanding the crucial role of the psychological contract in what has traditionally been a cash-based industry is likely to be difficult. In construction, excessive job demands and difficult or unpleasant work conditions have become part of the industry's culture, and managers at all levels may need sensitivity training to raise their awareness of employees' expectations as parties to a psychological contract.

The SHRM mechanisms explored below are all crucial to allowing the organisation to deliver on their employment and psychological contract obligations. Consideration of these issues should therefore help managers to be sensitive to employees' needs.

Employee resourcing

Central to the activities that make up the day-to-day aspects of the SHRM cycle is *employee resourcing* – the process of ensuring that an organisation has a ready supply of appropriately skilled people. The aim of a resourcing strategy is to ensure that the organisation identifies its personnel needs both now and in the future, and then meets them through its internal and external recruitment and staff-development activities. It encompasses the human resource planning function (deciding on the resources required by the organisation and arranging for their employment), deployment decisions (deciding on where to utilise people within the organisation – for example which project to allocate them to), recruitment and selection processes, and succession planning (i.e. ensuring that the organisation has a ready supply of managers to displace others as they leave the organisation). Supporting the resourcing process should be a range of basic administrative functions such as salary and wage administration, together with employee-support services such as welfare and health and safety provision. To avoid burdening project managers, construction companies usually provide these services centrally from specialist departments based at head office.

Managed effectively, employee resourcing can help to achieve organisational flexibility by providing access to a full range of skills that can be utilised for longer-term strategic planning as well as for deciding immediate responses to unexpected problems or opportunities. When employee priorities and preferences are taken into account in this process, resourcing can also enhance job satisfaction, employee-development and career-management processes. Strategically considered, allocation decisions can offer professional-development opportunities for employees and valuable learning experiences. For example, the Institution of Engineers in Australia has a graduate-development programme in which participant companies move graduates around their organisation in such a way that they attain the competencies required for full membership of the institution. Graduates' competencies are documented in the form of experience chapters and recorded in a database. Data can be transferred from company to company if an employee moves, and the database can be searched to identify gaps in graduates' professional development. Even in the absence of such a sophisticated system, job rotation schemes which move employees between roles or projects can lead to improved teamwork, innovation and decision-making, which in turn facilitates high performance, productivity and, ultimately, profit. The process also contributes towards staff retention, allowing for more efficient HR planning and succession management, which also contributes towards the accumulation, sharing and retention of important company knowledge and experience.

For most construction companies the resourcing process is a 'balancing act', in which managers attempt to take into account the long-term strategic considerations of the HR planning function while providing immediate solutions for the shorter-term operational needs of individual projects (Beardwell and Holden 1997). As we discussed earlier in this book, the nomadic, transitory, project-based environment of the construction sector presents a particularly problematic context for effective employee resourcing. Projects tend to be won at short notice, requiring the rapid mobilisation of teams to distant locations. In the rush to respond to this situation, individual employee needs can be excluded from the resourcing decision-making process.

The complexity of the construction project-resourcing environment is shown in Figure 5.2, which reveals the many potentially competing objectives which the resourcing process must meet.

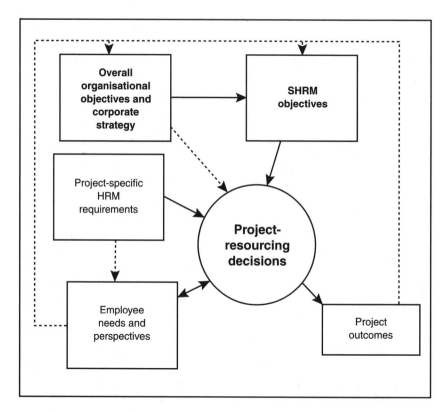

Figure 5.2   A model of the competing pressures and interdependent functions involved in the resourcing of project teams

Human resource planning

Any organisation must analyse current and future markets and ensure that it has the right number of people with the required skills to exploit them successfully. This is the essence of the human resource planning function. Assessing the available human resources and relating these to potential future market opportunities enables judgements to be made about resourcing requirements which will form the basis of recruitment and staff-development plans. These must be closely aligned with wider corporate planning and budgeting.

The competitive tendering process, which creates uncertain workloads, coupled with the cyclical nature of the construction market makes the HR planning process difficult. It does so by facing HR managers with the dilemma that laying off staff in response to a downturn may pose problems of recruitment when an upturn returns. It is for this reason that construction companies may decide to hold an apparently inefficient surplus of labour during recessionary periods. Human resource planning approaches used to mitigate some of these problems comprise four basic operations, which are outlined below.

Needs analysis

The first step in human resource planning is needs analysis, which is the assessment of current and future business needs. This is achieved through the careful analysis of an organisation's future goals and the market for its products or services. This applies as much to projects as it does to normal business organisations. In particular, organisations which are growing rapidly or operating within a very turbulent environment need to conduct frequent reviews of their human resource planning policies to avoid the possibility of labour or skills shortages restraining growth.

In detail, the stages of needs analysis involve:

1 Identifying what jobs need to be done, now and in the future.
2 Identifying the technology people will need to do these jobs effectively, and therefore the skills they will need.
3 Identifying the knowledge and qualifications required – qualifications will indicate knowledge levels and may be important to an organisation's profile.
4 Identifying personal requirements – types of personality that fit with the organisation's culture and ethos.
5 Identifying performance standards expected to assess individual suitability on the basis of past performance.

The strategic nature of needs analysis requires the involvement of senior managers.

The evaluation of current resources

The next step in human resource planning is to evaluate current human resources, in terms of people's skills, interests, abilities and experiences. When compared to the SHRM plans, this will identify skills 'gaps' that need filling and resource deficiencies that need addressing. However, the existence of deficiencies does not necessarily indicate the need for an external recruitment policy, since the necessary skills may be obtainable from existing human resources through retraining programmes or structural changes to the business. An adaptable multiskilled workforce is essential in enabling organisations to respond quickly to the highly dynamic reality of the modern business world. In this sense, the maintenance of an efficient database of existing employees is extremely important for any organisation. Such a database should contain information about employees' interests, experience, training and qualifications obtained before and during their period of employment. This will also allow the development of the individual to be monitored and action taken should they not be developing and contributing to the organisation's strategic objectives. For example, Telstra, Australia's largest telecommunications organisation, established such a database to monitor the organisation's implementation of strategic occupational health and safety plans. These plans contained stringent requirements for employees to undergo occupational safety and health training each year, and these could be tracked through a central database. Employees' training was audited during routine occupational health and safety management audits, and managers who had not released employees for training could be identified using the database.

Job analysis and design

The task of collecting information about existing human resources and the jobs they do and then analysing how an organisation can restructure itself to meet current and future business needs is called a job analysis. Job analysis should be seen as a process involving the following series of steps:

1 *Examine* the complete organisation and the *fit* of each person and job to future needs.
2 *Identify* deficiencies/misfits.
3 *Redesign* misfit jobs and people to better suit an organisation's future needs. This may involve modifying elements, duties and tasks associated with a particular position and retraining those who perform these tasks.

Data for job analysis is usually collected via questionnaires, interviews, observation and/or job dairies/logs. To enable the job analyst to develop an unbiased picture of a job it is important to collect this information from a variety of sources – operatives, supervisors, high and low performers, males and females, older and younger incumbents, etc. The emphasis should be on the identification of the inputs (i.e. the skills and competences required of the job-holder), the process (how they apply these skills and competences to the tasks at hand) and the outputs of these activities (particularly with regard to the value that they add to the organisation) (Armstrong 1991).

The analysis of internal availability

The above stages will lead to the identification of skills 'gaps' which need to be filled, internally or externally. Before committing to external recruitment, an organisation should check the current and future availability of internal staff. This involves analysing the future movements of people in terms of promotions, transfers, retirements, terminations and resignations. Some changes are easier to predict than others, and in this sense there will always be some reactive element to an organisation's SHRM policy. However, every attempt should be made to predict these trends so that an organisation does not suffer temporary periods of short staffing.

The human resource planning process will enable an organisation to compare future needs with the availability of internal resources. Following these analyses, the organisation will have ascertained its needs with regard to internal staff-development activities and external recruitment. It will also use the findings to help define its overall SHRM approach and to decide where to target its resources to ensure that it does not suffer from skills shortfalls.

It is important to note that the individualisation of the employment contract and restructuring have significantly affected career structures within many construction organisations. The traditional hierarchical structures have been replaced by more open systems, within which an individual is required to navigate his/her way with only minimal support from the organisation (Arthur *et al.* 1989). If HR professionals are to ensure employee commitment through balanced psychological contracts, they must incorporate the individual employees' varying needs and preferences into their planning and policy-making processes. Clearly, job design will only be effective as a mechanism for organisational change if it is carefully integrated with other organisation structures. Thus roles must also be viewed in relation to the impact that they have on new career and opportunity structures within the organisation, and checks made to ensure that these accord with employee needs and expectations.

Analysing employee turnover and throughput

Employee turnover should be analysed for two reasons: to assess the numbers of employees likely to be replaced in the future and to ascertain why people are leaving the organisation so that action can be taken to retain them. An appropriate level of annual staff turnover is deemed by many organisations to be around 8 per cent. This allows a sufficient influx of new people, who bring with them new ideas, innovative practices and energy, which in turn promote organisational development. Thus some degree of wastage is desirable. However, excessive wastage can be damaging to an organisation in terms of replacement costs and the loss of knowledge to potential competitors. At a project level such problems can be just as acute. For example, Loosemore (2000) showed how changes in project personnel during construction projects result in the loss of knowledge about potential problems, which then grow into major crises. Furthermore, Chapman (1999) identified the problems that changes in personnel could cause for the efficacy of the design process during a project. He noted that the amount of project information developed becomes so voluminous that it cannot be transferred in its entirety to an incoming team member. Moreover, it is important to note that much of the historical design and development of construction projects centres around responding to design issues and problems as they arise. Many of the steps and decisions taken during its development are not documented in a formal way, but are retained informally by the project participants. This informal knowledge is very important in running projects effectively, which means that the loss of core project members during a project can create major problems.

Most organisations calculate an annual turnover index by dividing the number of people leaving over the year by the number employed over the same period and multiplying by 100. Whilst this percentage measure is simplistic, it can be deepened by a range of other analyses which provide insights into the types of employee that are leaving (such as department, length of service, etc). Analysis of staff turnover should not just measure after-the-event departures, but also seek to identify predictors of turnover, such as diminished organisational commitment or employee burnout. Only if predictors are identified and monitored can proactive steps be taken to stem the turnover problem. Monitoring the predictors of turnover is also important in managing the performance of employees who may want to leave their job but who perceive the alternatives to be no better. Feelings of being 'trapped' in an unsatisfactory job are likely to have a negative impact on employees' performance, though turnover itself may not be high. Also, HR managers should be increasingly mindful of the need to consider not just the needs and expectations of the direct employees, but also those of their spouses or partners. Dual-career couples are on the increase and employees are more likely than ever to leave employment that is not compatible with the career of their partners or spouses. In addition, human resource planning also demands that an assessment is made of the impact of promoting and moving people around the organisation. The promotion of one employee to fill a position left vacant by someone leaving or a growth in the business can set off a chain reaction of promotions and transfers. This should trigger the organisation to revisit and revise its succession plans to ensure that its employees are moved around the business in a way which meets their psychological contract expectations. Bringing external people in to fill positions can lead to resentment from existing employees, who feel that their contribution is undervalued by the organisation.

Preparing job descriptions and specifications

Whether an organisation decides to fill skills gaps internally from existing employees or externally, job descriptions and specifications will need to be created and used as a basis for recruitment. A job description is an outline of the specific responsibilities and duties associated with a job, and a job specification is an outline of the educational experience and skills necessary to perform well on a job.

Whilst it is important that job descriptions and specifications are clear and unambiguous, the nature of construction activity is such that they should retain a degree of flexibility in order to cope with the fluctuating demands of projects. People should be made aware of this when they take up employment, so that resistance to redeployment and retention is minimised. For example, reporting relationships may need to be left fairly fluid, as project-team composition may vary depending on the stage of development. Similarly, the duties attached to a particular post may have to be amended to suit particular project types and/or client demands. However, this type of change can be minimised by seeking long-term relationships with clients via partnering arrangements.

Recruitment and selection

Once shortfalls in the skills available after restructuring, retraining and redeployment have been identified, the recruitment and selection process must then be used to attract the people who will meet these shortfalls. These people may be internal or external to the organisation.

Recruitment

Recruitment is the process by which managers attempt to locate people to fill positions identified. This is often a problematical process in construction projects, even in times of high unemployment, since both applicant and advertiser will have specific needs and expectations which will have to be matched. There is often a temptation for employers to see the process as one-way, a dangerous mistake, particularly in times of high demand. In such an environment good employees are a scarce resource and it is a seller's market, where employers need to impress potential candidates. For example, within the UK declining numbers of construction graduates have led to a shortage in certain skills such as quantity surveying. Construction employers have had to offer increasingly attractive remunerative and career-development packages in order to meet their recruitment needs.

Recruitment can take place either internally or externally. With changes in the structure of construction companies away from directly employed labour, most recruitment for projects is of an external nature and normally revolves around competitive bidding processes for specific work packages by subcontractors. Internal recruitment tends to be restricted to managerial and administrative support functions, although with the growth of part-time contracts in professional staff the distinction between internal and external employees is becoming more blurred.

The identification of internal candidates is normally facilitated by simple job postings in newsletters or on bulletin boards, etc. Alternatively, and more ideally, an information technology (IT) based human resource information system can produce a list of potential candidates who can be headhunted to avoid the inconvenience of a full recruitment process. The role of ICT tools in facilitating this process is explained in more detail later in this chapter. Internal recruitment has several distinct advantages over external recruitment. For example, it is considerably cheaper and quicker, particularly in the case of searching for managers. Also, internal recruits know how the organisation works and can enter a new job without having to suffer a costly and delaying learning curve. Furthermore, internal recruiting can provide a motivating force for the existing workforce by providing opportunities for progression within the firm. However, on the negative side internal recruitment starves an organisation of the opportunity to expand its experience and expertise base. In the rapidly changing world of business it is debatable whether organisations can afford to miss this opportunity, since adaptability will be the key to future prosperity. This is one important warning against the long-term use of partnering as a procurement practice. The freshness of ideas which variety, change and conflict bring is as important in construction projects as it is in any company.

External recruitment involves using direct advertising, recruitment agencies or informal means such as word of mouth and approaches to suitable candidates. The cost of repeated advertising and the competition for staff in a shrinking labour market have led many companies to increase their use of agency recruitment in recent years. This is fairly expensive in that a commission has to be paid for the agency to find suitable candidates, and it can lead the organisation to lose some degree of control over the process. However, it saves time and resources in advertising and vetting large numbers of applications. At site level, informal methods of recruitment predominate in construction, largely because of the propensity for site-based line managers to manage recruitment to their own projects (Druker and White 1996a). This also extends to senior positions, where companies may identify managers and professionals working for rival companies, whom they entice to join them with monetary rewards and other incentives.

Selection

Having identified a list of suitable candidates for a position, the critical task becomes the selection process. Selection is the task of deciding which people best suit the current and future needs of the organisation and offering them a position. Unfortunately there is no way to know for certain how a person will perform in a position, and in many companies, including those in construction, the process is no more than a lottery (Cooper and Robertson 1995). However, psychologists have developed a variety of methods for attempting to measure a candidate's suitability, and these include interviews, peer assessment, astronomy, graphology (analysis of handwriting), bio-data analysis, personality tests, trials, etc. The following is a summary:

Method	Variants
Interviews	Used in 90 per cent of selection decisions and the most common selection method in companies. However, at project level it is uncommon outside the design-and-build context, particularly for contractor appointments. Research indicates that interviews can be misleading, since the most polished performers, rather than the most able candidates, often get the job. Therefore it is critical to balance it with other techniques and to formulate questions carefully so that they expose the attributes which are important to the position. In this sense, the key to successful interviewing is preparation in identifying job requirements and in studying the candidates' strengths and weaknesses. The interview environment is also critical, in that it must be relaxed and undisturbed. When these conditions exist, research has found that interviews can be a useful way of assessing levels of motivation, intelligence and interpersonal skills. However, they are not very effective in assessing practical skills. There are a number of types of interview, which include:

Structured: the interviewer uses a structured approach based on a series of preconceived questions which facilitate direct and easy comparison between different candidates.

Unstructured: the interviewer acts as a facilitator, allowing discussion to take its natural course; the interview occurs in a relaxed setting and requires a considerable amount of expertise from the interviewer.

(*continued over*)

Method	Variants

Method

Variants

Panel: using a group of interviewers, which minimises bias in selection and facilitates greater discussion, but can be intimidating for candidates.

Individual: a one-to-one situation, which is useful when two people have to work closely together.

Employment tests

The candidate is given a problem to solve in a work context. Useful for testing people's performance in a workplace setting. Not widely used in the construction industry at a project level. There is a range of tests which can be used to assess the following:

General ability: focus on measuring intelligence and often used in preselection.

Aptitude: measures special practical abilities such as typing, bricklaying, drafting, etc.

Cognitive ability: tests of intelligence, intellect, numerical and verbal ability. Although subject to debate, widely accepted as giving consistent and reliable results.

Personality: measures dispositions to behave in certain ways in certain situations. Considers attributes such as emotional stability, honesty, motivation, leadership abilities, adaptability, curiosity, open-mindedness, extroversion etc. Widely accepted as reliable and, when used carefully, can play a useful part in selection processes.

Integrity: used for jobs in which honesty is essential, such as banking, insurance, government security/defence positions, etc. One method is the polygraph (lie detector), which works by detecting changes in electrical resistance on the skin and breathing rates, which are a result of anxiety when someone is asked leading questions such as 'Have you ever stolen money?'. The problem is that nervousness can cause the same response, and the technique has been banned in many parts of the US. The most respected method is to use a specific type of personality test by questionnaire.

Others

Self-assessment: when candidates are asked to make a personal assessment of their own abilities, skills, knowledge, etc. This is usually based upon comparisons with others. Research shows that people are usually very accu-

Method   **Variants**

rate at predicting their own abilities in simple tasks, but for uncertain, complex tasks, this is less so. Furthermore, people tend to overestimate their own abilities.

Peer assessment: there is much evidence to indicate that peer assessment (from colleagues of equal status) is valuable for assessing internal candidates, but not for external candidates. This is because of the impracticality of obtaining information from peer groups. However, internally peers have been found to provide very rigorous and reliable assessments, even though they often feel uncomfortable doing so.

Graphology: the study of handwriting. Used by 77 per cent of French companies in managerial selection, but seldom in other countries. Based on the assumption that handwriting style and content reflect personality attributes such as honesty, motivation, emotional stability, social skills, disposition and loyalty. For example, it is contested that right-hand slant indicates a more emotional person, that pressure on paper indicates energy levels and that the size of script indicates the ability to concentrate, etc. There is considerable scepticism about this technique's validity and reliability.

Bio data: based upon the assumption that a person's life history (education, age, marital status, number of children, employment history, etc.) shapes their attitudes and behaviours (work performance, reliability, stability, etc.). Information is usually collected by questionnaire. Care must be taken in making discriminatory decisions.

Astrology: in the UK less than 1 per cent of applicants are selected in this way. However, some people believe that the position of the stars and planets when we were born and subsequent movements determine our personality and behaviour.

Assessment centres: many companies outsource the selection process. Testing centres sit down with an organisation and construct an elaborate cocktail of tests to suit a particular position. Testing centres are effective but not cheap.

Whichever technique is used, the aim is the same – to measure the degree of fit between the 'gap' in the organisation which needs filling for the achievement of its goals, and the capabilities and interests of the candidate.

Although sophisticated selection systems which focus on a range of criteria have been developed in construction, most selection processes remain unscientific (Druker *et al.* 1996). Whilst some companies have begun to use psychometric tests, assessment centres and behavioural event interviews, such practices are not widespread. In many cases projects are staffed on the basis of availability alone, and subcontractors are normally employed on the basis of price (after pre-qualifying). This is despite considerable evidence that team structure has a significant influence on organisational success (Barbara 1997) and that personality can influence people's ability to cope with work demands (Dolan 1995). Until companies begin to lose tenders on the basis of not being able to staff a project appropriately this is unlikely to change. At the same time, this widespread ignorance of team-building research is a source of potential competitive advantage to more insightful, knowledgeable and imaginative companies.

HRD

HRD represents the developmental side of the SHRM cycle, in improving career management and the performance of the individual. The rationale behind investing in HRD is that investing in people in the right ways will ensure that they continue to contribute to the direction in which the business wants to go. Failing to address HRD needs inevitably leads to the reopening of skills gaps, which the above processes have been designed to fill.

Despite the importance of HRD, there is evidence that it is under-utilised within the construction industry. For example, Hancock *et al.* (1996) found that, whilst large construction companies generally understood the concepts of HRD, only around half actually practised it. B. A. Young (1988) found that 75 per cent of construction companies had no career-development policy to allow individuals to compare their personal career needs with those of the organisation, and Mphake (1989) found that only 17 per cent of large construction companies had formal management-development policies. This under-reliance on HRD reflects the industry's reliance on 'hard' systems approaches, or those commonly attributed to 'personnel-management' practices. In the light of the industry's economic revival in the late 1990s Knutt (1997a) made a series of recommendations to promote loyalty and motivation among construction employees, including extended training schemes, career-review systems and performance-management systems. Together these mechanisms can ensure that the organisation meets its succession needs, and that it has people moving through the organisation with the requisite skills and abilities to fill vacancies and drive the business forward. Each of these aspects is explored in more detail below.

Staff training and development

Employee training and development should closely interact with staffing and performance activities. It is a vehicle for facilitating organisational and individual learning through training and development (Beardwell and Holden 1997; Sisson and Storey 2000). Systematic as well as ad-hoc development programmes help to ensure that staff have the skills required for their current roles and can develop those required for future posts. It can also work as a motivating factor: significant training indicates commitment to people and the recipients are more likely to feel valued (Sisson and Storey 2000). A construction organisation must consider training both to integrate employees into the organisation and then to facilitate their development and retention, as outlined below.

Orientation programmes

With increasing understanding of organisational culture, learning processes and the important role of knowledge within organisations, issues of orientation and staff development have been attracting considerable attention. It is now recognised that organisations have distinct cultures and that new employees must go through a process of orientation/socialisation to avoid the inefficient effects of culture shock when they join a new organisation. Within construction firms the process of orienting and inducting new employees is also important for health and safety reasons. Explaining how systems and procedures work in order to protect employees is key in ensuring that everyone working on a project complies with the health and safety ethos.

There is considerable evidence to indicate that those who go through an effective orientation or induction programme are considerably more effective than those that do not (Feldman 1980). However, in reality most organisations give new employees no more than a quick tour around the workplace, with cursory introductions to their co-workers. Certainly, at construction project level there is very little evidence of comprehensive orientation programmes other than in the most innovative projects. Those orientation programmes that exist are typically superficial and factual, being restricted to mandatory requirements which have been imposed upon the employer by legislation relating to health and safety, etc. Often they are delivered by union officials. Such programmes completely miss the point of orientation, which is to acclimatise people to their new job environment, to an organisation's goals and objectives, to its culture and to the expectations associated with their roles. Essentially, orientation should be a process of socialisation and acclimatisation, which needs to occur gradually over a period of time, particularly in large organisations where there is much information to absorb.

Noe *et al.* (2000) describe the basic principles of a well-designed orientation programme:

1   It should involve a series of incremental stages, the first beginning with the most job-specific, relevant and immediate information and the latter with more general policies about the organisation as a whole.
2   It should occur at a pace with which the employee is comfortable. Four sessions over twelve weeks is widely accepted as ideal.
3   The most significant aspect of orientation is the human side – giving new employees information about their supervisors and co-workers, telling them how long it should take to reach effective work standards, and encouraging them to seek help and advice when they need it.
4   New employees should be mentored by an experienced and accessible worker during the induction period.
5   Introductions to co-workers should be gradual. A superficial introduction on the first day is insufficient.
6   There should be a period of lower work commitment to allow people to get their feet on the ground.
7   Supervisors should continue orientation after the twelve-week formal process is complete. Channels of communication to the SHRM department should be kept open to prevent isolation.
8   There should be a follow-up after six months to collect feedback from the new recruit to identify and redress any problems.

The importance of the follow-up cannot be overstated. If problems are dealt with early on there is far less likelihood that the employee will be lost due to misunderstandings or through being unable to conform with the organisation's procedures or processes. Follow-ups can also be used to establish the efficacy of the recruitment and selection procedures. If people are being recruited who find it difficult to integrate into the organisation, then this suggests that there is something wrong with the recruitment criteria and/or selection mechanisms, which can then be amended accordingly.

Training and development programmes

If employee orientation involves getting the employee started in the right direction, personnel development is about keeping them there and ensuring that they contribute and add value to the business. The continuous development and training of staff is essential to maintaining a healthy, motivated and adaptable workforce. It is critical to ensure that people have the skills that they need to perform their job and advance along their career path. Indeed, with the construction industry's need to recruit from an ever-wider range of disciplines, fewer and fewer employees starting a new job will have the necessary skills to do their job effectively, which makes training a necessity from the very beginning of a new job, particularly for new or recent graduates.

Essentially, the object of training is to alter permanently the behaviour of employees in a way which will bring improvements in the achievement of organisational goals. It should provide opportunities for an employee to learn job-related skills, attitudes and knowledge. Since training is a form of learning, to be successful it is essential that employees are motivated to learn, are able to learn, are able to transfer their learning to the job, and have their learnt behaviour encouraged and reinforced in the workplace. Arguably the most important purpose of training in the modern dynamic business environment is to bring about a learning culture. Keep and Rainbird (2000) suggest that there are three different states of learning within an organisation:

- individuals learning things;
- organisational learning, where the organisation collectively develops ways in which it can learn collectively;
- the learning organisation, where the central organisational goal is systemic learning.

Senge defined the learning organisation as one where 'people can continually expand their capacity to create the results they truly desire, where new and expansive patterns of thinking are nurtured, where collective aspiration is set free and where people are continually learning how to learn together' (1990: 3). Learning organisations seem to be particularly proficient at problem-solving, developing new ideas, learning from their experiences and transferring new ideas to the workplace. This idea has begun to change the traditional concept of *training* from an information-providing role to one of *education*, where people are helped to 'learn how to learn' and have an open and enquiring mind, rather than being stuffed with facts. In essence, the emphasis is on developing thinking skills and a new work ethic rather than on spoonfeeding information. This result is a much more flexible and creative workforce. The learning organisation is one in which this level of learning is embraced and a high level of dialogue exists, enabling employees to explore organisational issues freely in search of creative solutions to problems. Thus the concept of the learning organisation effectively questions the individualisation of learning, and suggests that the social and systemic dimensions of learning are key in determining how organisations acquire, develop and deploy their skills (Keep and Rainbird 2000).

Unfortunately, Druker *et al.* (1996) found that most construction companies are far from learning organisations, and there is no evidence to suggest that the same is not true for projects. The challenge of learning from project to project faces most construction firms, and one useful mechanism to capture and share experiences in projects is to conduct focus groups with participants as part of a post-project review (or post mortem). The data collected can then be compiled and shared with other project teams within the company. Unfortunately, this is often difficult, since most projects do not end suddenly, but with a gradual dissipation of staff on to other projects. The end of a project is therefore difficult to define, and learning would involve extracting people from other projects for a day or so. However, all the evidence indicates that the inconvenience is worth it.

Druker *et al.* (1996) also discovered that training was in decline within the construction industry and found little evidence of old training initiatives being replaced with new ones. Indeed, many companies were closing down their management-development centres. They also argued that the severe skills shortages, which predictably emerge in every construction boom, were largely the result of growing self-employment. The small subcontractors, which employ over 85 per cent of the industry, are so highly geared that long-term investments in training have been impossible in the traditional short-term boom–bust cycle of construction activity.

Nevertheless, training is still a fundamental requirement for improving organisational performance and filling skills gaps. Much of this information will be available from the job-analysis process, and these skills gaps will identify the needs and objectives of the programme in the short, medium and long term. They will also identify the type of training which specific employees will need.

An important part of developing a training programme is the decision of whether to use on-the-job or off-the-job training. In terms of avoiding the potential problems of persuading project mangers to release people from a project, on-the-job training offers a significant advantage over off-the-job training and is used for more than 60 per cent of training provision in the construction industry (Noe *et al.* 2000). Probably the biggest advantage is that people are still available to deal with problems occurring in the workplace. However, the close proximity to work also has the potential to reduce training effectiveness significantly by distracting those who attend. Furthermore, remaining on site can make the training seem less attractive and important than a day away from the project environment. However, one unobtrusive and stimulating method of on-the-job training is *job rotation* or *transfers*, which involve employees swapping jobs or sites for periods of time to give them a range of skills and experiences. Job rotation is particularly useful for providing people with insights into interdependencies with working colleagues and for placing their own project in context. Another unobtrusive method is *mentoring* or *coaching*, where a supervisor is given formal responsibility to train an employee. If managed well, this is an effective means of learning, although it depends upon the quality of the mentor and their ability to impart knowledge and form personal relationships with individual members of staff. In this sense it is crucial that potential mentors are given the proper training, resources and time to carry out this task. Simply asking a busy project manager to mentor his or her staff on top of existing workloads will be unlikely to succeed. Finally, one method of on-the-job training which is now becoming more popular relies upon company *intranets*. This is an internal company or project Internet site which can deliver programmes which have been customised for a particular site or company. Intranets can facilitate multimedia delivery, virtual reality, online discussion groups and can support interactive training software. Employees can complete training modules at their own pace, and progress can be monitored centrally.

In terms of off-the-job training, the choice is lectures/discussion groups run by training companies, distance-learning programmes, programmed learning or external mentoring/coaching. The most frequently used approach is for a trainer to give a lecture and involve the trainee in a discussion about the material to be learnt. Alternatively, one could organise a series of courses and arrange some kind of accreditation through a private accrediting body or university. Choosing a respected partner and capable deliverer is critical to the effectiveness of this approach in terms of knowledge, reputation, experience, track record and presentation technique. As with all training material, it is essential that the organisation undertaking the training be closely involved in developing and scrutinising the material to be delivered. It is essential that it is seen as relevant, useful and interesting to the people involved. Innovative delivery mechanisms should also be considered to facilitate the training process. Distance learning is set to expand dramatically over the next decade, with companies creating strategic alliances with educational institutions to form virtual universities where students can 'pick and mix' individual modules to suit their own needs. Clearly, the main advantage of distance learning is its flexibility and the opportunity to study without regular attendance at a remote educational establishment. Training and development approaches are discussed in greater depth in Chapter 10.

Performance management

Performance management aims to generate better results from individuals, teams and the organisation as a whole. Essentially, it is about planning goals, targets and standards, continually monitoring progress towards achieving them and providing support where necessary. According to Armstrong (1991), the process of performance management begins with a performance agreement between the manager and subordinate. This comprises a set of achievable objectives, together with the developmental needs required to achieve these objectives. During the ensuing assessment period (usually 6 to 12 months), performance against these targets is monitored and measured and action is taken should any of the objectives prove difficult to achieve. Poor performance can be dealt with through simple motivational techniques such as coaching and/or counselling, whilst effective performance is reinforced with praise, recognition and increased responsibility. The core principle of this approach is that problems are dealt with proactively during the monitoring process and not reactively at the end of the review period. A formal review is carried out at the end of the process, where both the individual and their manager compare the employee's achievements with the original targets and set new targets for the coming year.

The emphasis of performance-management systems is clearly to provide flexibility in approach, supported by high-quality feedback. The ultimate form of this type of performance-management system is known as 360-degree feedback, where a variety of internal and external parties all provide an assessment of an individual's performance. This process provides multiple stakeholder perspectives on performance, and allows the process to explore broader elements such as teamwork and relationships, as well as traditional criteria. Those asked to provide feedback on the performance of an individual come from both within and outside the organisation, and include their peers, subordinates, managers and clients. However, the very breadth of the appraisal process forms one of the key disadvantages of the 360-degree feedback approach – that it is costly to implement and difficult to manage. It usually requires a considerable staff resource to manage the data and control the processes attached to it. Another criticism is that subordinates of the individual being assessed are unlikely to provide honest feedback if they fear retribution from the manager in the future. Thus it relies on a culture of openness and honesty if it is to be implemented successfully.

Management development and career management

Management development is the organisation's primary tool for ensuring that it has the managers available to meet the organisation's present and future needs. The goal is to ensure that the organisation provides *management succession*, that managers move through the organisation continually replacing others, filling vacancies as they arise. Whereas in the past this may have been left to a process of natural selection, contemporary SHRM thinking suggests that this is a process that demands careful management if the organisation is to benefit.

Langford *et al.* (1995) discuss the practice of management development within the construction industry. They suggest that different firms adopt different practices depending upon the sector of the industry in which they operate. Moreover, they suggest that different professions and disciplines have different management-development needs. It is clear that management development is not a single technique or programme, but a series of interrelated activities which occur together. In reviewing the literature on the function and considering how it relates to construction organisations, Langford *et al.* distilled the following features of the function:

- it relates to the overall organisational strategy;
- its success is measured in terms of the extent to which the organisational goals are achieved;
- it tries to harmonise the development of a manager with the organisational context as far as is possible;
- it is an integral aspect of the structure and culture of an organisation;
- it is essential to the present and future effectiveness of the organisation;
- whilst some acknowledgement is given to developing the individual manager, the emphasis is on ensuring that people fit well into the organisation.

This list of defining features suggests that management-development approaches must be flexible in order to cope with the changing socio-economic context of the organisation.

Career management

Traditionally, the ways in which organisations function have depended upon the formal hierarchy of positions that define their structure. Accordingly, the thinking underpinning the personnel function has traditionally relied on hierarchical assumptions of the employee – that they are striving for vertical promotion within the organisation, or understand their position and role by virtue of who they are responsible to and for. However, contemporary thinking and research have questioned this traditional view of employee priorities, particularly in the context of flattening organisational structures. Under contemporary approaches to organisational design, the requirements from individuals are now much broader and the opportunities for vertical development much narrower than ever before. Thus, effective career-management and development policies must increase the likelihood that the right people will be available to meet the changing HR requirements whilst reducing frustration stemming from the downsizing and flattening of organisational hierarchies (Thompson and Mabey 1994).

The basis of career management begins with developing an understanding of the nature of career-progression dynamics within an organisation. This is the process of establishing how people rise through the organisation in order to map career trends and patterns for those in different roles and functions. From this analysis the extent to which the organisation is growing its own talent or 'buying in' expertise will become obvious. In addition, it will highlight where certain individuals have stagnated or reached a plateau in their development. The organisation can then tailor career plans for each of its managers, identifying training and development needs and opportunities which should help them to achieve their goals whilst contributing to the organisation's objectives.

In construction, like any sector, career dynamics and individual priorities vary considerably from manager to manager. For example, for some project managers site-based work is a mere stepping stone on the way to a senior head-office position and a transition which must be endured in order to reach a position of authority. For others, site-based roles provide them with innate job satisfaction that they could not obtain from office-based roles and so they do not wish to progress beyond this level. Generally, people working within flatter, matrix-structured organisations are far less likely to crave vertical promotion than those working in hierarchical line-and-staff organisations. Furthermore, as people's expectations concerning the balance between their work and non-work lives change, employees may legitimately adopt a 'middle-laner' mentality. The 'middle-laner' is someone who has, by choice, taken a step back from the 'fast-lane' career path. It is important to recognise the legitimacy of this decision, and if employees are happy to continue to perform their job well this decision should be respected. The career-management process must be able to respond to any of these priorities and create appropriate opportunities for individuals to utilise their skills and abilities in pursuance of their own career and personal goals.

Performance appraisal systems

Performance appraisal is a sensitive process which involves measuring the contribution of every employee towards the accomplishment of an organisation's objectives, now and in the future, for the purpose of identifying rewards and penalties. The underlying objective of performance appraisal, as with so many of the SHRM functions, is to improve the performance of individuals in such a way as to contribute to the performance of the organisation as a whole (Mullins 1999). Accordingly, in construction projects appraisal systems should operate at both individual and organisational levels, appraising the performance of subcontractors and consultants as well as individual employees. However, performance appraisal is not a one-way process and it should also serve as an important input of information to human resource planning and development. For example, it should help to identify skills gaps and whether employees need training, and it encourages supervisors to monitor their workforce in a systematic way.

In recent years there has been increased recognition of the difficulties inherent in the appraisal process, particularly with regard to ensuring that it contributes to effective SHRM practice. The contemporary focus on integrating SHRM with business strategies has led to most organisations seeing performance appraisal as just one part of a more systematic approach towards performance management (Bach 2000). Thus there has been a shift away from performance appraisal towards performance management. However, performance appraisal still remains an integral aspect of performance management and an important tool for managing reward within many construction organisations.

The importance of performance appraisals to the lives of individuals and the prosperity of organisations who are assessed has ensured that the process has been surrounded by controversy. For example, the quality expert Deming (1994) believed that performance appraisals were subjective, open to bias, unreliable, unmeaningful, short term in focus, divisive, destructive, and that they introduced fear and politics into the workplace. Indeed, in a survey of a range of industries Vinson (1996) found that only 10 per cent of employees are happy with their appraisal system and that, ironically, more often than not these systems are counterproductive to performance. The problems of appraisal are particularly acute at managerial level with self-motivated professionals who value autonomy. They tend to set their own goals and react badly to being asked to specify goals to which they will subsequently be tied. However, problems can also occur at operative level, especially if appraisals are seen as a non-genuine and cynical attempt by managers to gather some ammunition to justify staff lay-offs. However, the most controversial aspect of performance appraisal is its relationship with the law and issues of discrimination. The law relating to equal opportunities is highly complex and it is essential that organisations work within it to make their decisions legally defensible. To avoid potential problems Bernardin and Cascio (1987) suggest that an appraisal system should adhere to the following rules:

1 Procedures for personnel decisions must not differ as a function of the race, sex, national origin, religion or age of those affected by such decisions.
2 Objective, non-rated and uncontaminated data should be used whenever it is available.
3 A formal system of review or appeal should be available for disagreements over appraisals.
4 More than one independent evaluator of performance should be used.
5 A formal, standardised system for the personnel decision should be used.
6 Evaluators should have ample opportunity to observe ratees' performance (if ratings must be made).
7 Ratings on traits such as dependability, drive, aptitude and attitude should be avoided.
8 Performance appraisal data should be empirically validated.
9 Specific performance standards should be communicated to employees.
10 Raters should be provided with written instructions on how to complete the performance evaluations.

11 Employees should be evaluated on specific work dimensions rather than a single overall or global measure.
12 Behavioural documentation should be required for extreme ratings (e.g. critical incidents).
13 The content of the appraisal form should be based on a job analysis.
14 Employees should be provided with an opportunity to review their appraisals.
15 Personnel decision-makers should be trained on laws regarding discrimination.

Equity and diversity issues are covered in more detail in Chapter 8. However, the three most important steps in establishing an appraisal system are to establish the performance criteria, to establish who evaluates, when and how often, and to establish evaluation techniques. These are discussed in more detail below.

Establishing the performance criteria

There are three dimensions of performance that should be measured by an appraisal system, namely *behaviour*, *activities* and *outputs*. Behaviour measures reflect the characteristics a person brings to the job, such as reliability, positive attitudes, enthusiasm, initiative, knowledge and expertise. Activity measures reflect the efficiency with which people go about their work and interact with others. Finally, output measures focus on the results actually achieved by an employee, in terms of quality of work and productivity. All three can be applied to organisations and individuals in a project setting. However, in construction projects, appraisal systems are rarely based on anything other than processes and, especially, outputs. This is understandable in the light of the industry's engineering values, which have traditionally seen construction projects as technical systems. However, with the increasing emphasis upon trust which seems to be emerging within the industry (Latham 1994) there is a growing recognition that the industry needs to perceive projects as socio-technical systems and assess softer performance factors relating to traits such as trust, commitment, etc.

Who, when and how often

At an *individual level*, most organisations conduct appraisals annually, although about 16 per cent evaluate twice a year and 4 per cent evaluate three times per year (B. N. Smith *et al.* 1996). However, in construction projects most appraisals are based on processes and outputs and take place at an *organisational level* in relation to issues such as payments, work quality and progress. These normally occur at regular intervals such as once a month, but are increasingly based around milestone dates linked to a project programme.

Whether appraisals occur at individual or organisational levels, the process is unavoidably personal and has to be undertaken between the individual and his or her supervisor on a one-to-one basis. The problem with such a system is its susceptibility to bias due to personal differences between managers and employees, intimidation, stereotyping and discrimination. Consequently, problems are often experienced in the acceptance of such appraisals and there is a danger that the process becomes more harmful than useful. Indeed, in construction projects many conflicts revolve around personal disagreements between individuals who have differing opinions of appraisals that have been made. For this reason, it might be worth considering other methods of appraisal than those traditionally used within construction projects (Rowe 1995).

The first alternative is to use a *committee* of several supervisors (with some external independent representatives) to make assessments. This offsets the effect of bias and introduces more information into the process. Another option is to use *specialist outsiders* who are not stakeholders in an organisation or project. External performance-management specialists can take a more objective view, which avoids accusations of bias. However, this system can suffer from the assessor having a lack of information compared to internal assessors. Furthermore, it can slow down the assessment process. Finally, there is *self-evaluation*, which is often met with scepticism because of the self-interest of the employee. However, research shows that self-evaluations are highly reliable, correlating closely with supervisors' ratings (Rowe 1995). Despite this, it is difficult to see self-evaluations having any part to play in the traditionally divided and confrontational environment of the construction industry. However, they might have a role in cementing partnering relationships, since there can be no greater gesture of goodwill than to trust someone.

Performance appraisal in the construction industry

There are two broad types of technique used to conduct assessments, namely those based on direct comparisons with other employees (multi-person/company evaluations) and those conducted in isolation (single person/company evaluations). In construction projects and at organisational level, appraisals are normally undertaken in isolation and by measuring progress against objectives. Management by objectives (MBO) is by far the most common method of evaluation found in traditional job-oriented organisations and is based wholly upon *output* and *process* criteria. Other techniques for *traits* criteria are the graphic rating scale, the forced rating scale, the essay technique and the critical-incident technique. The graphic rating scale technique is essentially based on a checklist with criteria which can be graded on a scale of 1 to 5. The tendency of people to tick 3 and provide an indifferent evaluation has led to the forced rating scale technique, which simply requires assessors to rate a person's relative performance across a range of criteria. The essay technique requires the assessor to write a verbal exposition on an individual's performance, and the critical-incident technique requires assessors to maintain a log of behavioural incidents which give some insight into a person's performance at work. There is no reason why any of these could not be used on a construction project to assess the traits of both individuals and organisations. Furthermore, there is no reason why rewards and penalties cannot be based upon the assessment results.

The problem facing construction personnel managers concerns their capacity to deliver performance-related pay systems which link organisational and managerial rewards to broader SHRM objectives (Druker and White 1996a). Competence-based or skills-based systems are one way of achieving this goal, as are incentive schemes rewarding collective working and project-team cohesion for improved project performance. A distinction between the various methods of appraisal available concerns the nature of the measures that are used. Measurement can be made in either a quantitative way (for example using rating scales of 1 to 5 which relate to various aspects of performance) or qualitatively (for example using descriptive reports of employees' performance). One approach, known as behaviourally anchored rating scales (BARS), is designed to allow a measurement to be made of performance in the actual job being appraised. Essentially, the process involves defining a benchmark of appropriate practice based on the collective perceptions of managers, who are asked to decide what the key behavioural requirements for performance are and to attach a scale for poor, average and high levels of performance. Although costly to construct, BARS do offer the advantage that the performance measures are based on a consensus of managers' opinions of what is required for a particular function. BARS can also include more general performance metrics such as teamwork.

Reward management

All managers have to decide upon rates of pay, pay structures and how performance will be rewarded. Reward management is about the design, implementation and maintenance of these systems, which support the improvement of organisational performance (Armstrong 1998). Reward is the reason why the majority of individuals seek employment, and those who perform well expect to be rewarded in some way, particularly in the case of the highly commercialised construction industry. The objective of a reward-management strategy should be to create a system of rewards that motivates people to work towards an organisation's goals. In essence, a manager can either provide *monetary* or *non-monetary* rewards. Monetary rewards include pay, bonuses, stock options, profit shares, etc. and non-monetary rewards can include prizes, awards and promotions to jobs with

greater status, responsibility and authority. We can classify these induce-
ments under two headings:

- *Extrinsic rewards*: tangible rewards such as salary, bonuses, commission payments, working conditions, company cars and pensions. These
rewards are effectively the 'hygiene' factors identified by Herzberg.
- *Intrinsic rewards*: the opportunity to satisfy other goals such as
lifestyle, comfort, a sense of achievement, status, acclaim, challenge,
etc. These are the factors that lead to job satisfaction and motivation.

The balance between the monetary and non-monetary rewards offered
by an organisation is likely to depend on a range of internal and external
factors, especially the financial state of a company, competition within the
labour market and the policies of its rivals. For example, during boom
times there is likely to be a greater emphasis on monetary reward, particularly if rival employers are offering relatively high wages. In many
construction projects the emphasis is almost entirely upon monetary
rewards, but it is debatable whether this strategy achieves the aim of motivating people and organisations to perform to their highest levels. In
Chapter 2 we discussed how human relations theorists recognised that
all kinds of cognitive processes affect the relationship between pay and
motivation. In particular, Maslow (1943) argued that all human behaviour stems from needs and drives which are innately biological in origin.
He proposed a hierarchy of needs, which in descending order were self-
actualisation, ego/esteem, social, safety and physiological. Only lower-level
needs (physiological and safety) motivate people to earn money to buy
shelter etc., which means that the motivational impact of money is limited
and soon lost. Herzberg's (1959) hygiene theory identified two sets of
factors involved in motivation: *hygiene factors* such as pay, working conditions etc., which do not motivate but only act upon dissatisfaction; and
motivators such as achievement, recognition, responsibility and work
itself, which act directly upon motivation levels. In essence, Herzberg
argued that good pay would not motivate but merely reduce dissatisfaction. More recently, *social comparison (equity) theorists* have argued that
motivation is affected by perceptions of fairness and equity in terms of
comparing pay levels to others. Furthermore, there is an *expectancy* force
at play which ensures that motivation from pay will also depend upon
how closely it approximates to people's *expectations* of what is reasonable
for their efforts. Thus, the link between pay and motivation is far from
clear, and so it is important that managers have a reward system which has
monetary *and* non-monetary elements attached to it.

Developing effective reward systems in construction

Patton (1977) argued that to be effective, compensation systems must be:

- *adequate*: provide basic security for employees/organisations;
- *equitable*: provide rewards which are in line with efforts and contribution;
- *balanced*: between monetary and non-monetary rewards;
- *cost effective*: affordable to the employer;
- *incentive providing*: employees should see the benefits of diligence and of seeking opportunities to improve performance;
- *acceptable*: the employee should be happy with them.

Construction companies would benefit greatly from adhering to these basic principles, since it is debatable whether the current reward systems employed in construction projects do so. For example, there is a wealth of literature going back many years which points to the inequities contained within construction contracts, particularly towards subcontractors. More recently, with the abolition of fee scales, professionals are also complaining of inequities in the construction market. This inequity is of particular concern because it has been linked with poor performance. For example, Loosemore (1999a) found that it is a significant source of conflict between subcontractors and principal contractors in the construction industry, and a recent study of civil engineers in Australia (Lingard, in press) found that engineers compare their input–output ratio unfavourably with ratios enjoyed by other professions. This has led to cynicism and dissatisfaction with their profession. According to equity theory an individual perceives a situation as fair when their own ratio between outcomes and inputs is the same as that of a comparison other. Ensuring that this ratio is perceived to be fair within organisations is important, as failure to do so is likely to increase employees' turnover rate.

In construction, reward structures are particularly problematical for outsourced work and subcontracted labour, where, although systems are undoubtedly affordable to contractors, there is considerable evidence to suggest that organisations do not perceive the system to be equitable, adequate, acceptable or to provide incentives. Certainly, the emphasis is very much on penalties rather than rewards, and those rewards that are provided are almost exclusively monetary in nature. Finally, it is important to point out that this is not always the fault of principal contractors. Like any organisation's, their compensation systems are merely a response to and a reflection of the compensation environment imposed on them. If there is to be more imagination in compensation systems at construction-project level, it is clear that some initiative must be taken by clients. However, the majority of clients are in the hands of those who draft standard forms of contract, and ultimately it is their responsibility to make contracts more equitable, more encouraging of innovation, less penal and threatening, more balanced, more sensitive to good management practices and simpler to understand. This is the contractual challenge of the future.

ICT support for the SHRM function

One aspect not yet explored within this book is how ICT can be used to support the administration of the SHRM function. There is considerable evidence to indicate that ICT can bring significant benefits to the efficient management of the SHRM function (Eddy *et al.* 1999). The key advantages of using ICT in managing construction SHRM include:

- easier provision of information to line managers, thereby enabling rapid resourcing decisions during projects;
- easier processing and control of employee records and performance data linked to reward systems (i.e. removing the need for managers to maintain unwieldy paper-based systems);
- a reduction in the workload of the personnel function, thereby lowering the head-office overhead associated with the SHRM function.

In response to the needs of the modern SHRM function, increasingly sophisticated ICT packages are being developed. The latest generation includes those which interface with other organisational systems in order that senior managers can access information which allows them to make SHRM decisions that respond to business priorities. However, these systems represent the latest evolution in a line of computer systems that have been used to support the personnel and SHRM functions. Initially, they were little more than databases used for storing personnel records and related information. This type of systems helped to manage SHRM transactions and thereby supported routine high-volume HR decisions. In essence, they automated daily administrative HR tasks, integrated cross-departmental activities, and ensured the accuracy and consistency of employee records. The next generation of software incorporated greater functionality and the ability to facilitate the administration of the function. These systems enabled managers rapidly to prepare reports on workforce profiles or even to correlate levels of pay and positions with employee groupings. This allowed the rapid assessment of particular SHRM/HRD requirements or analyses of employee trends, such as relationships between particular employee groups and their rates of pay or grades. Known as expert systems, they effectively allowed the user to improve decisions through rules derived from careful analysis of expert decisions over time. The latest generation of ICT systems are known as human resource information systems (HRISs) and have the capability to act as decision-support tools to SHRM decisions. These allow the user to model trends in employee profiles and to try what-if scenarios based on particular demand-and-supply projections. Recent developments have led to HRISs having the potential to hold comprehensive, almost endless, databases of employee skills and competences, and to produce complex reports mapping employee abilities and preferences against forthcoming vacancies/projects.

The latest generation of web-enabled HRISs now also allows employees to update their own personnel records, submit timesheet data, review benefits, request holidays and enrol on training courses. This integration of so many key SHRM activities can facilitate both the recruitment *and* retention of staff by delivering automated recruitment features and quantifying the value of total compensation packages (McLeod 2001). Thus HRISs are revolutionising the HR function by providing up-to-date information, services to employees, return on investment, and strategic analysis and partnership.

Regardless of how powerful computer applications become, it is unlikely that they will ever develop to such an extent that they negate the need for a human element in strategic and operational decision-making. There will always be a need for subjective decisions based on dynamic requirements from an organisation's SHRM capability. Employees cannot be treated as a homogeneous group, and bespoke decisions will need to be made for each individual if their contribution, development and retention are to be effectively managed by the organisation. Thus, the key benefit of automating SHRM processes is that it leaves HR professionals with more time to focus on strategic activities and provides them with information to turn their employee assets into a source of competitive advantage.

Case study

Shepherd Construction Ltd and Birse Construction Ltd: an innovative approach to performance management in construction

Competence-based performance management has received considerable attention in recent years due to a widespread acknowledgement of the centrality of human resource development to the achievement of strategic business objectives. It is based on a cyclical and continuous process of performance planning, employee assessment and corrective action. Managed effectively, the process will ensure that employees meet the levels of competence required for improved business performance. Shepherd Construction Ltd and Birse PLC are two of the UK's leading construction contracting organisations. Conscious of the need continually to improve the performance of their managers, they are currently engaged in a research project in collaboration with Loughborough University and funded by the Engineering and Physical Sciences Research Council, in which they are developing an innovative, competence-based approach to managing the performance of their key project-based staff. They are adopting leading-edge tools and approaches used in other project-based environments such as the aerospace technology and marine engineering sectors. By developing a generic competence framework, this will eventually facilitate decision-making on issues such as recruitment, selection, training, professional development and succession planning within these organisations.

Behavioural competences are usually 'criterion validated', which means that they are established through the analysis of the behaviour of superior performers in a particular occupational role. Accordingly, Shepherd and Birse are establishing competency profiles of what defines an effective manager in *their own* corporate context. This involves establishing the knowledge, skills and other qualities required for the successful achievement of project objectives. The companies are using the McBer job-competency assessment process to provide a systematic methodology to underpin this performance improvement project (see Spencer and Spencer 1993). This involves:

- defining performance-effectiveness criteria;
- identifying a criterion sample of superior and average performers;
- collecting data on their competences;
- identifying job tasks and competency requirements;
- validating the competency model;
- applying the model within selection, training, professional-development, succession-planning and performance-management contexts.

Initially, the companies have identified a range of exemplary performers against a range of performance criteria that were derived through focus groups and workshops held with their project-team members. A range of superior performers were then identified whose skills, knowledge and behaviour were systematically mapped using *behavioural-event* interviewing, observation and activity-sampling techniques. This process has allowed the organisation to ascertain the generic competency requirements of superior performance. The next phase was to translate these competency frameworks into a performance-management tool that will encourage their managers to develop the skills that will drive their businesses forward in the future. Although this approach required detailed job-task and competency requirements to be established, the resulting competency frameworks should be robust and generic enough to be applied across the organisation in order that the performance of all managerial staff can be improved in the longer term.

Shepherd and Birse's work towards competence-based performance tools is particularly timely, as it addresses one of the key underdeveloped themes of Egan's *Rethinking Construction* report (Egan 1998), which defined new levels of performance for the UK construction sector. The report called for better training of top management and project mangers to enable productivity improvements to be made. It advocated programmes to develop leadership skills, and better career structures to develop more leaders of excellence. This project is developing one of the tools necessary for achieving these targets and for ensuring that people contribute to the process of continually improving construction business performance.

Conclusions

A successful project manager must harness the efforts of the human resources at their disposal. People are an essential ingredient in every organisation's success, and one of the manager's greatest challenges is in finding good people, utilising them to their full potential, guiding them towards the accomplishment of organisational objectives, integrating their efforts into the organisation, training and developing them, promoting/demoting them and retaining/terminating them. The challenge of SHRM for project managers is therefore significant, and requires a belief in and knowledge of effective HRM tools and techniques. However, the fragmentation of SHRM functions within construction companies and the separation of SHRM from those with specialist knowledge in this area have meant that project-level HRM has been largely ignored. This neglect has also been compounded by the task-oriented culture of construction, which devalues people issues. More often than not, time and cost pressures and an ignorance of SHRM have precluded a thoughtful approach to resourcing, HRD and reward management.

In this chapter we have reviewed some of the basic mechanisms that comprise the HRM function and have applied them in a construction context. It would be easy to dismiss these aspects as being of low importance because of their apparent distance from the high-level SHRM decision-making that underpins much of the thinking within this book. However, the effective management of these aspects is essential for translating broad strategic objectives into practical improvements in organisational performance. Moreover, they are the tools the organisation can use to prevent breaches of the psychological contract, which can ultimately undermine the commitment of employees to the organisation.

Considering the apparent under-utilisation of the HRD function within the construction industry, it seems reasonable to hypothesise that an incongruity may exist between organisational policy and the career needs of the individual within many construction companies. If this is the case, then this may explain why employees seek to develop their careers within competitor companies, or even outside the sector. In construction people are increasingly employed on what are effectively contracts for performance – an expectation of the performance of the individual in return for performance/profit-related pay or the achievement of particular objectives. This may have contributed to the culture of performance orientation that now pervades the industry and the long-hours culture which prevails, in which employees are unlikely to take lunch breaks, begin early and stay late, working harder but not necessarily smarter. The resultant person–job mismatch can lead to burnout and ultimately to withdrawal through turnover. If people are to be retained and developed so that they add greater value to the organisation in the long term, project managers must begin to understand the importance of psychological expectations and adopt appropriate techniques for managing this aspect of the employment relationship. This demands that more careful consideration be given to resourcing, HRD and reward systems, supported by the appropriate use of ICT as a delivery mechanism.

Discussion and review questions

1 Many people believe that the management of people within large construction organisations should be a centrally, head-office managed activity in order to reduce the burden on construction project managers. Discuss the validity of this view given the contemporary view of the SHRM function outlined within this book.

2 Discuss the aspects of construction work which provide you with intrinsic satisfaction, and compare and contrast these with the extrinsic factors used by your employer to encourage your motivation and commitment.

3 Consider the aspects of the SHRM function which you have been exposed to whilst working for construction organisations. Identify aspects where ICT could facilitate these tasks and encourage your commitment to the organisation.

6 Employee relations

'Employee relations' (traditionally known as industrial relations) refers to the relationship between operatives and managers and the system that attempts to control this. It has informal and formal dimensions that define and control the nature of the employment relationship. It is important to understand that good industrial relations are an automatic outcome of managing the topics covered in this book, such as occupational health and safety, equity and cultural diversity, compensation, reward, termination, etc. Therefore this entire book could be interpreted as a text on how to achieve good industrial relations. However, this proactive approach has been largely neglected in the industrial relations debate, which has traditionally focused on the reactive mechanisms used to manage the effects of poor management and on the relationship between employee and employer representatives. Of course, management *is* difficult and disputes *are* inevitable, and this chapter focuses on how this relationship can be managed to positive effect. It is hoped that readers will acquire the knowledge and skills to avoid many of these avoidable disputes in the other sections of this book.

Employee relations must be understood within the economic, cultural, social and political context of specific countries. Consequently, this chapter focuses on general trends in industrialised countries, where employee relations have traditionally been negotiated through collective agreements with the workforce, often brokered through workforce representatives and trade unions. While unionism has declined since its heyday of the 1970s, largely because of changes to established patterns of work, this has not diminished the significance of employee relations in the construction industry. Instead, the emphasis has changed from confrontational relationships between employers and employees towards attempts to achieve greater cooperation to maintain market share in the face of greater international competition. This chapter examines the theory and practice of employee relations in relation to formal collective agreements, which are traditionally negotiated through unions, and less formal arrangements, which often exist at an organisational or project level. Both types of arrangement are now embodied under the employee relations concept. We will explore the nature of employee relations within the construction industry, examining the roles of unions, managers and employees in forming and maintaining an effective working relationship.

Introduction: unitary and pluralistic perspectives on employee relations

The 'employment relationship' is a key aspect of managing human resources because it brings together the rights, power, legitimacy and obligations that both employers and employees seek from each other (Beardwell and Holden 1997). The traditional view of industrial relations was to associate it with the rules and procedures relating to employment. Under this model, trade unions were seen as protecting the interests of their members, for whom they negotiated through collective bargaining to gain improved pay and conditions from employers (see Dunlop 1958). This rather narrow and somewhat outdated view of the employment relationship fails to take account of the informal social relations that are now acknowledged as influencing the effectiveness of how people work within organisations. Moreover, it fails to account for the role of the individual in employment relations (Armstrong 1996). For this reason, the term employee relations is now more commonly used, and it takes account of the motives, ideologies and perspectives of both the organisation and its employees. The effective management of employee relations is the prerequisite of positive psychological contracts and to ensuring that employees feel involved in decision-making processes. These are the foundations of improved levels of performance.

Before considering the intricacies of managing the employee relations function in construction, it is first necessary to explore some basic theoretical perspectives on the nature of work and organisations. In particular, it is important to distinguish between the unitary or a pluralistic perspective, two radically differing views of work relations which will influence the nature of HRM within them (Fox 1966):

- The *unitary* perspective views the organisation as a harmonious environment in which management and employees share common goals. Conflict is seen as unnatural and trade unions are seen as unnecessary because there are accepted leaders and a single focus for effort.
- The *pluralistic* perspective sees the organisation as a collection of distinct and competing subgroups that vie for power, frequently conflicting with each other. Conflict is not seen as wholly undesirable, but as a force which can act as an agent for change and innovation. Pluralist theories focus on conflict resolution and problem-solving in order to develop a state of dynamic equilibrium where the needs of groups are held in a state of balance. The pluralistic perspective implied by the collective bargaining process regards the workforce as

being represented by 'an opposition that does not seek to govern' (Clegg 1976, quoted in Armstrong 1996: 717). This implies that the organisation must form a relationship with employees as a collective body which allows for a balance of power between them, but which does not allow one side to dominate so as to destabilise the operation of the organisation. In the past this has been a relationship managed between trade union representatives and industrial relations specialists within the organisation.

In the construction industry both perspectives exist – managerial white-collar workers being bound by individual contracts committing them to working towards organisational goals, and blue-collar workers being served by collective bargaining. Construction is unusual in that it maintains a relatively strong and continuing reliance on collective bargaining that has long since ended in other sectors (Croucher and Druker 2001). However, major social changes have occurred in the last decade which have affected industrial relations. Higher levels of unemployment, self-employment and other economic changes have affected employment patterns and management strategies, leading to a decline in trade union membership. For example, in Australia the number of union members as a proportion of all employees has declined, standing at 24.5 per cent in 2001. However, it is interesting to note that, while union membership is on the decline in the traditional strongholds of mining, manufacturing and construction, it is increasing in non-traditional service sectors (*Australian Financial Review*, 1 March 2002). In Australia total union membership is around 2.1 million, which still makes unions the largest independent social organisation in Australia.

The structure of the modern construction industry poses problems for union-driven collective bargaining in that the vast majority of construction firms, particularly in the residential construction sector, are small businesses. For example, in Australia 97 per cent of general construction businesses employ fewer than 20 employees and 85 per cent employ fewer than 5 employees (ABS 1998a). A similar situation exists within the UK industry, where only 1.6 per cent of construction firms employed 25 people or more in 1994 (Druker and White 1996a), a situation which is unlikely to have changed greatly in the past few years. Small-firm employees are less likely to be union members and to engage in industrial action. For example, an *Australian Workplace Industrial Relations Survey* (Moorehead *et al.* 1997) revealed that 71 per cent of workplaces of between 5 and 19 employees had no union members. In contrast, 98 per cent of employees in workplaces of 500 or more are union members. Furthermore, only 2 per cent of businesses in the 5–19 employees group had experienced an industrial dispute in the preceding 12 months, compared to 49 per cent in the 500-plus employees group. However, the figures vary from sector to sector. For example, Barrett

(1998) found that, regardless of size, employees are more likely to join a union in traditional industry groups such as mining and construction. In the mining and construction sector 44 per cent of businesses in the 5–19 employees range had experienced industrial action in the preceding 12 months. Barrett also points out that, while employees in small businesses seem less likely to be involved in a union, this does not mean that employee relations are more harmonious in this sector. Low levels of unionisation merely indicate that group consciousness may be harder to foster among employees in small businesses and that employees in such firms may not benefit from collectively negotiated standardised terms and conditions. However, even in large companies legislative changes in many countries are making collective agreements more difficult. For example, in Australia the Liberal–National Party Coalition introduced a Bill in which individual contracts were the favoured means of regulating the employment relationship and the coverage of nationally agreed awards was stripped back to 18 allowable provisions (Teicher and Svensen 1997). The labour-market flexibility introduced by new legislative models such as these has led to a greater reliance on enterprise-level agreements between employers and their workforces (i.e. those negotiated within a single organisation). For example, in the construction industry it is now not uncommon for all aspects of the employment relations on a major project to be established a priori in a project-specific agreement. This, in turn, has placed considerable responsibility on project managers for managing employee relations.

Approaches to managing employee relations

There is no one best way of managing employee relations – organisations must develop policies and strategies which suit their individual circumstances. Many factors are likely to play a part in determining this, such as the type of product and labour markets in which an organisation operates, the type of technology in operation, the internal structure and organisation culture, and the degree and nature of trade union representation. Inevitably, industrial relations policies and practices will vary from industry to industry and from organisation to organisation – and they should. Having said this, it is useful to classify generic approaches to managing employee relations, and Purcell and Sisson (1983) have identified four main strategies: traditionalist, sophisticated paternalist, standard modern and sophisticated modern.

- *traditionalists* adopt a policy of 'forceful opposition' to trade unions and an overtly exploitative approach towards their employees;
- *sophisticated paternalists*, a category held to include companies like Hewlett-Packard, Marks & Spencer and Kodak, also tend to be non-union, but differ from traditionalists in that they spend

considerable time and money developing personnel policies which ensure that unions are seen as unnecessary and inappropriate by their employees;

- *standard modern* is considered to be the most common category and is characterised by an approach to industrial relations which is essentially pragmatic and opportunistic, and while this approach recognises that the idea of unions as a 'fact of life' depends upon specialists to handle industrial relations, they tend to view it as a 'firefighting' activity;

- *sophisticated modern* adopts a long-term and more positive approach to trade unions, legitimising their role in certain areas on the grounds that they can help maintain stability, promote consent, assist management–employee communications and help the process of change.

In arguing that the standard modern category is the most common, Purcell and Sisson highlight the fact that most employers do not tend to approach industrial relations in a particularly strategic and positive manner. Indeed, a study of UK practice in 175 large multi-establishment enterprises found that, while over 80 per cent of firms claimed to have an overall industrial relations philosophy, only half said this was written in a formal document and even fewer that a copy of this was given to employees (Marginson and Sisson 1988). The authors concluded that the general weight of evidence seemed to confirm that most enterprises remain pragmatic in their approach to the management of employees. Indeed, the presence of a particular philosophy or policy at corporate level did not necessarily mean that it was actually implemented. Managers within companies tend to differ in terms of the work they do, and in their attitudes, philosophies and personal experiences. These differences inevitably influence how they personally approach the management of their employees. However, as we will see later in this chapter, forward-thinking approaches to the management of the employment relationship and the attainment of mutually beneficial partnerships between employees, unions and organisations require a sophisticated modern approach towards the management of employee relations.

Offering an alternative model, Bernardin and Russell (1998) have identified four main approaches to the management of employee relations. In contrast to Purcell and Sisson's model, it focuses on power relationships and participation in decision-making activities:

- *adversarial* managers do what they want and employees are expected to fit in;
- *traditional* managers propose strategies and employees react through their elected representatives;
- *partnerships* managers involve employees in making decisions but ultimately make the decisions themselves;

- *power-sharing* managers directly and genuinely involve employees in day-to-day and strategic decision-making.

These approaches relate closely to Purcell and Sisson's categories and range from retaining decision-making autonomy to jointly sharing responsibility with employees. The reality of the modern employment climate in most developed countries makes the adversarial approach inappropriate, outdated, ineffective and possibly illegal. However, management approaches have not progressed to the point where genuine and complete power sharing with employees is the predominant employment relations strategy. In reality, most managers rely on there being some degree of representation from employees, but retain the power to make the ultimate decisions concerning employment practices. Making the shift to a relationship based on genuine trust and collective responsibility is the secret of an effective SHRM policy.

The changing role of trade unions

The nature and role of trade unions varies from country to country. For example, in some countries, such as Korea, unions are affiliated to individual companies – for example the Hyundai Workers' Union. In other countries, such as China and Singapore, unions are strongly affiliated to the government, which sets their agenda and selects their leaders. In contrast, in countries like the UK and Australia unions are more independent, in that their leaders are selected by members, who pay their wages through annual membership fees. However, despite claims to independence, UK and Australian unions have strong connections with the Labour Party in their country, which was originally established as the political wing of the union movement.

Construction trade unions represent their members across a wide range of issues, including wages, conditions, health and safety, and increasingly in dealing with disputes, disciplinary cases and grievances with employees. However, the ability of unions to provide these improvements varies from country to country and largely depends on the legislative framework which exists. For example, in some countries employers have no obligation to recognise trade unions and participation in union-led activities such as collective bargaining is voluntary. In other countries legislative provisions protect employees' freedom of association and require organisations to recognise trade unions. For example, in Australia an employee's right to belong to a union was upheld by the Australian Industrial Relations Commission (AIRC) when the Australian mining company CRA put pressure on employees at the Comalco Aluminium installation at Weipa to sign individual contracts. CRA's attempt to de-unionise its workforce was based on the premise that employees work to their full potential when they believe that 'their leaders are committed to a safer and fair workplace; and when they know that their efforts will be individually recognised' (CRA management; quoted in Hearn-Mackinnon 1997: 57). In an attempt to render unions irrelevant, CRA had adopted a sophisticated paternalist approach to employee relations and offered workers signing individual contracts a higher rate of pay than union members, who remained covered by an award. The majority (approximately 375 workers) signed individual contracts and those who refused became the centre of a bitter national campaign of industrial action (Hearn-Mackinnon 1997). The AIRC upheld the right of 75 workers to remain represented by their union and ruled that unionised employees should be subject to the same terms and conditions as staff-contract employees.

In Britain quantitative data on union membership point towards a developing crisis facing British trade unions since the early 1980s (Tailby and Winchester 2000). Whereas in the 1970s and early 1980s British unions wielded a great deal of power through their large membership, the late 1980 and 1990s saw a gradual erosion of their influence. In Australia the union movement faces a less bleak picture. The left-wing Construction, Forestry, Mining and Energy Union (CFMEU) is still highly influential in the Australian construction industry, as is demonstrated by its successful negotiation of a 36-hour week for Victorian construction workers. The CFMEU has approximately 120,000 members and, while it is only the third largest union in Australia, it is generally considered to be the most powerful and influential, representing the interests of workers in Australia's most dangerous industries – timber, construction, mining and energy. However, changes in Australian federal labour-market regulations have made the organising environment for trade unions more difficult, meaning that in the future unions will find it harder to retain members and exert influence (*Australian Financial Review*, 1 March 2002). This is unlikely to change in the future as the Australian Labour Party seeks to distance itself from the union movement which founded it, in the same way as New Labour did in the UK in the 1990s.

Despite a general trend towards the erosion of trade unionism in post-industrial economies, they still have an important role to play in the context of the modern construction industry. Due to negative media coverage which has associated unionism with left-wing militancy and industrial conflict, it is all too easy to neglect the huge contribution which they have made to the advancement of well-being in society and to the productivity of industry. If managed in a constructive manner, conflict is a healthy part of organisational life and unions continue to play an extremely important balancing role in the employee relations arena. It is undoubtedly largely due to union activities that most developed countries have legislation governing wages and hours, equal opportunities, occupational health and safety, medical leave, holidays, etc. Unions have set the benchmark for safety, wages and working conditions in the construction industry, and this has also led to improved working conditions and wages for managers, whose working conditions are inevitably linked in some ratio to those of operatives. Thanks to union activities, it is now widely accepted that there is little to be gained from worker exploitation.

While unions are still primarily concerned with improving working conditions for their members, many modern-day unionists have adopted a more commercially astute and less confrontational role than their predecessors. In the UK during the peak of union membership in the 1970s unionists led their members and clashed with employers in protracted and bitter campaigns of industrial action. Subtler tactics now prevail, which include lobbying for legislative change through the political process and achieving better terms and conditions by directly negotiating with employers and employer organisations. Unions have also responded to the changing nature of work, where the service industry now provides a substantial proportion of employment. In this context many workers face mental rather than physical strains, and unions are beginning to address these issues through their negotiations and collective bargaining with employers. The result of this amended role for unions is that, despite the problems caused by confrontational relationships between trade unions and employers in the past, many employers now choose to cooperate with unions because they provide an important channel for communications when dealing with employment issues. The alternative is to deal with employees directly, which can be costly and problematic to implement effectively (as is discussed in Chapter 7). Thus, building a climate of trust and mutual respect with trade union representatives can lead to a more harmonious workplace environment where organisational objectives can be aligned with employee goals.

The effect of unions

Two perspectives dominate the literature on the effects of labour unions: the *monopoly* and *collective-voice* perspectives (Freeman and Medoff 1984). The monopoly perspective is based upon a belief that unions behave as labour-market monopolies, raising wages and causing economic inefficiencies and adverse effects for workers and firms. In contrast, the collective-voice perspective views unions more positively by concentrating on their beneficial economic and political aspects. These two perspectives are described in more detail below.

The monopoly perspective

The monopoly view of unions starts with the premise that unions raise wages above competitive levels. In the UK overall, estimates of union and non-union wage differentials were around 15 per cent during the height of union power (Lewis 1986). However, these estimates vary by demographic group, and studies show that younger workers, non-whites, people living in economically deprived areas and blue-collar workers seem to gain the most from unionisation, perhaps because members of these groups possess relatively low levels of individual influence. Interestingly, little apparent difference exists between the wage gains from unionisation for males and females. Thus, although unions undoubtedly have some impact on wages and benefits, this varies across different employee groups. Research indicates that this influence depends on two main factors, namely the elasticity of labour demand and the extent of unionisation.

Elasticity of labour demand

Union power largely depends on the union's ability to take wages out of competition. Wages can be taken out of competition if labour demand is relatively insensitive to wage changes (inelastic) – that is, if consumers will absorb increased labour costs without offsetting employment effects. Factors that contribute to labour-demand inelasticity include labour costs being a small proportion of total costs; insensitivity of product demand to changes in prices; and an inability to substitute labour for capital, either through technology or through markets. In construction labour costs are generally accepted as accounting for about 45 per cent of the total costs of a building, making the demand relatively elastic and making it more difficult to negotiate wage increases without affecting job security (*Australian Financial Review*, 5 May 2000).

The extent of unionisation

The extent of union organisation in a particular market also affects union monopoly power: more unionised markets have greater union/non-union wage differentials because of less non-union wage competition. The extent of bargaining coverage further augments this effect, maximum influence being exerted when one union bargains for the entire market, meaning that all firms in the industry are covered by the same terms and conditions. For example, in Australia an arbitral model of industrial relations existed at the federal level from the beginning of the twentieth century until the early 1990s. Under this model, prescriptive conditions of employment were regulated by awards endorsed or arbitrated by a Federal Tribunal (Teicher and Svensen 1998). Trade unions were actively involved, through collective bargaining, in establishing the content of these awards, which regulated employees across entire industries. However, although nationally agreed terms and conditions have traditionally governed employment terms and conditions in the construction industry, there is now a trend towards individual enterprise agreements. It is believed that enterprise bargaining introduces more competition into the marketplace, making industry more competitive. Initially the trend towards enterprise bargaining was resisted vehemently by unionists, who perceive their bargaining power to be fragmented and reduced by such developments (*Australian Financial Review*, 5 May 2000). However, unions have found that they are still heavily involved in bargaining enterprise agreements on behalf of their members.

The collective-voice perspective

Workers have several choices when they feel dissatisfied with their jobs: they can do nothing, they can leave their organisation or industry, or they can complain and try to improve the conditions around them. In work settings the voice of one employee is rarely effective in bringing about change. In addition, many workers fear termination and victimisation for revealing their true feelings to management. For these reasons, most workers find it easier to fight for work improvements when in a union. Banding together and creating a 'collective voice' offers protection from the fear of raising concerns with management.

Union advocates maintain that the collective voice reduces staff turnover, thereby leading to retention of experienced and loyal workers, lowering a firm's training costs and raising its productivity. Another benefit attributed to collective employment agreements is the fact that management is forced to become more efficient when faced with the necessity of providing higher wages to unionised employees. It is also argued that in large organisations the transaction costs associated with negotiating collective agreements are less than those of negotiating multiple individual contracts. Thus some proponents of the collective-voice argument suggest that unions may actually have positive effects on organisational effectiveness. However, many managers would argue the opposite and there is considerable controversy surrounding the productivity impact of unionisation (Hirsch and Addison 1986). Many managers argue that unions reduce their decision-making autonomy and slow down decision-making at a time when speed of response is becoming more important in taking full advantage of business opportunities. Others argue that productivity will decrease in unionised firms because unions create resource misallocation, demand restrictive work rules and thereby stifle innovation. For example, in Australia the managing director of a large resource company recently stated that undue union influence had added $1.5 million in labour costs to a maintenance and field services joint venture his company had formed the previous year, despite several rulings by the AIRC in the company's favour (*Australian Financial Review*, 1 March 2002).

Unions are said to have a particularly negative effect in conditions of poor labour–management relations, where the focus is on win–lose outcomes. Finally, managers argue that union productivity effects do not sufficiently outweigh the negative impact of unions on stock prices and shareholder wealth. Thus, while unions may be able to increase productivity and even profits, they adversely affect the capitalisation value of an organisation and redistribute profits from the stockholders to the workers. Studies investigating the effects of unions on business conclude that unions are associated with lower profits. Kleiner (2001) argues that this has led shareholders in the US to exert considerable pressure on managers to oppose unionisation in their workplaces. Such tactics, according to Freeman and Kleiner (1990), include threatening managers' jobs and career enhancement by dramatically reducing their chance of being promoted if unions attempt to organise and obtain a collective bargaining agreement in their establishment. Although construction is not the most market-driven industry, no organisation can afford to ignore its share price, which will ultimately determine its ability to finance future developments.

A final area of contention is whether unions improve worker satisfaction. For example, one would expect that better wages, benefits and improved working conditions would result in union workers being more satisfied than non-union workers. Indeed, research in the US has found that occupational health and safety performance is significantly better on unionised construction sites than on non-unionised sites. However, Schwochau (1987) found that supervision, co-workers and job content created more dissatisfaction for union workers than for non-union workers, with only pay providing more union satisfaction. One possible reason for this is that unions encourage members to voice their dissatisfaction rather than to leave, which means that voluntary turnover rates are substantially lower under unions and that the workforce is more dissatisfied. In contrast to these findings, more recent research indicates that union membership has no effect on either job satisfaction or intention to resign (Gordon and De Nisi 1995). Thus links between trade unionism and worker satisfaction are far from clear.

Collective bargaining

Collective bargaining occurs when workers group together to negotiate employees' wages and benefits, to create or revise work rules, and to resolve disputes or violations of the labour contract with management representatives. Traditionally, collective bargaining was carried out under the auspices of the trade unions. However, more recently non-union forms of collective bargaining have emerged in line with legislative changes that have diminished the role of unions. For example, when faced with the abolition of individual contracts under the Western Australian industrial relations legislation, mining giant Rio Tinto opted to adopt a non-union collective agreement under the Federal Workplace Relations Act (*Australian Financial Review*, 7 March 2002). However, the Rio Tinto situation is not typical, and of 3,856 certified agreements formalised in the period 1 January–15 October 1997 only 136 were non-union agreements, indicating the important role still played by Australian trade unions in the collective bargaining process (Teicher and Svensen 1998). For many blue-collar workers in particular, collective bargaining represents the primary process for securing fair wages, benefits and working conditions. Despite the decline in union membership in post-industrial countries in recent years, it is unlikely, and arguably undesirable, that the union role in collective bargaining will disappear. This means that both organisations and unions will need to maintain knowledge of bargaining strategies and guidelines to represent their interests successfully.

In the UK collective bargaining is a largely voluntary process where employee representatives aim to reach agreements with employer representatives over terms and conditions of employment (Druker and White 1996a). It is important to note that in many countries collective agreements do not become legally binding unless they are built into an individual's contract of employment. In Australia the Workplace Relations Act requires enterprise agreements to be certified by the AIRC, which must be satisfied that the agreement has the approval of a valid majority of employees. Once certified, the agreements are binding.

Despite the move away from collective bargaining arrangements in many sectors, they have remained in construction, albeit at a less significant level, given today's reliance on subcontracting and self-employment. Nevertheless, for those in the field of HRM, knowledge of labour relations and collective bargaining is critically important. Indeed, it is often difficult to separate labour relations as an HRM function from any other HRM function. For example, labour relations are closely tied to HR planning because the labour contract generally stipulates policies and procedures related to promotions, transfers, job security and lay-offs. Recruitment and selection may also be tied to labour relations, particularly if those hired belong to or are likely to join a union. Finally, the area of HRM where knowledge of collective bargaining is probably most critical is compensation and benefits, since almost all aspects of wages and benefits are subject to negotiation.

Collective bargaining should be viewed as a two-way process in order that the basic interests of both parties are protected. It is important to appreciate that both sides (employee and employer) have a responsibility to each other and are ultimately dependent on each other for survival. For example, workers should not expect management to concede on issues that would ultimately impair the company's ability to stay in business. Likewise, management must recognise the rights of employees to form unions and to argue for improved wages and working conditions. The end result of the collective bargaining process will be a formal agreement between workers and management that specifies the conditions of employment and the employee–management relationship over a mutually agreed period of time (typically two to three years, but sometimes up to five years). The major issues covered by the contract and therefore by the process of collective bargaining fall under the following four categories:

- *wage-related issues* such as basic wage rates, cost-of-living adjustments, wage differentials, overtime rates, wage adjustments and two-tier wage systems;
- *supplementary economic benefits*, which include pension plans, paid holidays, health insurance plans, dismissal pay, reporting pay and supplementary unemployment benefits;
- *institutional issues* such as the rights and duties of employers, employees and unions, including union security (i.e. union membership as a condition of employment), employee stock ownership plans and quality of work–life programmes.
- *administrative issues* such as seniority, employee discipline and discharge procedures, employee health and safety, technological changes, work rules, job security, worker privacy issues and training.

Typically, wage and benefit issues receive the greatest amount of attention at the bargaining table, but in recent years issues of job security and employee benefits have become increasingly important as bargaining items. In addition, the unions have adapted to a variety of workplace changes and have played an important role in defining non-traditional employment policies. For example, they have been active in negotiating family-friendly contract provisions such as childcare, elder care, domestic partnership benefits and paternity leave. Teicher and Svensen (1998) note that an almost universal feature of certified agreements is the 'flexibilisation' of hours of work. Through their involvement in collective bargaining, unions have also been involved in promoting health and safety protection for their members and, most recently, have even become involved in immigration, cultural diversity and environmental debates. Indeed, the Australian CFMEU has been involved in environmental issues for some time. For example, in the 1970s the CFMEU developed the concept of the 'Green Ban', which involved banning their members from working on projects which destroyed the environment. It is acknowledged that this has saved some of Australia's most important heritage buildings and natural environments, and it is still being used today to prevent environmentally insensitive developments. For example, it is widely acknowledged that in the 1970s Green Bans prevented the redevelopment of The Rocks in Sydney, an area containing Australia's oldest buildings (96 are now heritage listed) and which has become a major tourist attraction and amenity for local residents. It seems that as the traditional role of unions has declined they have identified new challenges and issues on which to focus.

Levels of bargaining

Terms and conditions for the sale of labour can be negotiated at an individual or a collective level. The main difference between industry-wide (collective) and individual (enterprise) bargaining is that the latter results in individual workplace contracts which differ from one employer to the next, whereas collective bargaining results in standardised national agreements across an industry. Collective bargaining appeared at the early stages of the industrial revolution and by the mid-twentieth century was accepted by most Western governments. Most industrial relations theorists argue that collective bargaining is the only means by which the power imbalance between employers and workers can be overcome to prevent exploitation of workers.

If this argument is to be accepted, then the deregulation of the labour market that has occurred in post-industrial economies, with its encouragement of individual contracts, could represent a return to nineteenth-century conditions, under which management is able to dictate workers' terms and conditions. For example, Teicher and Svensen (1998) argue that individual agreements permitted under the revised Australian Workplace Relations Act, known as Australian Workplace Agreements (AWAs), are shrouded in secrecy and there is little scope for third-party intervention, either in agreeing terms and conditions or in checking compliance with provisions. Teicher and Svensen suggest that the Workplace Relations Act has failed to meet its objective of fairness, since AWAs have been approved which reduce the pay of employees by as much as 34 per cent. However, supporters of individual contracts argue that workers' rights are protected because AWAs must be lodged with the Office of the Employment Advocate, who determines whether they satisfy the 'no-disadvantage test'. This ensures that agreements do not result in overall terms and conditions inferior to those provided in the relevant award. Where the test is not satisfied, the Employment Advocate can request that amendments be made to the agreement.

Proponents of individual contracts also argue that by making employees responsible for getting the job done, and rewarding them via merit-based pay systems, they foster a sense of common purpose and commitment. This approach is in keeping with that of sophisticated paternalists, who expend time and resources on employment policies which ensure that unions are seen as irrelevant by employees. Employers in traditional union strongholds, such as the resources sector, have successfully implemented individual contracts and there is evidence to suggest that productivity gains are possible. For example, during the previously mentioned Weipa dispute hearings, Comalco presented data to the AIRC to the effect that the ratio of output of metal to powder use had risen from 91 per cent average in the two years before the introduction of individual staff contracts to about 92.5 per cent following their introduction (Hamberger 1995). The growth of individual contracts will inevitably be encouraged by such experiences and also by the shifting balance of employment in construction towards professional and managerial staff (Druker and White 1996a). These groups do not tend to have formal union representation and so agreements tend to be made at an individual or an enterprise level. Furthermore, very large construction contracts often require bespoke working agreements, particularly those involving working in especially remote or hazardous locations, where staff may prove hard to recruit or require particular training or skills. Project-specific agreements can have other benefits. For example, Druker and White (1996a) discuss the example of long-duration projects, where collective bargaining can allow the long-term variation in labour costs to be planned at the outset.

The decline in the importance of multi-employer bargaining has been compounded in recent years by developments stemming from government privatisation policies in many countries, which have prompted the demise of the most powerful public-sector unions and their established systems of national bargaining. Thus, in practice many companies have been moving away from highly centralised bargaining arrangements and seeking to negotiate at divisional or workplace level. The trend also appears to be associated with a change in corporate strategy and business policy towards local profit centres, and the granting of greater autonomy at project level.

Thus, despite the union movement preferring industry-wide agreements (because fragmentation makes it far more difficult to monitor and control employees' rights), industry-wide bargaining will probably continue to decline in importance in the construction industry. However, to date in countries like Australia individual contracts have not been adopted and powerful unions remain vehemently opposed to their adoption. Thus it is likely that union-driven collective bargaining will remain a key feature of employee relations for the foreseeable future.

Managing the collective bargaining process

The collective bargaining process needs to be carefully managed, and this is best achieved by understanding the context in which negotiations occur, negotiation tactics, the nature of negotiation as a process, and disputes procedures should negotiations break down.

The context of negotiations

Negotiations are inevitably influenced by both the issues under discussion and the particular circumstances surrounding those discussions. For example, in the case of pay negotiations union claims will often be supported by reference to one or more of the following: rises in the cost of living, improvements in productivity, trends in company profitability, the level of settlements negotiated elsewhere in the industry or the economy, and the earnings of other employee groups, either inside or outside the company. Notions of equity, fairness and justice will frequently underpin, either implicitly or explicitly, union arguments put forward concerning pay and other issues. The context of negotiations varies from country to country. For example, Australia has a unique system which is designed to avoid disputes over basic rates of pay and working conditions. In this system, employers, unions and government officials meet to negotiate in the AIRC, a legal court which has the authority to conciliate and, if necessary, arbitrate in matters relating to such issues.

Negotiation tactics

Negotiations involve a combination of coercive and persuasive strategies. The latter encompass threats and arguments; the former, the application of direct sanctions intended to inflict harm on the opposing side and prompt it to shift its negotiation position. Threats involve one party promising to impose sanctions on the other unless a more favourable settlement is offered. For example, management may threaten to lock the workforce out, cease negotiations and impose a pay settlement unilaterally, dismiss any workers taking industrial action, or close the workplace down. Union-initiated sanctions can include strikes, go-slows, work to rules, overtime bans and the withdrawal of cooperation. Sometimes these threats may be genuine, while on other occasions they may involve an element of bluff.

Most industrial disputes that arise are accidental and are due to tactical miscalculations and misunderstandings by one or both parties. It follows, therefore, that a crucial part of the process is intelligence gathering about one's opponent. For example, in dealing with unions, managers will have to gather intelligence data and make calculations about the ability of union members to take effective action, since this will influence the ability of unions to impose sanctions on a particular issue. Issues to be considered might include the degree of support among members, union resources available to fight a dispute and the capacity of those members to disrupt their employer's operations. Furthermore, the nature of the employer's products, the position the workers occupy in the work process, the immediacy and extent to which they can affect the supply of goods and services, and the ability of the employer to find alternative means of meeting customer demand (such as through the use of alternative labour or production from another site) are some of the more important factors influencing the balance of power between employer and employee groups.

The process of negotiation

Negotiation is like a game of chess, where parties try to use a range of tactics to encourage or force their opponent into a position which is closer to their own interests. The process involves one or more representatives of two or more parties interacting in an explicit attempt to reach a jointly acceptable position on one or more divisive issues. Negotiation only truly stops when this equilibrium is achieved. Various models have been put forward to analyse management–union negotiations, but the most enduring has been that developed by Walton and McKersie (1965), who argue in their model that negotiations consist of four subprocesses: distributive bargaining, integrative bargaining, attitudinal structuring and intra-organisational bargaining.

Distributive bargaining exists where the function of the negotiations is to resolve conflicts between the parties, the resolution of which by definition requires one party to win, the other to lose. Typically, distributive bargaining involves rigid, directed and controlled information processes and little exploration outside the preconceived solutions brought to the table by the respective parties. Walton and McKersie argue that this form of bargaining is the dominant activity in management–union negotiations.

In contrast, *integrative bargaining* refers to negotiations over issues which are not marked by fundamental conflicts of interest and hence are capable of solutions which benefit both sides to some degree, or at least do not result in the gains of one side representing equal losses for the other. The main characteristic of integrative bargaining is its tentative and exploratory nature and its open communication processes, which facilitate a supportive and trusting climate. Productivity bargaining, where employees agree to accept more efficient working practices in return for improved terms and conditions, is a good example of this type of bargaining.

Attitudinal structuring is more concerned with the process by which one party to the negotiation attempts to influence the attitudes of the other in a way favourable to itself. Attitudes can be influenced both during negotiations and before they start. For example, management and unions may try to influence the expectations of the other side with regard to forthcoming negotiations during discussions in forums like joint consultative committees. Management may also try to influence attitudes in the build-up to negotiations by providing information on financial performance, trading prospects and likely future developments in the enterprise. Indeed, in many countries all parties to a bargaining encounter have the right, on request, to be provided with information without which they would be materially impeded from carrying out collective bargaining.

Intra-organisational bargaining reflects the fact that those directly involved in negotiations are acting as representatives of their respective parties. This is particularly relevant to construction, where people in projects will be acting as representatives of a wide variety of organisations. In this situation an important aspect of negotiations is the politics and negotiations that go on *within* each of the parties involved in order to reach consensus on matters like the offer which should be made to the other side, the tactics to be employed during negotiations and the acceptability of a proposed settlement. In effect, two agreements are needed to achieve a negotiated settlement between the parties: an agreement within each party (intra-organisational) and one between them (inter-organisational). Indeed, intra-organisational negotiations can often be more complex than inter-organisational negotiations and end up being the main barrier to settlement. This is likely to be the case in the public sector, where the government might try, either formally or informally, to influence the policies and strategies adopted by management. Moreover, management negotiators need to appreciate that even if their union counterparts find an offer acceptable they may have to reject it because of the views of their members. Thus a sensitivity to, and awareness of, the internal political pressures which opposing negotiators are under can be helpful in the search for the compromises and packages necessary to resolve disputes.

Responsibilities for negotiation

Negotiations may involve full-time union officials, shop stewards (company delegates) or a combination of the two. Union officials are paid employees of a union and have a range of responsibilities, for example organising media communications, coordinating industrial action, representing employees in disputes, arranging general union activities and meeting shop stewards on a regular basis. Shop stewards are part-time representatives who are elected by union members in a particular location, department or section to represent their interests. They are not paid with money but 'in kind', in that their employer gives them time off work to pursue union activities. While the position is voluntary, they play a crucial role in union affairs since they are normally the first point of contact for a disaffected employee. They also make up the bulk of a union's workforce. For example, the Australian CFMEU has over 4,000 shop stewards (delegates) representing its interests across Australia. Traditionally they were associated with manufacturing industry and manual workers, their widespread appointment developing first in the skilled engineering trades during the nineteenth century. However, today it is estimated that there are more shop stewards in the public sector than in the private, and that there are nearly as many representing non-manual employee grades as manual. Shop stewards are found in the majority of workplaces where unions are recognised, and where two or more stewards are present they may appoint senior stewards, and/or conveners. Committees representing stewards from one or more unions may also exist. Consequently, when negotiating, stewards have to take account not only of the policies of the national union, and the expectations and aspirations of their members, but also of the views of their fellow stewards.

The nature of the relationship between stewards and the members they represent will reflect the composition of the workforce and their own personality and opinions. A study of shop stewards in a car factory distinguished four different types of steward based on the extent to which they espoused trade union principles and the nature of the relationship they had with their members. The most important distinction was that drawn between what were termed *leader* and *populist* stewards (Batstone *et al.*

1977). Leader stewards adopted an initiator role in decision-making, actively shaping the issues raised on behalf of members, and they sought to achieve objectives supportive of wider trade union principles. On the other hand, populists saw themselves as more akin to a delegate, whose role was merely to carry out the wishes of the membership, whatever these might be. Interestingly, Batstone found populists to be more common among white-collar groups and leaders among manual workers.

Negotiation roles and the mechanics of negotiations

Negotiations can range from relatively informal discussions between a supervisor and a shop steward to a formal meeting involving a number of representatives from each side. In this second situation it is important to clarify the roles to be played by each member of the negotiating team before the start of negotiations. Roles commonly distinguished are those of the *chief negotiator* (the person who does most of the talking), the *secretary*, (who is expected to keep notes and look for any verbal or non-verbal signals from the other side) and the *analyst* (whose role is to scrutinise and analyse what is being said and to summarise the issues, if and when this is appropriate). Negotiating teams may also include one or more specialists with technical knowledge and experience relevant to the issues under discussion.

In conducting effective negotiations it is vital to ensure that team members do not give contradictory messages, since this is not only confusing and may lead to tactical miscalculations, but may give the other side an opportunity to exploit any differences of opinion that exist. Disagreements within a negotiating team should be discussed during adjournments, an important part of negotiations. Adjournments give negotiators time to consider the arguments and offers put forward by the other side, to reappraise their bargaining tactics and to seek clearance from senior management if a change in bargaining objectives is thought desirable.

Good preparation is an essential prerequisite of effective negotiations. Each side needs to consider carefully what its negotiating objectives will be, including fallback positions and the arguments that will be used to support them. Where a number of issues are to be discussed it may be useful to allocate the various objectives to one of three categories: *essential, desirable* or *optimistic*. Negotiators should also give thought to the objectives likely to be pursued by opponents and the arguments that they may use to support them. Collectively, this information should lead to a planned bargaining strategy with agreed tactics. For example, management may choose to make a relatively generous opening offer which leaves little subsequent room for manoeuvre, or it may adopt the opposite approach.

Once negotiations start, the convention is that the party seeking the negotiation should open the encounter. For example, in the case of a pay claim the union will start the process by outlining its claim and the rationale underlying it. At this stage, management should listen carefully to what is being said, seek any necessary clarification and make sure that the union's claims are fully and properly understood. Research indicates that skilled negotiators tend to seek more information and more frequently summarise and test understanding than average negotiators. They also depersonalise their negotiations, focusing on issues rather than the people involved in the negotiation process. Exchanges that could be perceived as personal attacks by the other side's negotiators should be avoided at all costs, since this is unlikely to be helpful to the discussions and could have adverse implications for future negotiations. It is important to remember that collective bargaining is part of an *ongoing* relationship between management and workers. This needs to be borne in mind during negotiations, care being taken to ensure that relatively minor short-term gains are not obtained at long-term harm to working relationships.

When negotiating, research also indicates that it is important to think creatively about how the various issues on the bargaining table can be combined or packaged in a manner acceptable to both sides. This is difficult and ultimately agreement is possible only if one or both sides are willing to make concessions and move from their opening positions. However, where communications are poor between negotiators both sides may be wary of offering concessions for fear that they will not be reciprocated. In this situation a technique called *signalling* can be used to try and 'break out' of circular and non-conclusive argument in a way that minimises this danger. Signals are qualifications placed on a statement of position indicating a willingness to consider alternative proposals if some reciprocal concession is made. For example, a negotiator may add the phrase 'in its present form' to a statement that 'we will never agree to what you are proposing'. Where a party makes a concession like this it is sensible to help them do this without losing face.

Finally, once agreement has been reached it is important to clarify exactly what has been agreed. Often, in the emotion and complexity of negotiations it is easy to lose track of what issues have been resolved. One way of doing this is periodically to prepare a detailed summary that both sides then agree on. An important point is that negotiators will frequently have to refer agreements to others for approval before they can agree to them formally. This applies to managers and unions. In this situation each side should seek an undertaking that the other side will recommend acceptance.

Disputes procedures

Disagreements will inevitably occur which will prove difficult, if not impossible, to resolve in a way that is reasonably acceptable to both managers and union officials. When negotiations become deadlocked the parties will usually register a 'failure to agree' and refer the issue to the next stage of the disputes procedure. It is normally understood, if not explicitly stated, that both sides will refrain from taking any industrial action until all stages of the procedure have been exhausted. Nevertheless, *unconstitutional action*, which is action in breach of procedures, can occur.

Disputes procedures are intended to aid the resolution of disputes by enabling them to proceed through a number of hierarchical stages, each involving the introduction of more senior personnel who are less directly involved with the issues under discussion. To avoid dispute procedures becoming unduly cumbersome and time consuming, only those levels of management able to play an effective role in resolving disputes need to be involved. Another important issue is whether individual grievances and collective disputes should be covered by the same procedure. Whatever decision is made, procedures vary considerably both in relation to the number of stages they contain and in the identity of those to be involved at each stage. Nevertheless, three main levels or stages can be distinguished, namely *department*, *establishment* and *external*. An additional *corporate* level stage may be distinguished in multidivisional companies. Clearly, external intervention should be a last resort, only to be used when all else fails.

Where provision is made for issues to be referred outside the organisation, it may take the form of reference to an industry-wide disputes procedure if the employer concerned is a party to the relevant set of industry-level negotiations. Alternatively, the matter may be referred to an independent third party who may conduct conciliation, mediation, adjudication or arbitration depending on the seriousness of the dispute. Procedures that include provision for third-party intervention differ in terms of how the process is to be triggered. In some cases this can only occur with the joint agreement of the two parties, which may itself be difficult if relationships have broken down. In others, one party may have the right to refer an issue to a third party. A third possibility is an automatic trigger in certain circumstances, such as where a 'failure to agree' has been formally registered. Moreover, it is possible for procedures to specify more than one type of intervention, but to lay down different procedures regarding how they are to be initiated. For example, an agreement may provide for joint reference to conciliation, but unilateral reference to arbitration. Essentially, the possibilities are endless and the important point is to agree them in advance of negotiations.

Of the various intervention strategies mentioned above, *conciliation* is the least formal. With conciliation, the third party supports the negotiating process by assisting the parties to identify themselves, the nature of their differences and possible ways of resolving them. *Mediation* permits a rather more interventionist stance in that the third party puts forward recommendations for settlement, although the parties are free to accept, reject or amend the terms proposed. In *adjudication* the third party's decision is binding for a period of time, when reference to arbitration can be made. In construction projects this is normally for the duration of the project. Adjudicators are often supported by contractual powers and used within such project situations as are commonly found in the construction industry. Indeed, in some countries like the UK and Australia adjudication has now become a compulsory aspect of the bargaining process. However, a burning debate currently revolves around the effectiveness of this process and, in particular, the qualifications of adjudicators, their powers and their knowledge of the law. Finally, *arbitration* involves an even greater degree of intervention in that the third party has to make an award which is binding on the parties. In general this is more expensive, formal and time consuming than the previous methods, although it is not as costly and public as litigation, which is the ultimate and most formal form of redress.

While third-party intervention can provide a valuable means of resolving disputes without costly industrial action, it does have several potential drawbacks. In particular, it involves management allowing a third party to decide issues which could have considerable implications for internal costs, efficiency and public relations. It may also act to undermine the collective bargaining process in two important ways: first, because it provides the parties with a way of 'getting off the hook' without having to work out their own solutions; second, because compromise solutions often result. These potential drawbacks have prompted considerable interest in the concept of 'pendulum' or 'last-offer' intervention. Used widely in the US to resolve public-sector disputes, this type of intervention differs from that traditionally used. In pendulum intervention the third party cannot make an award somewhere between the final negotiating positions of managers and unions, but has to opt for one side or the other. It is argued that this serves to support rather than undermine collective bargaining since it encourages the parties to narrow their differences as far as possible before embarking on the process. However, it also introduces difficulties surrounding what constitutes the final negotiating positions of managers and unions. Furthermore, it is a rather crude way of resolving a complex set of negotiations covering several different issues such as pay, holidays and hours of work. Finally, it is unlikely to identify possible solutions which may suit both parties' interests and result in more amicable long-term industrial relations. Inevitably, there will always be a winner and a loser in this type of resolution.

Future developments in employee relations

The future of industrial relations is inexorably linked to developments in the union movement, the process of globalisation, and the growing awareness of the need to engage employees in a spirit of partnership and mutual commitment. The influence of these factors on employee relations is discussed below.

The future of the trade union movement

A variety of economic and structural factors are contributing to a general decline in union membership. These include relatively high levels of unemployment combined with generally rising real incomes; a shift of employment away from highly unionised industries, such as mining, to the less well-organised private services sector; an increase in the number of white-collar and female workers; and a growth in self-employment and temporary and part-time work. There has also been a significant move towards outsourcing and subcontracting and away from direct employment, which has stretched union recruitment resources to the limit. The cumulative effects of these developments have been compounded by changes in employer and government attitudes towards trade unions. Governments around the world have introduced a host of legislative changes which have made it harder for unions both to secure recognition and to take industrial action without legal action. For example, in the UK secondary (sympathetic) action and the enforcement of closed-shop arrangements have been made unlawful; the inclusion of union-only clauses in commercial contracts has been prohibited; and employers have been given the right to seek damages from unions in respect of industrial action which has not been approved by a membership ballot. At the same time, and partly as a result of these legislative developments, employers have been considerably less willing to cooperate with unions, and in some cases have decided to withdraw recognition, either completely or in respect of particular groups of staff.

The situation is similar in Australia, where the conduct of the construction unions is under particular scrutiny with the commencement of a Royal Commission into the Building and Construction Industry. This Royal Commission was established by the federal government to inquire into and report on matters relating to corrupt employment practices, and its terms of reference include:

- practice and conduct relating to the Workplace Relations Act, occupational health and safety laws or other laws relating to workplace relations;
- fraud, corruption, collusion, anti-competitive behaviour, coercion, violence or inappropriate payments, receipts or benefits;
- dictating, limiting or interfering with decisions on whether or not to employ or engage persons, or relating to the terms on which they be employed or engaged.

The Commission, which has been criticised as a political stunt designed to smear the unions, has thus far uncovered evidence of the involvement of underworld figures and payments of large amounts of money to secure industrial agreements amidst a climate of fear and retribution (*Australian Financial Review*, 5 March 2002). The Commission is ongoing but its findings are likely to harm the reputation of unions, and indeed some employers, in the Australian construction industry.

One response of unions to the loss of subscription income has been to merge to build greater institutional security and create economies of scale. For example, the number of unions in the UK dropped from 454 in 1979 to 268 in 1992. Another union response has been to improve the organisation and resourcing of their recruitment activities. For example, new legal and financial services have been made available to members, and increased efforts are being made to recruit among part-time, temporary and even self-employed workers. There is considerable debate as to how far activities of this type can succeed in securing an expansion of union membership. It would seem questionable whether they can significantly reverse recent membership trends in the absence of an economic and political environment more conducive to union organisation.

The other inevitable result of a decline in union representation has been a growth in non-unionised workplaces, which seems set to continue in the future. For example, Cully and Woodland (1996) estimate that over half of all workplaces in the UK are already non-unionised. However, companies who have sought to engender a non-unionised workplace have faced enormous challenges in effectively replacing the perceived fairness of the collective bargaining process. Some have decided to do nothing, thereby avoiding expense and union involvement. This approach has been termed the 'Bleak House' strategy (Bacon 2001) and inevitably results in lower employee commitment. Clearly, organisations which choose a non-unionised route must be prepared for the additional expense of setting up communications channels and actively managing their employee relations. However, replacing the role of unions has proved much more difficult than was once thought and, if engaged positively, they still perform an important function in securing efficiency in the modern-day workplace.

Globalisation

The process of globalisation has had a major impact on industrial relations in all industries and will continue to do so. While the construction industry remains a domestic industry in most countries, the time will soon come when it will face the same competitive forces as other industries. If construction companies are to compete in the global market they will need to be mindful of differences between the employee relations climate in different countries. Employee relations, perhaps more than other aspects of HRM, is a highly context-specific area, depending on local cultural, social, political and economic forces. The following examples show how these variables have interacted to impact on employee relations in South Korea, Canada, Europe and America. These examples are particularly interesting because of the significant social, economic and political changes which have influenced unionisation and HRM practices in recent years.

South Korea

Lee (2001) argues that South Korean firms have always adopted a patriarchal stance to employee relations. It is argued that this stance is firmly rooted in the Korean heritage, based upon the strict, well-established Confucian culture, an extended family system and a rice-agricultural production system. According to Lee, the effect of the socio-cultural environment in establishing a set of social norms and moral values is stronger in a uni-cultural society, such as Korea, than in a multicultural society, like Australia or the United States. Korean Confucianism focuses on piety to parents, benevolence to children, loyalty and allegiance to the nation, and justice and faith among peers. The traditional extended family system, based on blood relations and male lineage, also contributes to a norm of familism, collectivism and paternalism, and a stable society where family members cooperate with one another, fostering a sense of collectivism. According to Lee, these three socio-cultural features have contributed to the development of patriarchal human resource practices in Korea. This patriarchism was initially expressed in an authoritarian style of people management in which employees were expected to obey management in much the same way as a child was expected to obey his or her father. Employers had complete discretion in making decisions affecting employees' terms and conditions of employment and had never experienced resistance from organised labour. However, during the 1980s several variables intervened to change the balance of power between employers and employees in South Korea. First, there was a serious labour shortage as a result of rapid economic growth. Second, the authoritarian military government succumbed to public pressure for democratisation and signed a declaration of democratisation in 1987. This led to the implementation of a raft of democratic changes, which changed the political status of the South Korean people and triggered a great increase in industrial disputes. For example, in 1987 alone there were 3,749 disputes. Union density also rose within the workforce, from 12.4 per cent in 1985 to 18.6 per cent in 1989. The government responded by amending labour laws in November 1987, reducing state intervention in union activities and establishing a system of collective bargaining to determine wages and working conditions. Lee (2001) argues that after this period of labour disputes the government and employers adopted a more pragmatic approach to employee relations. Realising that the authoritarian style was no longer tenable, they began to adopt paternalistic HRM practices, based on mutual understanding, shared information, respect and improved working conditions, remuneration and benefits. It is argued that this approach fostered spontaneous and genuine employer–employee cooperation, which is far different from the transactional-exchange climate fostered by American-style contractual HRM. Using this approach, Korean companies avoid industrial conflict by establishing a sense of familism. For example, following the unrest of the late 1980s, in 1995 employees and employers in 2,700 Korean companies jointly declared a cooperative movement called *ga-sa-bul-i*, meaning that home and company are no different from one another. Lee (2001) argues that this shift did not abolish the patriarchal employee relations system rooted in Korean heritage but represented a move away from authoritarian HRM to paternalistic HRM within the patriarchal system.

Canada

Reshef (2001) also draws on political and economic characteristics to explain the quiescence of the labour movement in the Canadian province of Alberta. In the face of budget cuts and public-sector restructuring under the Klein government, trade union leaders vowed to unite workers against the government and overturn government policies designed to balance the budget. However, Reshef highlights the unions' failure to achieve this labour 'revolution', and he attributes this failure to a number of characteristics of the unions' internal and external political and economic environments. First, Reshef notes Alberta's tradition of conservatism, in which labour laws are more employer-friendly than those of other Canadian provinces and workers' right to strike is substantially restricted. Economically, Alberta has also lacked a significant manufacturing base. For example, in January 1999 the manufacturing sector employed only 8.5 per cent of the labour force. Even the oil and gas industry has been characterised by a paternalism and laissez-faire ideology not conducive to effective labour resistance. In this context, trade unions have received only modest support, and in a recent survey 68 per cent of respondents reported that they would not join a union if one existed in their workplace. Between 1983 and 1996 private-sector union membership fell from 16.0 to 13.8 per cent, whereas public-administration unionisation fell from 79.3 to 58.6 per cent. Coupled with a massive reduction in the civil-service workforce through lay-offs, voluntary severances and early retirements, declining unionisation meant that even the larger public-sector unions were substantially weakened. In this weakened position unions complied with government requests to negotiate wage rollbacks with their members. In the restructuring of public-service departments healthcare was regionalised, meaning that the boundaries of long-established bargaining units were redrawn and rules governing union membership and recognition meant that some unions lost their bargaining rights. The government also used the mass media and community consultation to popularise and legitimise its activities, with the result that the public accepted the government's argument that public-sector spending must be curtailed. Internally, Reshef (2001) argues that the Albertan union movement had adopted an apolitical 'business unionism' approach, in which all it did was negotiate collective agreements between employers and employees. Being devoid of a socio-political dimension, the union movement was not able to mobilise for collective action. Furthermore, the union movement was fragmented, with workers in similar occupations represented by different unions, and this fragmentation was associated with entrenched union rivalries. When their survival was threatened by government economic policies, unions were not able to cooperate and the rivalries and competition were increased. Thus Reshef suggests that the nature and structure of the economy and the political forces at play were key features in the inability of the labour movement to mobilise.

The United States of America and Europe

Blanchflower and Freeman (1992) state that the union/non-union wage gap in the US is about 20 per cent, compared to 10 per cent in the UK and 7 per cent in West Germany, illustrating the increased incentive for managers to oppose unionisation in the US. Indeed, Kleiner (2001) identifies a range of economic incentives for managers in the United States to resist unionisation and argues that these incentives are a major factor in the rapid decline of US unions. US firms now spend more than US$200 million annually on direct payments to consultants and lawyers hired specifically to stop union-organising drives, and between the 1950s and 1980s this was the fastest growing area of management consulting. The success of these anti-union consultants has meant that the cost of violating employee relations legislation (the National Labor Relations Act – NLRA) is now so low that the economically rational firm would actually choose to violate rather than comply. This contrasts with the legal situation prevailing in European Union countries, in which employee relations are more heavily regulated. In most EU countries bargaining takes place at a national or industry level and there is little economic incentive for managers to try to stop union organising drives. Furthermore, Kleiner (2001) argues that the social contract in the EU has fostered a greater social acceptance of unions than exists in the US. While EU penalties for violating the social contract determined by member countries tend to be low, there are still few benefits to be gained because incurring the anger of large powerful unions is likely to harm rather than benefit employers. However, Kleiner (2001) notes that the UK's Employment Relations Act, passed in 1999, may lead to outcomes similar to the US experience, including the extensive use of management consultants to advise firms how to remain non-union.

Employee involvement and partnership

In recent years UK firms have shown considerable interest in employee involvement and empowerment in defining conditions of employment (Marchington and Wilkinson 2000). This has profound implications for the future of unionism, since it emphasises the involvement of the individual via direct communication rather than collective involvement through union representatives. This is increasingly reflected in UK legislative changes relating to employee relations. The election of the Labour government in 1997 has led to a significant change in labour relations in the form of a new act (the Employment Relations Act 1999), a 'Fairness at Work' programme and a closer engagement with the social policies of the European Union (Bacon 2001). This has provided routes for union recognition *and* greater rights for individual employees. The new rhetoric surrounding these changes within the employee relations field concerns the need to build *partnerships* with employees in order to enhance business performance. However, the term 'partnership' is extremely ambiguous and has been subjected to considerable interpretation. On the one hand, it has been used by managers to increase the power and identity of the individual within an organisation, while, on the other, unions see it as having a 'respect for union' influence. The precise impact of this movement is discussed in depth in Chapter 7, but it is clear that it has major implications for the future of union–manager relationships in the future.

Conclusions

The purpose of this chapter was to discuss industrial relations management as a central aspect of SHRM. As seen by the employer, the ultimate test of effectiveness should be whether their relations with unions or other representative groups enable the organisation to operate in a way that is conducive to the achievement of both its short-term and long-term objectives. The days of powerful trade unions locked away in conflict with industrial relations directors have now largely disappeared and the function of industrial relations has now broadened to encompass other aspects of the employment relationship. Relevant questions to consider under employee relations are whether strategies are helping or hindering the achievement of high levels of output and productivity; satisfactory labour recruitment and retention; the maintenance of an atmosphere supportive of change; and the presence of a committed and motivated workforce. Organised conflict can dramatically harm an employer's operations, and the ultimate objective must be to help managers and unions resolve disagreements without recourse to costly industrial action. On the other hand, a low level of conflict is not necessarily a sign that relationships are satisfactory. It is quite possible for low levels of conflict to exist alongside high levels of absenteeism and labour turnover, inflexible working practices, widespread resistance to change, and low levels of innovation and competitiveness. The key is how differences of interests are managed. If managed well, they can be a source of strength rather than weakness to an organisation.

As a result of workplace changes and the shifts towards maintaining employee relations rather than imposing industrial relations frameworks, strike activity has declined significantly within modern industry. Even in countries such as Australia, where the construction unions are considered to be militant, innovative collective agreements have been established with companies incorporating 'no-strike' provisions, closer links between pay and performance, extensive labour flexibility, and harmonised terms and conditions of employment for all employee grades. This flexibility has been brought about in response to pressures from increased competition, which has united the interests of workers and unions in a need for continuous improvement. While construction remains a relatively domestic industry in most countries, industrial relations will eventually have to change dramatically as international competition increases to threaten hitherto safe home markets. Thus employee relations are set to become more important in the future. The next logical step in the development of employee relations will be to engender the full participation and involvement of employees in SHRM issues in order to increase their commitment to the organisation. Approaches for securing this involvement are discussed in the next chapter.

Discussion and review questions

1 Discuss the likely future significance of collective bargaining and trade union involvement in the context of your own construction industry.
2 Collective bargaining negotiations can discuss one or more of the following issues: wage-related issues (wage rates, cost-of-living adjustments, overtime rates, etc.); supplementary economic benefits (pension plans, paid holidays, health insurance plans, etc.); and institutional issues (rights and duties of employers and employees, unions, employee stock ownership plans and quality of work–life programmes, etc.). Discuss the extent to which the parties' objectives under each of these remain mutually exclusive in the context of the modern working environment.
3 Consider the viability of a system of individual contracts for construction workers. What are the potential benefits or disadvantages for employers and employees?

7 Employee participation, involvement and empowerment in construction

This chapter explores the value of employee participation, involvement and empowerment in creating a motivated and productive workforce. These concepts are becoming central to the development of SHRM because they help to breach the gap between the aims of organisations and those of employees. They are also central to many managerial initiatives that are currently being advocated as leading-edge practice within the construction industry – for example partnering, teamworking, value engineering, TQM, risk management and an improved customer focus. While the project-based nature of construction activity inherently demands the devolvement of decision-making responsibility from central decision-makers, the level of true participation, involvement and empowerment is limited, particularly at project level. This chapter explores these concepts in detail and argues that they are an important part of any effective SHRM strategy.

Introduction

The UK construction industry has undergone significant change over the last 50 years, which has put considerable pressure on its organisations to respond appropriately and swiftly to their business environment. However, much of this change has been externally driven, and the construction industry has gained a reputation for being relatively unproductive, confrontational and resistant to change compared with other industries such as manufacturing. In particular, the last decade has seen unprecedented demands on the industry from leading client groups, which are demanding a fundamental examination and redefinition of working practices in order to improve its efficiency. Within the UK, for example, a series of government-commissioned reports have questioned whether the industry's structure, culture and operation are capable of meeting the needs of its clients (see Latham 1994; Egan 1998). These concerns are reflected in other countries. For example, in Singapore the Construction 21 Report *Re-inventing Construction* (MMS 1999) echoed similar concerns and recommendations, and in Australia numerous government reports have proposed strategies for implementing change in the construction industry (see AFCC 1988; DPWS 1996). More recently the debate has shifted towards questioning whether the industry's lamentable performance with regard to managing and respecting its human resources could explain why the sector has so far failed significantly to improve its performance (Rethinking Construction 2000).

Unfortunately, despite governmental and institutional pressure to improve employee participation within the construction industry the use of empowerment to enhance performance has not yet been universally embraced by construction organisations. The experience of other sectors has shown that, used appropriately, empowerment could play an important part in helping construction organisations to address increasing performance demands whilst mitigating the negative effects of the fragmented project-delivery process. However, there remain many barriers to individual and team-based empowerment strategies that must be overcome before the industry can benefit from their implementation. In this chapter we examine the evolution of the empowerment concept, consider its relevance to the construction sector and discuss ways in which construction companies could manage its implementation. We pay particular attention to the significant communication barriers that have plagued the industry for so long.

The evolution of empowerment within HRM

As noted in the previous chapter, in recent decades union-led collective bargaining processes have been undermined by legislative and social changes in Western economies such as the UK, US and Australia. These economies have adopted more flexible hybrid systems of regulating industrial relations that recognise the legitimacy of individual employment agreements and non-union representation. These changes have been accompanied by increases in employees' involvement and participation within the industry, and enhanced by considerable management rhetoric surrounding the benefits of mutual commitment, empowerment and direct communication (Armstrong 1996: 774). Terms such as 'employee participation', 'employee involvement' and 'employee empowerment' are now commonplace in the HRM literature. Essentially, all of these concepts are about deliberately engaging with employees and encouraging them to exert an influence on decisions that impact on their own interests within organisations. The argument is that increased commitment to organisational goals can be engendered by increasing the amount of information provided to employees and by encouraging their involvement in decision-making.

The idea that employee participation enhances performance is not new and was a significant aspect of the organisational behaviour literature throughout the latter half of the twentieth century. More recently *empowerment* has become the latest stage in the evolution of this concept and its

development can be traced to the trend amongst postmodernist organis
tions to flatten structures and devolve decision-making responsibility to
lower organisational levels than they did previously (Sykes *et al.* 1997). In
relation to management theory, empowerment can be seen as the antithesis
of the traditional classical models of management discussed in Chapter 2,
which advocated compliance and productivity at the expense of commitment and good employee relations. It is far more closely linked to the
human relations school, where theorists such as Vroom and Maslow
suggested that motivational approaches such as employee involvement can
develop a more interested and hence productive workforce. By enriching
jobs through the addition of responsibilities and decision-making
autonomy, they argued that employee involvement could promote more
meaningful work for employees, which in turn encouraged commitment to
the organisation. This requires a democratic culture where a workforce is
given freedom and flexibility at the expense of control, hierarchy and
rigidity.

In construction companies empowered working is inherent in the way
projects are run as autonomous profit-centres. Nevertheless, the concept of
empowerment is much deeper than this and involves a far greater commitment to human resource management than the construction industry has
so far demonstrated. To assist this process, we will now discuss the evolution of the empowerment concept in more detail.

Employee participation

Employee participation first emerged in the 1970s, when employees began
to exert influence on decision-making through their leaders or representatives. Employee participation is a collective process where employees exert
influence through their representatives rather than as individuals. It
provides employees with an opportunity to exert influence over decision-
making *and* to share in the rewards through mechanisms such as profit
sharing and share ownership. Employee participation reflects contemporary management thinking and proponents argue that organisations which
develop systems that value the opinions of their employees are more likely
to be effective than those which do not. This arises from lower levels of
misunderstanding, dispute and confrontation, higher levels of motivation
and self-esteem, lower levels of frustration and greater use of people's
skills, knowledge and experience. Apart from greater efficiencies, there are
also other potential benefits, such as improved occupational health and
safety (OHS). Aside from the moral argument for the involvement of
people in decisions affecting their welfare, an important principle underpinning contemporary OHS legislation is that the management of
workplace risk is most effectively achieved through collaboration between
the people who manage and those who work with the risk on a daily basis.
This conclusion was first drawn by the Robens Committee following an

inquiry into OHS in Britain in the early 1970s. Since then, the committee's report has informed preventive OHS legislation in countries such as the UK, Australia and Hong Kong. However, critics of the Robens philosophy have argued that used in its purest form employee participation could be more damaging than useful if implemented without consideration of an industry's culture. For example, in construction the confrontational nature of the industry could undermine any initiatives to devolve responsibility for decision-making. Some have argued that conflict has become so institutionalised in the industry that any attempt at a softer approach without broader cultural change is likely to be met by exploitation and even higher levels of conflict. Indeed, this has always been one of the problems encountered by those attempting to introduce procurement systems and contracts which more closely reflect contemporary management thought (Loosemore and Hughes 1998). It is clear that any strategy for greater employee participation in construction must be a long-term one which is clearly thought out and negotiated with relevant stakeholder groups.

Unfortunately, as with other concepts in HRM, there is no general consensus as to exactly what the term 'employee participation' should encompass to be effective. Examples of participative practices include profit-sharing schemes, quality circles, job enrichment, problem-solving groups, team briefing and employee share ownership. However, whether all of these concepts can be encompassed under a single term is debatable. What is clear is that participation can take various forms depending on the level at which it is focused within the organisation. For example, it could take place at a job level between individual supervisors and their immediate subordinates. In this case it could involve an informal interchange between employees and supervisors as to how a job could be done more effectively. Job-safety analyses are one form of workgroup participation, involving group members and first-line supervisors discussing the OHS issues involved in each task before it commences. At the other end of the scale, employee representatives could be allowed to share in strategic decisions concerning the goals, objectives and direction of a business. This can be achieved by making them shareholders in the business. Clearly, there will be a range of levels of participation between these two extremes, but ultimately the degree of influence employees enjoy will be largely determined by the willingness and enthusiasm of both employers *and* employees to embark upon a process of meaningful change (Armstrong 1991: 715).

There is a variety of formal and informal mechanisms for implementing employee participation, and according to Armstrong (1991) the degree of formality should depend on the level of influence that employees are to have. For example, in terms of influencing working at the job level, informal group meetings are suitable, but where employees are to influence policy at a senior level, formal representations are likely to be more appropriate. Some of the most popular formal approaches include joint consultation (where managers exchange views in a frank and open manner on key issues), consultative committees (where representatives elected by their respective employee groups come together to discuss issues through a formal committee structure) and quality circles (where groups working on related aspects of the production function regularly come together to discuss ways of improving performance and productivity). Regardless of the form of employee participation adopted, plans for increasing employee participation should be prepared and methods for implementation should be discussed with employees, who should also be kept fully informed of progress against the mutually agreed targets.

Employee involvement

The 1980s witnessed a new agenda for the employee participation movement and an extension of the concept into 'employee involvement' (Holden 1997). In the UK the concept of employee involvement grew throughout the 1980s and 1990s, largely in response to reduced union power and government policy, which encouraged firms to evolve arrangements which suited their needs (A. Wilkinson 2001). These trends have been mirrored in other countries such as Australia, where some large employers have implemented employee involvement in an attempt to de-unionise their workforce (see Chapter 6). The main difference between employee involvement and employee participation is that involvement stresses *individualism* rather than collectivism and therefore requires a focus on direct communications with the employee rather than with employees' nominated representatives. In addition, employee involvement is management initiated rather than set within the domain of industrial relations specialists, largely because it is geared towards achieving economic efficiency. Employee involvement links closely to the Harvard Model of HRM, discussed in Chapter 3, where employee influence is firmly set within the centre of the process (see Beer *et al.* 1984). The Harvard Model defines employees as a key stakeholder and so inherently values their influence in decision-making processes.

Management initiatives used to engender an involvement culture include team briefing, teamworking (giving teams autonomy and responsibility for their own tasks) and quality circles. These initiatives have in turn become connected to cultural change mechanisms such as TQM, customer-service improvement schemes and business-process re-engineering (BPR). These have all become popular in the construction industry in recent years. However, in order to achieve these new modes of working a range of communication and participation mechanisms are required which may be difficult in construction, an industry with a stubborn history of communication problems between its culturally diverse and organisationally fragmented occupational groups (Crichton 1966; NEDO 1988; Latham 1994). Nevertheless, the communication and participation mechanisms required for an effective involvement culture have been classified by Holden (1997) under four headings:

- *downwards communication from managers to employees*: newsletters, briefings, journals, etc. to keep employees informed of organisational priorities and directions and the results of involvement polices;
- *upwards communications from employees to managers*: quality circles, suggestion schemes, customer-care programmes, etc. to ensure that employees' opinions and perspectives influence the direction and policies of the organisation;
- *financial participation*; profit sharing, share ownership, etc. in order that employees have a genuine stake in and commitment to the business;
- *representative participation*: works councils, consultative committees, collective bargaining, etc. in order that employees have a real voice in decisions affecting their role and employment.

Proponents of employee involvement argue that if companies demonstrate respect for individual employees and foster a culture of mutual trust and commitment, employees are more likely to derive job satisfaction and commit themselves to an organisation's objectives. Other espoused benefits of employee involvement include enhanced performance and productivity derived from utilising the resources of all employees more effectively, improved job satisfaction and greater employee influence (IPD 1993). Opponents, on the other hand, argue that involvement does not necessarily guarantee commitment from employees and that the cost of its implementation means that it inevitably gets squeezed in times of recession. Marchington (1995) warned that in some organisations employee involvement may become little more than a one-way communication channel, rather than a meaningful process of two-way exchange and influence.

Employee empowerment

The most recent stage in the evolution of the employee involvement concept is 'employee empowerment'. This takes the involvement concept to another level, providing employees with responsibility for their own tasks, deciding how they will be achieved and who will lead them. Thus, rather than attempting to control employees, proponents of empowerment seek to give employees the responsibility for improving their own performance (Wilkinson 2001). Theorists and practitioners espouse the benefits of empowerment to individual and team performance. For example, Dufficy (1998) presents comparative benchmarking data from over 100 British companies which shows that more empowered businesses perform better than 'traditional' companies in terms of a range of business performance indicators, including turnover, return on sales (ROS), return on capital employed (ROCE) and profit per employee. In particular, firms recording over 50 per cent on a specially developed empowerment scale recorded sales increases of 30 per cent, ROS increases of at least 7 per cent, ROCE increases of at least 12 per cent, and profit per employee increases of more than £5,000. In comparison, Dufficy (1998) claims that only one-quarter of the less highly empowered firms matched this rate of progress. At an individual level empowerment has been linked to reduced psychological strain, enhanced job satisfaction, and decreased absenteeism and turnover (Koberg *et al.* 1999).

However, despite compelling arguments in favour of empowerment, the concept remains poorly defined, with the literature providing a wide range of competing definitions. This may be explained by the tendency to attach empowerment to popular management ideas and developments such as TQM, BPR, teamworking and the 'learning organisation' (A. Wilkinson 2001). These movements have all required the increased participation and autonomy of employees as a necessary component of improving the efficiency and effectiveness of organisations (see Cook 1994). However, they have also demanded their own approaches towards ensuring the inclusion, involvement and participation of people. It is clear that empowerment should not be treated as a homogeneous concept, but as embodying a range of concepts.

Theorists have tended to adopt one of two main approaches in conceptualising empowerment, both of which are important to its successful implementation. Some theorists describe empowerment as a set of *power-sharing* managerial strategies, practices and techniques. For these theorists, empowerment occurs when structural changes are made which give employees greater latitude to make decisions regarding their work, through which they can exert greater influence in their workplace (Eylon and Bamberger 2000). Other theorists have adopted a *psychological approach* to employee empowerment. This group of theorists defines empowerment in terms of the perceptions of job incumbents rather than the actions of their managers. Thus, Spreitzer (1995) suggests that management actions will not necessarily empower employees and therefore employees' perceptions of empowerment are more important than management practices designed to enhance employees' influence. This cognitive approach is endorsed by Siegall and Gardner (2000), who suggest that the benefits of empowerment will not emerge unless people perceive themselves to be empowered. Thus employees may be given 'permission' by their organisation to act autonomously, but unless they also believe that they have the capability to perform effectively, enhanced individual or organisational performance will not result. Eylon and Bamberger (2000) also express the view that empowerment cannot be regarded as a cognitive experience independent of managerial action, because managerial actions have such an important impact on employee perceptions of power relations in the workplace. They argue that the empowerment experience cannot be separated from workplace structures and practices. For example, Thorlakson and Murray (1996) evaluated the impact of an empowerment programme in Canadian businesses and found that employees exposed to the programme did not view their organisation more favourably than a control group of employees. They suggest that this finding could be explained by the disruptive corporate downsizing going on in the organisation at the time of the experiment and conclude that the timing of empowerment interventions must be considered with respect to structural changes occurring in the organisational context. Thus empowerment appears to be a multifaceted concept, comprising both sociological (structural) and psychological (cognitive) elements, which are inextricably linked. These are discussed in more detail below.

The experience of empowerment

Within the psychological approach to employee empowerment it is also acknowledged that the value and effect of an empowerment exercise is multidimensional. For example, Thomas and Velthouse (1990) identified four psychological dimensions of empowerment as impact, competence, meaningfulness and choice. *Impact* is defined as the degree to which an employee feels his or her behaviour 'makes a difference' in accomplishing a purpose; *competence* relates to the degree to which an employee believes that he or she can skilfully perform a task if they try; *meaningfulness* relates to employees' intrinsic caring about a given task; and *choice* relates to individuals' perception of their control over the outcome of their actions. Spreitzer (1995) built on this model and developed a measure of empowerment based on these four dimensions, suggesting that, together, these four dimensions represent a proactive individual frame of mind (Spreitzer 1996).

Empirical examination of individual and organisational characteristics associated with employees' psychological empowerment are relatively few, so that making recommendations for effective psychological empowerment strategies is difficult. However, some recent studies have shed light on the requisites of psychological empowerment within organisations. For example, Spreitzer (1995) found that individual factors such as self-esteem and organisational factors such as employees' access to information about the company mission were related to psychological empowerment. In a later study Spreitzer (1996) also identified work-unit level variables impacting upon empowerment. These included strong socio-political support from subordinates, work groups, peers and supervisor/managers. Spreitzer (1996) also found a positive link between low levels of role ambiguity and the effectiveness of an employee's psychological empowerment. More recently, Koberg *et al.* (1999) studied the antecedents of psychological empowerment and found that group and organisational variables were more important influences than individual differences. Thus they conclude that employees are more likely to feel empowered if they work in a group with an approachable leader who encourages a sense of group worth and effectiveness. These studies should inform managers eager to implement an empowerment programme.

It also seems that the immediate relationship between employees and their supervisors or managers is crucial to feeling empowered. This may be a problem in the organisationally fragmented construction industry, where some employees may be separated by numerous organisational interfaces from project managers. Gomez and Rosen (2001) explain the impact of the quality of the manager–employee relationship on empowerment in terms of leader–member exchange theory. This theory holds that managers develop different types of relationships with employees, and that employees who feel they belong to an in-group of favoured employees will enjoy greater latitude, more access to information, involvement, and confidence and concern from their managers than employees belonging to an out-group. Gomez and Rosen (2001) provide empirical evidence to show that the quality of the leader–member exchange is positively associated with employees' feelings of empowerment, and emphasise the importance of immediate interpersonal relationships between managers and employees in designing effective empowerment strategies. The leader–member exchange theory also suggests that care must be exercised to assure equity in the development of empowerment strategies because some employees may be systematically assigned to out-groups by managers. The risk of ethnic minorities, female employees or employees with disabilities being treated as out-groups in the construction industry may be particularly acute given their under-representation (see also Chapter 8).

Siegall and Gardner (2000) also identify communication with a supervisor and the quality of interdepartmental relations (teamwork) as contextual factors influencing employees' psychological empowerment, although perhaps the most interesting finding from their study is that different dimensions of empowerment in Spreitzer's model were predicted by different contextual variables. This suggests that multiple empowerment strategies are required and that there is no single 'fix-all' solution to fostering psychological empowerment.

The act of empowerment

Empowerment techniques can be individually or team focused, and can involve the devolution of varying levels of decision-making power (see Honold 1997; Nykodym *et al.* 1994; Pastor 1997). Advantages of each type of empowerment can include the following:

- *Individual empowerment* can lead to better utilisation of skills and innovative capabilities, as well as greater job satisfaction, motivation and organisational loyalty (Mullins and Peacock 1991). Employees gain the power to be heard, to contribute to plans and decisions, and to use their expertise to improve their performance for the benefit of the organisation (Foy 1994: 5). Individually focused empowerment techniques are important for improving the self-efficacy of employees and their personal motivation and satisfaction.
- *Team empowerment* emphasises the importance of self-management and group decision-making autonomy. This requires a shift from task-oriented working to an output-focused approach, where the team has collective responsibility for both the utilisation of resources and the division of tasks necessary for performance (see Belbin 1996). Team empowerment helps groups to cope more easily with turbulent business environments, where there is a demand for informed, responsive and adaptive workforces (Swenson 1997), and it is more effective in harnessing and integrating knowledge (Nykodym *et al.* 1994).

The variety and interaction of the individual and teamwork concepts is complex, but recently they have converged in the form of the *high performance work system*, which offers a form of organisation and performance well in advance of traditional teams (Huczynski and Buchanan 2001: 388). Perhaps unsurprisingly given the historical reliance on classical models of production management, empowerment has also come to be associated with the process of culture change within organisations. This is because the delegation of responsibilities and power is more appropriate in modern, de-layered organisations and in project-based environments, where dedicated project-management teams control discrete work packages relatively independently. Interestingly, empowerment practices have been found to offset the effects of downsizing in reducing employee loyalty, suggesting that empowerment can help employees to cope with the financial realities of working in a modern firm (Niehoff *et al.* 2001). However, this effect is not direct but occurs through job enrichment. Thus, Niehoff *et al.* (2001) suggest that employee loyalty and support behaviours are more to do with the characteristics of a job, including autonomy and responsibility, than with the empowering style of a job incumbent's manager. This finding supports the view that power sharing must be genuine and not illusory to yield the benefits of empowerment. The use of 'empowerment' strategies to offset the effects of downsizing on employees' loyalty and trust may also be dangerous, in that care is required not to overload employees who 'survive' such structural changes. Where empowerment strategies result in fewer people taking on more work and responsibility, burnout and work–life imbalance can occur (see also Chapters 8 and 11).

A cautionary note

It should be noted that empowerment has also been widely criticised. A. Wilkinson (2001) identifies a number of problems within the empowerment literature:

- the term 'empowerment' is elastic and so it is not always clear what it means;
- it is often treated as an entirely new concept with no historical context;
- little research is available on the effective implementation of empowerment strategies;
- the context within which empowerment takes place is largely ignored when implementing strategies.

Empirical research is beginning to deal with some of these shortcomings in the empowerment literature. For example, attempts to conceptualise and develop valid measures of empowerment allow for clearer definitions of the concept, and studies exploring the contextual variables related to employees' empowerment are emerging (Siegall and Gardner 2000). However, the link between management acts designed to empower employees and psychological empowerment is still not well understood. It is possible that in some contexts managerial acts designed to empower employees may have a negative impact on employees' psychological empowerment. For example, Eylon and Bamberger (2000) suggest that when individuals are unprepared to take on higher levels of responsibility, or do not have the experience, knowledge, training or capacity to do so, empowerment strategies may have a negative impact on individual and organisational outcomes. This might be the case in some sectors of the construction industry due to relatively low levels of education, historical practices and a culture which exclude employees, and cynicism towards management activities. Eylon and Bamberger suggest that structural changes designed to empower employees which are not supported by them often increase uncertainty and role ambiguity in jobs, and they postulate that employees' tolerance for uncertainty, autonomy and responsibility will moderate the effect of management power sharing on job performance and satisfaction. Thus, where employees have a low tolerance for uncertainty or ambiguity, managerial empowerment acts will actually result in diminished job performance and/or satisfaction (Eylon and Bamberger 2000). There is a need to better understand the relationship between empowerment acts, perpetrated by an organisational elite, and empowerment experiences, felt by job incumbents, if effective empowerment strategies are to be designed.

Others writers have highlighted the common structural barriers to the implementation of empowerment, such as a lack of management support (Wysocki 1990; cf. Swenson 1997), linkages with rationalisation or organisational downsizing (Adler 1993) or the development of inappropriate competition between teams and departments, which can often detract from its objectives (O'Conner 1990; cf. Swenson 1997). Lack of support from middle management is a particular structural concern, since empowerment may be seen to threaten the role of middle managers in that not only are tasks devolved, but bottom-up communication channels may also bypass middle management. For example, Denham *et al.* (1997) presented data collected in public- and private-sector organisations undergoing empowerment initiatives and revealed middle managers overtly supporting empowerment strategies but covertly demonstrating no real commitment to, and in some cases actively blocking, the implementation of these initiatives. Thus, although it has a high degree of theoretical attractiveness, the complexities inherent within the implementation of empowerment strategies must not be oversimplified. There may be significant resistance from both managers and employees to its effective implementation, and this necessitates a senior champion of some kind.

Claydon and Doyle (1996) identify a more fundamental problem with the way the concept of empowerment is often described and justified in modern management literature. They suggest that tensions and contradictions are inherent in the empowerment concept and explain these in terms of the inconsistent ethical frameworks underpinning it. On the one hand, they suggest that empowerment is often justified on the grounds that the involvement of employees is virtuous in its own right and justified in terms of individuals' rights to self-determination and personal growth in work. On the other hand, however, empowerment is often said to be good because it leads to enhanced business performance. Claydon and Doyle argue that these two ethical positions are fundamentally at odds since one is founded on self-interest and the other requires that self-interest be disregarded. From this ethical standpoint, the motivation for managerial adoption of empowerment strategies comes under scrutiny. Arguably, the prime motivation is not to create a humanising work environment, as is often stated, but to help managers to manage and, in doing so, to encourage employees to engage in beyond-contract work (A. Wilkinson 1998). Furthermore, the individualistic nature of empowerment strategies, which blocks the collective organisation of labour, is also paradoxical, since it reduces employees' power base in an organisation or industry (A. Wilkinson 1998). Furthermore, when combined with strategies such as BPR, which often result in organisational downsizing (see Chapter 11), employees' 'self-empowerment' is threatened rather than enhanced. In such circumstances the element of choice, a fundamental component of psychological empowerment (Thomas and Velthouse 1990), might be absent.

These contradictions in the empowerment debate are rarely highlighted in the popularised management literature, but remain important issues. They indicate that unless empowerment strategies are carefully implemented they are likely to result in a cynical response from parties whose economic position leads them to perceive no real power sharing to exist, despite the rhetoric. Evidence has shown that poorly managed empowerment strategies can erode trust within an organisational system and be counterproductive rather than helpful (Robinson 1997).

The application of empowerment to construction

The recent focus on the effective integration and management of supply chains within the construction sector has emphasised the need for all key stakeholders to contribute fully to the achievement of project objectives. This emphasis is evident in the increased application of partnering and strategic alliances between organisations (Barlow *et al.* 1997). Effective alliances can provide a seamless supply network offering better business relationships, integrated solutions, reduced costs, higher quality and shorter delivery times. However, considering the shift in relationships and responsibilities that such changes require it is surprising that the process of empowering organisations and individuals has not been an extensively researched topic within the construction literature. It is even more surprising when one considers that the level of fragmentation and autonomy in the industry provides an ideal climate for the systematic implementation of performance-enhancing empowerment techniques. In addition, empowerment also offers the opportunity to create a workplace culture that is more responsive to employee needs. This is essential to retaining and developing a workforce that will generate sustained performance improvement in an increasingly demanding business environment.

The role and importance of empowerment as a potential panacea for the industry's poor performance in people-related issues has been emphasised by recent government-formulated reports and initiatives. For example, in his *Rethinking Construction* report Egan (1998) identified commitment to people as a key requirement for performance improvement within the UK construction industry. A follow-up report, which focused on the 'respect for people' theme as a key driver for change, was then released, entitled *A Commitment to People: 'Our Biggest Asset'* (Rethinking Construction 2000). This report set out to identify practical ways for the industry to improve its performance on people issues. It presented a powerful business case for improving respect for people working within the industry, based on the premise that failure to improve the current situation will result in falling profits, an inability to achieve effective teamworking and partnering, and the industry falling behind leaders in process productivity and innovation. The report identified several 'cross-cutting themes' which underpin key performance measures for improving respect for people. One such theme concerned the issue of workforce involvement and empowerment. Specifically, the report stated that:

> Respect for people means that all workers need to be consulted, involved, engaged and ultimately empowered in a spirit of partnership – not just management. The workforce is a rich source of ideas to improve the way work is carried out. And involving the workforce will not only demonstrate that they are respected and valued, but will improve productivity and quality.
>
> (Rethinking Construction 2000: 13)

The empowerment concept is based on the premise that imposed values, culture, training and information do not necessarily release people's store of creativity and innovation. Rather, it recommends the creation of a climate and structure in which employees willingly take on *responsibility* for the achievement of organisational objectives. It suggests that if they are treated as partners, trusted and given responsibility within a clearly defined framework, employees will then be encouraged to work for the good of the organisation. Thus, if empowerment is seen as the *process* of enabling employees to make workplace decisions for which they are accountable and responsible, within acceptable parameters, its utilisation should be positive in terms of improving business performance and enhancing innovation. In this context, empowerment can be seen as a process of *gaining influence* over events and outcomes of importance to an individual or group, rather than merely devolving power and control, as opponents argue (Foster-Fishman and Keys 1995). Applying this definition within a construction context, it could be viewed as the process of apportioning appropriate responsibilities for aspects of project delivery and overall strategic performance throughout the supply chain. Developing and applying appropriate empowerment strategies throughout the project-delivery system is therefore essential to exploiting the potential benefits of self-managed teams within construction.

On first examination, it would seem that empowerment strategies are ideally suited to the construction industry. Traditionally, the industry has neglected to take advantage of its workforce, which means that there is a considerable amount of untapped productive potential waiting to be released. However, determining how empowerment should be applied in practice, in a way which aligns with the industry's operational environment and culture, presents a significant challenge. As Cunningham and Hyman point out:

> The empowered organisation seems to assume a transition through cultural change to an organic unit in which all employees continuously learn to contribute to organisational operations in the context of an evolving and fluctuating product market. Whilst discussed extensively in managerial literature in recent years, few examples of this model have actually been investigated empirically, leading to questions as to

whether the empowered organisation exists more in the minds of management consultants than in actuality.

(Cunningham and Hyman 1999: 193)

It is clear, even from this brief critical review of the empowerment concept, that if it is seen as the process of enabling employees and teams to make workplace decisions for which they are accountable and responsible, then the first thing that must be established is the acceptable parameters of this decision-making. This will depend, in part, on the structural and cultural context in which the organisations are embedded. Thus, its utilisation should be viewed within the context of the specific application under consideration, and the strategies used must be sensitive to the particular business and operational requirements of the organisation. We will now explore the contextual factors which underpin the case for employee empowerment in construction, and isolate the requisite conditions for its successful implementation.

Contextual parameters of empowerment in construction

The task for all construction organisations, whether they are involved in the design and/or the construction processes, is successfully to complete a series of projects. Projects are the basic commercial building blocks that contribute to the success of the companies that manage them. In accordance with the needs of projects, large construction companies are made up of a number of small project teams comprising groups of specialist staff in project-related functions. These project-delivery teams are effectively self-managed, cross-functional groups, which combine the expertise of various designers, managers, cost-control specialists and other technical experts in a manner which is appropriate to the nature of the particular development under construction and the environment in which this occurs. As we discussed in Chapter 4, typically underpinning this configuration will be a 'matrix' organisational structure, in which functional departments form a permanent location for project-based staff working in temporary cross-functional teams (Langford *et al.* 1995: 69). Upon completion of the assigned project tasks, teams are usually disbanded and redeployed elsewhere within the organisation. The success of a project team therefore relies on them having clearly defined objectives and well-defined tasks, and on team members having been carefully selected (Mullins 1999).

A traditional assumption surrounding production teams is that they involve a number of individuals with shared production goals, with work status defined through their social roles, and supported by incentives and sanctions (Huczynski and Buchanan 2001). This view of teamwork implies that work will form a satisfying and meaningful activity for those involved, and that there will be little conflict between employees' functional and project-based goals. However, within construction organisations' matrix structures the competing demands of functional and production-oriented goals have been shown to provide a potential for tension within teams that can impact on the empowerment process. For example, Moore and Dainty (1999, 2001) found that the success of construction teams could be undermined by issues arising from the rigid professional cultures of individual participants within project workgroups. They identified cultural non-interoperability as a significant barrier to the management of change and the achievement of innovation within the design and construction processes. They argued that project responsibilities delineated along professional identity lines produce design and construction solutions that fail to fulfil the potential of the integrated team environment. Thus it is far from clear that common goals will necessarily result from the industry's project-based environment. Claydon and Doyle (1996) conducted a series of case studies, and they also identified tensions and contradictions in managers' perspectives on power sharing. They suggest that experts actively withheld cooperation from cross-functional project teams when suggestions made by these teams were seen as challenging their power.

The matrix structure of project organisations presents other significant challenges. First, teamwork and cooperation depend on each element of a system working in conjunction with every other element (Landes 1994). This is particularly difficult for those working within matrix structures, who find themselves torn between their project and functional loyalties. Second, it is essential that empowerment takes account of the transitory and fragmented delivery process, in which separate organisations often provide design, management, construction and supply functions, and at different times. Managing the empowerment of individuals and sub-teams which are constantly changing presents a considerable challenge for project-management teams. A third challenge stems from the temporary nature of construction teams and the fact that they are held together for very short periods. Thus the time needed to develop the trust and cohesiveness required for effective self-managed teamwork is often not available within construction project environments.

Although few would doubt the need for the industry to embrace leading-edge thinking on employee involvement and empowerment, the ingrained employment practices and time-honoured organisational delivery structures will not be easy to break down. Internal challenges to the implementation of empowerment in construction include a lack of management commitment, an underestimation of the extent of change required, a resistance to behavioural change, a failure to adopt continuous learning, too much bureaucracy and ineffective communication channels (Holt *et al.* 2000). Furthermore, encouraging people to buy into the performance-improvement process and add value to the industry's products and services is likely to be particularly problematic given the fragmented project-delivery structures, reliance on subcontracting and self-employment within the sector. These characteristics demand that to be implemented effectively the empowerment concept should be dynamic and should be applied throughout the entire project supply chain.

Swenson (1997) identified a range of measures to overcome barriers to effectively empowered teams. The barriers within the construction industry that may prevent their implementation are listed in Table 7.1.

It is evident that whilst in many environments these would offer a complete solution, they do not mitigate all of the challenges posed by the construction industry's project-based delivery structure. The list reveals that managers of construction companies should be selective in deciding who to empower and the extent to which various teams and individuals should be encouraged to self-manage to enhance project performance.

Another contextual challenge to empowerment concerns the apparent paradox that appears when applying the empowerment concept in a team-focused project-based environment – that promoting individual decision-making autonomy can move teams away from an integrated operation. This stands in marked contrast to the current industry emphasis on improving delivery through supply-chain integration and improved internal project relations. This demands that empowerment within the construction project environment should be aimed at achieving an optimal balance of individual and team decision-making. However, establishing a desirable position along the continuum bounded by team integration and individual employee empowerment for particular project types is complex, and has yet to be informed by empirical research.

Implementing empowerment strategies in construction

The challenges to the effective use of empowerment identified above need redressing if the construction industry is to take advantage of its hitherto untapped potential to improve its efficiency. Whilst many within the industry would argue that the empowerment of its workforce and project-delivery teams is the time-honoured way in which the sector has always managed the production function, the current inefficiency and lamentable performance of the sector with regard to respect for its workforce reflects an under-utilisation of the concept in current management approaches.

Taking advantage of the empowerment concept in construction demands that employees embrace the ethos behind its implementation. If it is seen as a mechanism to offload responsibility on to line managers, or merely as a consequence of outsourcing or downsizing initiatives, then a backlash is likely. Furthermore, if it is seen as an unnecessary formalization of the existing devolution of project-level responsibilities, then the value-adding benefits and finer points of empowerment are likely to be

Table 7.1   Barriers to the changes required for empowerment within the UK
construction sector

External change required for empowerment (Swenson, 1997)	Barriers to empowerment within the construction sector
Teams should be valued as part of the strategic plan and receive support from top management.	Teams are temporary, held together for very short periods of time and often include non-permanent members. Moreover, the composition of teams changes during the project lifetime to take account of different skills needs at different development phases. It is therefore difficult to integrate teams into the strategic planning process.
The use of teams should be congruent with larger organisational culture.	This is problematic when team culture is largely dependent upon the members themselves, many of whom will come from different functional parts of the organisation, from external organisations or even different parts of the industry.
Management should provide clear goals, parameters and resources to the teams.	Construction projects are often 'one-offs' with there being little resulting contribution to wider organisational goals. Resource requirements and allocation tend to remain fluid to allow for rapidly changing workloads and urgent requirements.
Care should be taken in the composition of teams to include interpersonally competent, motivated and complimentary diverse members whenever possible.	Team composition and deployment decisions are often made reactively depending upon the number of projects won through competitive tendering. External recruitment is often used for the selection of some team members.
A cooperative relationship should be formed with unions and workers by emphasising common goals and the benefits of teams to all stakeholders.	The matrix structures used by construction companies mean that project team members have different (and sometimes competing) project and functional goals. The use of sub-contracted labour and widespread self-employment militate against the formation of common goals.
Expectations for improvement should be realistic and based on time needed for training and transition.	Training and time allowed for team transition is often restricted by time pressures which result from increasingly demanding programme schedules and the necessity for quick project turnaround.
Employees should be trained in leadership and team skills and build a culture that fosters reflective learning and continuous improvement.	Training opportunities are often restricted during a project's development due to the demanding timescales and inherent pressures of the industry's project delivery structure.
Employees should be involved in formulating performance appraisals and recognition and reward structures for team work.	Performance management and reward systems are underdeveloped and the HRM/employee development function generally undervalued within construction companies.

missed. However, whilst enthusiasm for the principles would be welcomed, it is important that senior managers remain cognisant of the limitations of empowerment as a teamwork-enhancing approach. Like any management idea, it is not a wonder drug which will cure all of an organisation's ills. In particular, the inherent cross-functionality of construction teams presents a considerable challenge for team members attempting to reconcile functional authority structures with team-integration requirements. Furthermore, when the complexities and difficulties of implementing team-based empowerment are viewed alongside the inherent weaknesses of matrix structures (such as ambiguous reporting lines, competing demands on employees and problems with coordinating activities) its effective utilisation can be seen to be even more problematic. This raises the question of who should take responsibility for the implementation of empowerment within construction projects – does the real power to empower lie with the project manager or with the managers of companies which contribute to a project?

Together, these issues demand that a range of measures be put in place to enable the smooth transition to empowered individuals and teams within the industry:

1 There is little point in attempting to implement empowerment strategies within rigid, bureaucratic, hierarchical structures. These must be replaced with flatter structures which can more readily accommodate and complement concepts of employee ownership (see Holt *et al.* 2000).

2 Formal support networks must be put in place which provide empowered individuals and teams with training, assistance, guidance and leadership when required.

3 Construction companies at the head of the production effort must apportion appropriate responsibilities for aspects of project delivery and overall strategic performance *throughout* the supply chain. Although this is reliant on inter-organisational trust and cooperation, developing and applying appropriate empowerment strategies throughout the project-delivery system is essential to exploiting the potential benefits of self-managed teams, given the industry's reliance on a fragmented delivery structure.

4 The current drive for employee empowerment must be viewed within the context of the strategic management need to secure commitment from employees and to ensure their identification with organisational goals. Balancing this with the need to apportion responsibility and autonomy to project teams presents a significant SHRM challenge.

5 Construction contracts and professional contracts will need to be adapted to take account of the wider sharing and distribution of risks which empowerment implies. If more people are taking decisions they

must take responsibility for the outcomes of those decisions. However, it is also important that construction contracts become less penal, confrontational and inequitable. Trust is the foundation of any empowerment strategy.

As long as these factors are carefully considered, empowerment has the potential to play an important part in helping construction organisations address increasing performance demands whilst moderating the negative effects of their fragmented project-delivery structure. However, as with all forms of employee participation and involvement, a prerequisite of the implementation of empowerment needs to be the development of an appropriate and effective communications infrastructure to ensure that goals are understood and the performance opportunities associated with empowerment maximised. The role and importance of communication is explored below.

Communication requirements for employee empowerment

Organisations using empowerment effectively should comprise groups of people making individual and collective decisions which support common strategic objectives. However, without effective communication the decisions and actions taken by people may not be aligned, leading to key objectives not being achieved. It is appropriate that we consider the issue of communication within this chapter, as the existence of an effective communication network is essential to ensuring the participation, involvement and empowerment of individuals and teams. Indeed, communication and access to information have been found to be a key determinant of psychological empowerment in organisational research (Spreitzer 1996; Siegall and Gardner 2000). The need for well-developed communication mechanisms is particularly acute in construction, since the delivery of its products is managed in a devolved, dispersed and fragmented manner, thus making the process of communication more complex than in more stable manufacturing environments.

Communication forms an enabling factor for many aspects of the SHRM function. For example, communication is an important factor in motivating the workforce, since it is essential to keeping people feeling informed, happy, valued and involved. It is also essential to preventing simple misunderstandings, which are often the cause of industrial relations problems in many organisations. Furthermore, communication within a project team will influence general employee opinion of the levels of openness and involvement within the organisation. An open culture is likely to be a more egalitarian and trusting culture. Finally, good practice in many areas such as occupational health and safety, equal opportunities, training and development relies on good management systems and communication between key stakeholders.

These examples suggest the need for a communication system which combines both formal (structural) and informal (interpersonal) communication mechanisms if a culture is to be developed in which employees are to feel *directly* informed, motivated, involved and ultimately empowered. However, it is important to understand that establishing effective communications is not just about the ways in which information is transmitted from one person to another, but also about the ways in which it is interpreted, understood and then acted upon. This demands a directness and openness of communication that have not been a characteristic of many construction projects, which, as a result of conflicting interests and penal contracts, are traditionally characterised by secrecy, top-down information flows and manipulative behaviour. The importance of employers being able to communicate openly and directly with employees becomes more significant if empowerment is embraced. The decline in the significance of collective bargaining and the individualisation of the employment contract are likely to lead to further emphasis on direct communication (Emmott and Hutchinson 1998). Indeed, without an accompanying emphasis on empowerment the trend towards individual contracts and enterprise bargaining will merely result in the exploitation of workers, and will ultimately have a counterproductive effect on the construction industry's effectiveness.

Influences on communications effectiveness in construction

The effectiveness of communication is essentially dependent on three key factors:

- the effectiveness with which information in encoded and then transmitted through communication systems, channels and networks;
- the appropriateness of the communication medium used;
- how those receiving the communication decode, interpret and act upon it.

Each of these barriers is explored below in order to reveal the complexities and challenges of the empowerment concept in construction.

Encoding and transmission

The effective encoding of information has been a perennial problem in the construction industry. For example, drawings are often issued incomplete, with information missing and contradicting other project documents such as specialist engineers' drawings, specifications and bills of quantities. Many of these problems originate in the many different occupational cultures which characterise the industry, each having different educations, languages and customs which others find difficult to understand. Arguably the best example of this has been the problem of construction contracts, which for the uninitiated are almost impossible to understand due to their complex structure and legalistic language. This has historically been a major cause of conflict in the construction industry (Latham 1994). The divergent occupational cultures which give rise to such problems can be traced back to organisational changes in the construction industry between the Middle Ages and the nineteenth century, which led to the demise of individual artisans and to the domination of large capitalist contractors and a proletariat of waged labour (Musgrave 1994). These changes, fuelled by the process of industrialisation, led to the development of the professions and a hierarchical structure of social superiority with the architect at its pinnacle (Hindle and Muller 1996). Over time, professional institutions have developed vested interests in maintaining the roles which have provided their social status, cementing them in place by developing specific methods of communication which are reinforced by exerting control over the educational system and standard forms of construction contracts. Thus it is not surprising to find much contemporary evidence that hierarchical images of professional status remain deeply rooted within the modern construction industry's social fabric (Loosemore and Tan 2000). This has provided the foundations for the development of strong occupational stereotypes, which have been, and continue to be, a major source of misunderstanding in the industry. While new project procurement systems such as design and build provide a more egalitarian social environment, the influence of stereotypes may be strengthened by other factors like the industry's confrontational, macho and time-pressured culture.

Once a message has been encoded it has to be communicated through effective communication channels, something which is particularly challenging in construction because of the number of boundaries across which communication must take place. Furthermore, trends towards partnerships and improved client satisfaction are extending the need to communicate beyond organisational boundaries to embrace project-delivery structures and their supply chains. This is because employees within construction companies do not operate in isolation, but must liaise and interact with the suppliers and customers who surround them. For example, contracting organisations subcontract the vast majority of their work and buy in specialist design services where required. These service providers may be outsiders to the organisation, but they form integral members of the project-delivery team. Partnerships between supply-chain partners cannot work unless effective communications exist both within and between the organisations involved. This presents a real challenge for companies at the centre of the project delivery, which must communicate in a way that empowers the others involved in a spirit of mutuality.

Communication media

The medium through which information is transacted will also have an important bearing on the efficacy of the empowerment process. In terms of empowerment, a useful classification is to see media as either one way or two way. *One-way* media facilitate a linear and unidirectional flow of information from sender to receiver, with no opportunities for feedback. This might include letters, drawings, faxes, etc. In contrast, *two-way* media facilitate a reciprocal and multidirectional flow of information from sender to receiver, with numerous opportunities for feedback. This might include meetings, telephone calls and conferences, etc. If the aim is to secure trust, mutuality and a spirit of cooperation amongst the members of the project-delivery team, then two-way media are preferable. One-way media have the potential to undermine the atmosphere necessary for involvement and empowerment within the workplace. However, they are also cheaper and this is why they are increasingly relied upon in construction projects, often where meetings might have been more appropriate. As margins within the construction industry are further eroded through short-sighted tendering practices, the reliance on media which are not conducive to empowerment is likely to increase.

Interpretation and action

It is essential for any manager to be aware of the natural human tendencies with regard to the interpersonal communication process. Armstrong (1998) defines the key barriers to effective communication in this regard as follows: people hearing what they want to hear; ignoring conflicting information; having perceptions (and preconceptions) about the communicator; the influence of the group (people are more likely to listen to the opinions of their own group than those of outsiders); words meaning different things to different people (terminology and language barriers); non-verbal communication issues (misinterpretation of body language and other non-verbal signals); emotions (which can colour the reception of the message); noise (in both the literal sense and in terms of other distractions to the process of listening to and assimilating information); and the size of the organisation (the larger it is, the more difficult effective communication is to achieve). The likelihood that people will react in the desired manner can be enhanced by taking an *open* approach to communicating with people. This means being fair, equitable, honest and clear about information transfer and not manipulating it in one's own favour. This said, in reality those choosing to transmit the information must decide the degree to which people need to know it – it is important to recognise that in construction organisations, as in all others, information is power, particularly when it deals with commercially sensitive matters. Indeed, Loosemore (2000) found that, due to conflicts of interest within construction projects and the lack of risk sharing in construction contracts, it is common for information to be manipulated by project members in an attempt to force resource redistributions brought about by unexpected problems to go in their favour. The effect is that the problem's resolution is prolonged and made more costly. In these closed communication systems the real agenda behind the transfer of information is kept secret and hidden, and this erodes the trusting culture that is necessary for effective empowerment. The concept of empowerment is therefore a multifaceted issue and cannot be considered in isolation from commercial realities and other procurement issues, such as contract formulation and risk-distribution patterns.

Empowering people through effective communication mechanisms

It is evident that contemporary SHRM practice should pay particular attention to the influence that communication with the workforce has on the achievement of a climate of employee involvement. Developing effective communication systems is vital in ensuring that an organisation's favoured direction, approach and objectives are translated by the workforce as intended, and that employees contribute in a way which aligns their needs with the strategic needs of the business. This can be achieved by devising a communication strategy which combines several different methods to reinforce the messages needing to be transmitted within the organisation. For example, some construction companies use in-house publications such as magazines, newsletters or bulletin sheets. All can provide information on the overall performance of the business, interesting stories about a particular employee's achievements, or information on new priorities, systems or procedures that the organisation is adopting. Whilst these can be very effective mechanisms, simply placing notices on boards and assuming that people will read them is not adequate. Needless to say, avoiding jargon, using different languages where appropriate, reinforcing important messages and using feedback to check how well the receiver has understood the information being communicated are essential to ensuring the efficacy of these types of simple communication mechanisms (Armstrong 1998).

A more effective communication strategy is to use two-way communications media such as workshops, team meetings and briefing sessions to convey important information to employees. Training sessions can also be used to communicate important messages to employees in an interactive setting. Of course, these are expensive and time-consuming methods, but it is essential that two-way communication is facilitated which integrates all levels of the organisation, explaining policy decisions, future plans and progress in terms of how the organisation is performing (Armstrong 1996). In an empowerment culture it is essential that people know the effects their actions are having, since they will be responsible for them. In a construction project context this could extend to bringing together subcontractors and suppliers along with designers and client representatives to discuss the progress of the project against the original objectives. This will help to engender a spirit of mutual understanding, tolerance, cooperation and trust, helping to align the objectives of the participants in achieving their goals.

ICT is particularly important in facilitating communication, and the organisation of communication systems has been one area that has arguably seen the greatest change and development in recent years (Hamilton 1997: 5). ICT has radically transformed the communication process in construction projects, in that people can now communicate almost irrespective of their location, using mobile phones, satellite connections, the internet, etc. This has brought about easy access to a wealth of information, but has also brought its own problems related to the difficulties of information overload, confidentiality and people's tendency to over-rely on them as methods of communication. Thus, whilst ICT can help to facilitate the communication necessary for empowered workplaces, its use must be carefully managed. Like any method, it is best used in combination with other methods.

Case study

Empowering employees in Multiplex Asset Management

Multiplex Asset Management (MAM) is a subsidiary of Multiplex Constructions Pty Ltd, one of the largest construction companies in Australia and a major force in construction across Southeast Asia, the Middle East and, more recently, the UK. MAM was established in 1998 and has grown rapidly to offer developers, building owners and investors an integrated range of facilities and asset-management services. Being a relatively new company, MAM has had the rare opportunity to develop completely new systems that reflect best practice and complement its unique culture, which revolves around openness, trust and collective responsibility. Since MAM's core business is managing the risks and opportunities of its customers' facilities, the directors decided to develop a common framework for risk and opportunity management across the company, with consistent terminology, tools and techniques. The aim was to develop a new umbrella system, which would complement MAM's culture and empower employees to take responsibility for effectively managing risks and opportunities. In particular, the system had to be consultative and inclusive by recognising the needs of all stakeholders and encouraging collective responsibility, cooperation and teamwork for the management of risks and opportunities.

Initial discussions with key stakeholders soon revealed the complexity of the decision to develop a new risk- and opportunity-management system. While the creation of the guidelines was to be a major undertaking, so was changing people's existing expectations and behaviour. Only then would the system become an integral part of MAM's business culture, automatically determining the way that people operated, acted and behaved on a day-to-day basis. The directors knew that no matter how well designed the system was, people's acceptance of it would be critical to its success. Furthermore, while it is very easy to change people's behaviour in the short term, it is extremely difficult to keep it changed. This was the challenge MAM faced, which was equally important to the development of the system itself. In particular, MAM did not want to sterilise the decision-making process and eliminate the 'gut feeling' and intuition that had been at the heart of many good decisions made by Multiplex over the years. MAM was acutely aware of the dangers of 'paralysis by analysis', and of the importance of being decisive and risking being wrong rather than agonising at length and being right too late.

In MAM's enthusiasm to develop the system, the temptation was to rush in. Instead, the process was carefully planned over a one-year period. The first seven months involved writing the system and the remaining five months were used to implement it. To encourage people to 'buy in' to the system, MAM's directors championed it and were intimately involved in its development so that it realised their initial vision and MAM's corporate goals. Senior management commitment is widely recognised as being vital to the success of risk-management systems. There was also extensive consultation with other MAM stakeholders, external and internal to MAM, at all stages of the system's development. As a result, the final document evolved in a dynamic way and went through numerous changes before it was ready for use and everyone was happy with it.

An extensive training programme was also planned to bring people's confidence and skill levels to a point where they would feel comfortable with the system. This training was essential to overcome the common fear of risk management, and the perception that it is 'rocket science' and just another management fad that will eventually go away. An effective support infrastructure was also created which involved forming a risk-management team with specific responsibilities for implementing, managing, monitoring and updating the guidelines.

MAM also established a risk- and opportunity-management section on its existing intranet site, which allowed those working on MAM facilities to share, visualise and communicate information about risks with clients, staff, subcontractors, suppliers, consultants and authorities, anywhere in the world. Finally, to encourage people to use the new guidelines a monitoring and reward system was established, with the intention of linking people's risk-management performance to their annual performance appraisals.

Following the extensive consultation process, the final system is presented in a fifty-page document which specifies procedures to follow and risk-management tools to use. It was important to stakeholders not to have an intimidating system which was complicated to use and to have a document which explained how to do risk management. Furthermore, it is written in plain English, with minimal jargon, and provides numerous practical examples of how to use the tools specified. This was important for Australian users, but particularly important for non-English speakers on Australian and overseas projects. As requested by stakeholders, users are taken through the document by a series of simple questions, which they would be likely to ask as they encounter the system or the concept of risk management for the first time. Most questions are presented in the first person and examples include: What types of decisions are covered by these guidelines? When do I start risk and opportunity management? What is my first step? What do I do when I have identified a risk? Considerable thought also went into the design of the document, which is important from a psychological perspective. For instance, white space is maximised on each page to make it less daunting, colour and images are used to enliven the document and help readers navigate, simplicity of approach is maintained by restricting the number of steps to follow to no more than three in any one stage of the process, simple box diagrams are used to summarise processes, and the margin on each page is used to summarise the main points that have emerged.

The implementation of the system is currently being monitored for its effectiveness. This will be assessed in terms of how it has changed people's attitudes and working practices and how it has improved the management of MAM's facilities. Most importantly, the success will be measured against the impact on MAM's clients' business objectives rather than MAM's objectives. This is the most important attribute of a truly customer-focused system.

Conclusions

In this chapter we have advocated the use of employee involvement and empowerment strategies as an important part of an effective SHRM strategy. Used effectively, this could create a multi-organisational project environment founded on mutuality and common goals. However, if treated as a term to encompass the delegation or handing-off of responsibilities without due consideration to the process of managing the transition, empowerment is likely to lead to little more than abandonment and to be treated with cynicism and resentment. Similarly, it must not be seen as an act of management relinquishing control, but as a fundamental reorganisation of an organisation's operating practices. This presents a significant challenge for an organisation attempting to utilise empowerment strategies, since it is likely to require changes to both working practices and structural relationships. In particular, communication channels must be put in place which help to engender the two-way information flow that is a prerequisite of involving employees. Thus perhaps the greatest challenge for managers wishing to implement empowerment strategies is to ensure that the 'soft' rhetoric of empowerment does not mask the 'hard' reality of someone else taking risk and responsibility, which may in itself have implications for project costs and duration. Clearly, these are trade-offs that we would consider worthwhile in the short term for benefits in the long term. However, we also acknowledge that this ethos is at odds with the commercial culture of construction, which has traditionally placed short-term gain over investments in longer-term strategies. For empowerment to be properly embraced there are many deeply set attitudes, perceptions and behaviours to overcome, and this will take vision, determination, time and a considerable amount of goodwill on the part of those involved.

Discussion and review questions

1 Distinguish between employee participation, involvement and empowerment, and discuss how each might be applied in the construction environment to enhance project performance, teamwork and synergy between individual and employer objectives.

2 Effective communication channels and information communication networks (ICN) are prerequisites for the effective application of empowerment strategies. Map the internal ICN for a small construction project team, and identify the members through which relevant internal and external information should flow to ensure the motivation and self-managed autonomy of the workgroup.

3 Explain the difference between the cognitive (psychological) and behavioural (action-oriented) approaches to empowerment. Consider how these two approaches might affect one another.

8 Workforce diversity, equal opportunities and work–life balance in construction

The potential benefits of workforce diversity are enormous, but so are the challenges of harnessing it. In contrast to industries such as information technology (IT) and electronics, the construction sector is not widely regarded as an industry that effectively represents and uses the talents of all groups within society to its advantage. Ironically, however, it is one of the best-placed industries to do so. As the working population in most countries becomes increasingly diverse, the construction industry once again faces the prospect of losing an opportunity to enhance its creative and productive potential. The danger is that its poor record will make it a less attractive career option for potential new entrants from under-represented groups and will prevent it from benefiting from the skills, talents and varied perspectives of a balanced workforce.

The key to solving these problems lies in the way that construction activity is organised and in the development, communication and implementation of effective equal opportunities and diversity policies. As well as ensuring a fair and equitable workplace, these should include initiatives to facilitate the employees' attempts to combine their work and out-of-work lives effectively if people are to be retained. In this chapter we explore all of these issues and suggest ways in which project managers can address diversity issues effectively.

Introduction

The process of globalisation, demographic change and workplace reform are making equal opportunities and workforce diversity two of the most pressing issues in modern industrial relations. In countries like the UK, where workforce diversity in the construction industry is particularly poor, companies are becoming increasingly aware that in a highly competitive labour market restricting recruitment to under half of the population is likely to severely restrict their growth and future development. In other countries like Australia, where workforce diversity is particularly high, companies are recognising that this is a potential resource that has not been harnessed effectively.

One of the reasons why construction is facing so many challenges with regard to needing to diversify its recruitment base is that it is in the unfortunate position of having one of the worst public images of all industries. Most people come into contact with the industry at some point in their lives, and to the average person the industry is synonymous with stress, unreliability, high cost, low quality, chaotic working practices, and a dirty and dangerous work environment (Harris Research Centre 1988; Ball 1988). Therefore it should come as no surprise that in the 1999 edition of *Jobs Rated Almanac* civil engineering plunged from 18th to 70th position in expressed job preference, and 14 construction trades were among the least preferred of any occupation. This is reflected in the state of university education in most countries, where construction-related courses have among the lowest entrance requirements of any discipline. Faced with this problem, the industry in the UK has seen a recent push to improve its image through recruitment campaigns targeted at schools, colleges and minority groups, and has benefited from the success of the children's programme *Bob the Builder*. Professional institutions are also withdrawing accreditation from many courses to force the demand-and-supply curve in their favour. However, these strategies will not change the root cause of the problem, which lies in the way that the industry treats its employees and business partners. If the industry cannot portray itself as well paid, professional, safe, clean, caring, technologically advanced and innovative, then the problems will continue.

Effective education is the key to bringing about this change, yet very few construction books or journals have addressed the issue with any rigour. If people understand the value of diversity and the challenges of managing it, then the industry will have gone a long way towards solving its problems in this area. In the following sections we will begin to redress this problem by providing a background to the issue of diversity in the construction industry. In doing so, we restrict our discussion to issues of gender, culture, disability and age. These appear to be the main areas of concern facing the construction industry.

Differentiating between diversity and equal opportunities

The concept of diversity recognises that there are differences between people, which if harnessed can create a more productive, adaptable and creative work environment, where people feel valued and talents are fully utilised. Before commencing this chapter, it is important to distinguish between the concepts of diversity and equal opportunity, since they are often used synonymously. Indeed, the two concepts are quite different in the principles that drive them, as depicted below.

Table 8.1 Opportunity and diversity

Equal opportunities	Diversity
Externally driven from outside the organisation	Internally initiated from within the organisation
Driven by legislation	Driven by business needs
Focused on improving the numbers of underrepresented groups	Focused on improving the work environment
Concentrated on reactively addressing problems	Concentrated on proactively taking advantage of opportunities
Assumes assimilation	Assumes pluralism

Source: Adapted from Kandola and Fullerton 1998: 13

In particular, it is evident that a company must have an equal opportunity and diversity policy if it is to have a balanced approach to this increasingly important area of SHRM.

Minority groups in construction

In order to appreciate the importance of both equal opportunities and diversity issues for the construction project manager, it is important first to explore the current workforce profile of the sector in order to appreciate the scale of the problems that exist within the industry.

Women

A particularly pressing problem in the construction industry in most countries is gender balance (Dainty *et al.* 1998, 2000a). For example, in the UK women now make up over 45 per cent of the total working population but under 10 per cent of the construction workforce. This makes it the most male-dominated of all major UK industrial sectors. The underrepresentation of women in construction is the same in construction industries around the world – for example in Australia, America, Hong Kong and Singapore. This is known as *horizontal* labour-market segregation, meaning that women tend to work in certain industries and are either denied access or choose not to enter traditionally male-dominated sectors. Furthermore, in most developed countries *vertical* segregation is also apparent, where the women that do work in male-dominated sectors such as construction tend to occupy relatively junior positions or administrative and clerical support roles rather than being involved in leading the production function. In the UK, for example, women comprise under 4 per cent of the membership of the construction-related professional bodies, despite making up 10 per cent of the industry's workforce (Davey *et al.* 1998).

The current under-representation of women in construction is set against a historical backdrop of low representation in and deliberate exclusion by the industry and its institutions. Throughout their development in the nineteenth century the engineering professions and unions adhered to a policy of deliberately excluding women (Drake 1984). UK census data indicates that in 1871 there were only 171 women builders out of a total of 23,300 (0.75 per cent), and of 5,697 architects only 5 were women (Powell 1983). During the First World War the total number of women employed in construction and engineering increased, with almost all of them employed to replace directly, men called up to serve in the armed forces (Drake 1984). Although most returned to traditional occupations following the end of the war, by 1939 15,700 women were employed in the building industry, compared to 1.2 million men (Powell 1983). Since the Second World War there has been a steady increase in the number of women entering the industry in professional positions (Garner and McRandal 1995). Within the UK this has been boosted by targeted initiatives such as the Women into Science and Engineering (WISE) campaign, which has promoted science and engineering careers for women (see Shillito 1992). Other countries have also actively targeted women to attract them to the sector using similar campaigns in which schoolgirls are given an insight into the industry and the rewarding career possibilities available.

The poor image of construction, a lack of role models and knowledge, poor careers advice, biased recruitment literature, peer pressure and poor educational experiences have all been cited as problems which reduce women's entry into the industry (Wall 1997; Bronzini *et al.* 1995; Gale 1994). A survey commissioned by the Construction Industry Training Board (CITB) in 1988 showed that many women believed that if they chose a construction career they would not be treated as equals or would face harassment from their work colleagues. Other studies have shown that most women view the industry as a male-dominated, threatening environment, with an ingrained masculine culture characterised by conflict and crisis (Gale 1992). Women also perceive the industry to have inadequate facilities and poor training (see Sommerville *et al.* 1993). In a survey of schoolgirls only 17 per cent felt that the industry was suitable for women (EOC 1990). It is unsurprising, given this background, that women still view construction as an unattractive career option.

As well as the problems associated with the image of the industry, there may be other gender-determined factors which influence the number of women entering construction. S. J. Wilkinson (1992), for example, found that some 20 per cent of *employers* believed that construction work was 'unsuitable' for women. Greed (1997) also attributed women's underrepresentation to the construction subculture, which blocks the entrance of people and ideas that are seen as different and/or unsettling. Employer prejudices in this regard may manifest themselves through the recruitment process (Morgan 1992), particularly as recruitment in construction is often informal and through personal contacts (Druker and White 1996a). Other women are made to feel unwelcome and are not encouraged to develop towards the senior positions, from which they could contribute to shaping the built environment (Greed 1997). This is likely to affect their job satisfaction. For example, a recent study of professional employees of a large Australian construction organisation found that employees' satisfaction with their personal progress differed significantly between male and female employees, female employees reporting lower levels of satisfaction with their progress.

Despite these formidable barriers to entry, recent employment trends and membership data from the professional institutions show women's representation as having increased year on year since the mid-1980s (Dainty 1998). Women comprise around 18 per cent of the undergraduates on construction-related courses within the UK (Kirk-Walker and Isaiah 1996). However, women in the built-environment professions are neither evenly distributed by profession, nor proportionately represented at the middle and senior levels (Booth 1996; *New Builder* 1995). The most recent census data indicated that women were concentrated in clerical and secretarial positions, or other positions not directly involved with the construction process, with men occupying craft, operative and professional positions (Court and Moralee 1995). If the patterns of discrimination and harassment faced by women are not addressed, then there will be no substantial long-term gains in the number of women and underrepresented groups in the industry.

Ethnic minorities

In the UK ethnic minority groups make up only 1.9 per cent of the industry's work force, compared to 6.4 per cent of the general working population (CITB/Royal Holloway 1999). Indeed, of the seven major industrial classifications used in the UK, construction has the lowest representation of black and Asian workers. Further analysis reveals that ethnic-minority employees account for under 2 per cent of professionals in the industry. For example, the RICS membership comprises only 0.5 per cent ethnic minorities. These figures define the UK construction sector as a distinctly white-male bastion, and suggest that its historical development and culture have not been accepting of non-traditional entrants, whether they be female or from an ethnic minority. By failing to recruit a workforce that reflects the wider population, the UK construction industry is vulnerable to skills shortages and potentially to developing projects that do not meet the needs of the wider population.

In other countries the opposite problem exists. This is not surprising, because in most countries the construction industry employs a relatively large proportion of the working population and involves a relatively large number of menial and manual jobs, which makes it an attractive target for new immigrants. For example, in Saudi Arabia 30 per cent of the construction workforce is foreign, and in Australia and Singapore the figure is over 80 per cent, presenting managers with a range of potential problems, from communications, through conflict and safety to discrimination (CIDB 1998; MLSA 1998; ABS 1999b). Indeed, countries like Singapore and Australia have traditionally relied on a large immigrant population to meet their labour demands. For example, between 1976 and 2000 1.36 million people migrated to Australia from another country and became employed members of the Australian labour force (ABS 2001). Many of these workers have entered the construction industry, and statistics show that the ratio of construction employees born inside Australia (75 per cent) to those born outside Australia (24 per cent) is no different from that of the entire Australian workforce. Thus, in terms of employment participation it seems that the construction industry is not discriminating against immigrant workers. Furthermore, the proportion of employees born within and outside Australia does not differ substantially between different occupational groups: 73 per cent of managers and administrators were born inside and 22 per cent were born outside Australia; 73 per cent of professionals were born inside and 27 per cent outside Australia. The proportions are very similar for tradesmen and related workers (76 per cent born inside and 24 per cent born outside Australia), and for labourers and related workers (73 per cent born inside and 27 per cent born outside Australia). This suggests that barriers to the career advancement of immigrant workers do not exist in Australia. One possible reason for this is the fact that immigrants into Australia tend to be well educated (ABS 1999b). For example, an ABS survey of immigrants to Australia after 1980 reports that more than 55 per cent arrived with a post-school qualification. Accordingly, the vast majority (81.6 per cent) of professionals worked in the same occupation in Australia as they did prior to migration (ABS 1999b). However, caution must be exercised when interpreting these figures, because the figures are not industry specific and it is possible that the construction industry ratios could differ from those of Australian industry in general. Furthermore, statistics can be extremely deceiving and grossly oversimplify a highly complex problem. For example, the case study later in this chapter (pp. 188–92) on racism towards Asian workers in the Australian construction industry indicates that when one looks closely, racism and discrimination are common.

In recent years a particular source of concern in Australia has been the under-representation of Indigenous Australians (Aboriginal Australians and Torres Strait Islanders) in the workforce. The Australian government has come under significant political and public pressure to apologise for its policies of cultural assimilation during the 1950s, 1960s and 1970s. The result of ignoring the cultural identity and needs of this ethnic group has led to serious social problems for this group and significant under-representation within the working community. For example, in February

2000 the employment to population ratio for Indigenous people was 44 per cent, compared to 59 per cent among non-Indigenous people (ABS 2000). Furthermore, the unemployment rate was 17.6 per cent among Indigenous people and 7.35 per cent for non-Indigenous people; and the labour-force participation rate, which is the number in the labour force expressed as a percentage of the population aged 15 years and over, was 52.9 per cent for Indigenous people and 63.7 per cent for non-Indigenous people. While industry-specific figures are not available, it is likely that the labour-force participation rate of Indigenous people is lower in construction than other sectors of the economy because most construction work occurs in capital cities and urban areas, where indigenous people are under-represented. Only 34 per cent of the employed Indigenous population work in capital cities, compared to 66 per cent of the non-Indigenous population (ABS 2000).

The disabled

Gender and ethnicity are not the only issues in the equal opportunities debate. For example, the issue of disability has become of increasing concern in recent years, with 10 per cent of the potential UK workforce having some impairment or disability which affects their ability to gain employment. The problems for disabled people are not confined to difficulty in getting a job, but extend to the quality of the employment experience. For example, a recent Australian study (ABS 2001) found that 82 per cent of disabled employees said they were restricted in the type of job they could do, 71 per cent found it difficult to change jobs or get a better job, and 47 per cent felt restricted in the number of hours they could work. In the UK there is little data to indicate that disabled people are any more under-represented in the construction industry than in any other industry. However, anecdotal evidence suggests they are. In Australia the limited evidence available paints a similar picture. For example, according to ABS (2001) figures, the participation rate of working-age people who are actively seeking work is 50 per cent for disabled persons, compared to 80 per cent for able-bodied people. In 2001 there were 313,700 disabled people in paid employment in New South Wales and only 22,000 (7 per cent) were engaged in the construction industry. The ABS (2001) reports that people with a disability in New South Wales work mostly in manufacturing (19 per cent) and business services sectors (12 per cent). Whilst this might at first indicate a significant level of under-representation, many construction jobs are site based and present obvious problems for disabled people. One would therefore expect this employment group to be under-represented in construction. However, many disabled people can work effectively and safely on site with appropriate support, and there is no reason why disabled people could not work in office-based design and management roles.

The low participation rate of disabled people in general also highlights the potential importance of people with a disability in meeting future labour requirements. The ABS statistics indicate that in New South Wales some industrial sectors are utilising this labour source better than others. In manufacturing, for example, there exists a significant difference between the proportion of disabled persons in employment and that of employed people without a disability (18.6 per cent, compared with 13.0 per cent). Although many people become disabled through injuries sustained whilst working in the industry, most leave the industry, and it is not part of the culture of construction to see disability as a reason for taking people on (Greed 1997). The construction industry should examine ways in which jobs can be structured and workplaces designed to enable people with a disability who want to work to do so. The New South Wales study reports that in 1998 almost 28 per cent of working-age people with a disability stated that they were permanently unable to work. Of these, 5 per cent (10,200) said they would be able to work if their needs were met. In order to enable workforce participation, training, time off work, the provision of special or modified equipment, help with personal care and provision for work at home were cited as requirements.

The aged

Most developed countries have an ageing workforce, and employers will increasingly have to consider recruiting and training more employees from the over-45 age group. In Australia recent research indicates that over the next 10 years the over-45 population will grow at a rate five times faster than that of the total population (Hooper and Skeffington 1999). In Australia the construction industry is already reflecting this trend, with 57 per cent of employees 35 years of age or over, 30 per cent 45 years or over and only 16 per cent under 25 years.

In an ageing population it is important to recognise the value of older people and encourage their retention and participation within the industry. Advances in healthcare now mean that older people, who would have previously retired due to ill health, are more willing and able to lead a productive working life. These workers bring with them a wealth of experience and expertise that is extremely valuable in industries such as construction, where the 'shelf life' of knowledge is greater than in high-tech industries such as electronics and IT. This is often lost when people retire and needs to be more effectively transferred to younger employees. A second issue stemming from the ageing population is the increasing need for employees to care for elderly dependants. This is exacerbated by a shift from institutional to home-based aged care in many countries. Bourke (2000) estimates that between 1996 and 2041 the aged dependency ratio will double, rising from 18.1 to 34.8, meaning that for every 100 workers there will be 34 aged dependants. Francis and Lingard (2001) report that, currently, 70 per cent of all providers of personal care and home help for the aged, terminally ill or disabled persons are also in the workforce. Many of these carers may also bear responsibilities for childcare. Given this situation, companies should be cognisant of their employees' responsibilities in caring for aged dependants and assist them to fulfil these responsibilities wherever possible.

Arguments for diversity and equality in construction

Whilst there is a strong *ethical* reason for taking equal opportunities seriously, it will be the commercial argument that wins the day. The commercial issue for the construction industry is whether companies can compete successfully if all groups of people are not represented in the workforce and treated as equal. The challenge is to convince managers that in an increasingly competitive and global business environment a heterogeneous organisation made up of diverse groups containing a wide range of abilities, experience and skills is likely to be more productive than one which is more homogeneous. Companies should also recognise that they will suffer if they do not recognise the social concerns of the communities in which they operate (Ansari and Jackson 1996). However, the key driver for many construction companies is likely to be the lack of availability of appropriate skills in the construction industry. This makes traditionally under-utilised groups such as women, ethnic minorities, disabled or older workers an attractive pool of potential expertise. The construction industry is in competition with other industries for the best people and any perception that it does not provide an equal opportunities environment will encourage future cohorts of good ethnic-minority recruits to take their talents elsewhere. Thus, the construction industry continues to ignore the issue of diversity and equality at its peril. These arguments for diversity and ensuring parity of opportunity are discussed in more detail below.

Demographic arguments for workforce diversification in construction

In Chapter 1 we discussed the impact of projected labour shortfalls on the future competitiveness of the construction sector. In many countries the construction industry faces an impending 'demographic time bomb', where skills shortages are being caused by a declining population and an increasing tendency for younger people to stay on into higher education and/or choose a career in other industries, sectors and professions. The threat of skills shortages can also be attributed to other factors such as the traditional image of the industry, low pay and poor working conditions, the changing nature of skills requirements, extreme fluctuations in the construction market, the growth of self-employment, industry fragmentation and the decline in apprenticeships and training resources.

In the UK these problems were exacerbated in the late 1980s by a rapid rise in construction activity, which eventually led to skills shortages before the recession of the early 1990s (Agapiou *et al.* 1995b). The post-recessional growth of the late 1990s and early 2000s, coupled with an ageing population, terrible publicity, a continued ignorance of training needs and changing technologies, suggests that skills shortfalls are likely to be even more acute in the future. Therefore it is incumbent on all companies operating within the construction market to ensure that the industry is seen as an attractive place to work, that leakage of skilled employees to other sectors is minimised, and that employees feel that they are valued and treated accordingly.

In the UK an awareness of these potential skills shortages has stimulated a response strategy from both government sources and the industry itself. These include image-enhancing initiatives aimed at raising the profile and status of the sector, skills certification and employment schemes, and specific marketing campaigns aimed at broadening the industry's appeal beyond its traditional recruitment base. At a professional level, many construction companies are currently debating whether, in the face of declining entrance to degree courses, non-cognate graduates should be recruited to make up the shortfall of graduate entrants to the industry. For example, in 1998 around half of all graduate vacancies currently available are open to *all* graduates (Hall and Torrington 1998). Although there is a traditional reluctance to do this, it is common in other industries and would have the advantage of increasing the disciplinary diversity of the construction industry. Many of the industry's problems stem from the educational system, which in many cases instils and reinforces the preconceived occupational stereotypes which sustain the industry's confrontational culture (Loosemore and Tan 2000). Looking outside the traditional feeder courses might be a way of breaking down this culture and introducing greater innovation into the industry.

Another solution to the skills crisis is to diversify recruitment to the industry in response to changing demographics, which means targeting non-traditional groups such as women. The proportion of the population currently seeking work has followed a defined pattern in recent years, which is expected to continue; whilst male activity rates will continue to decline, female rates will increase. Indeed, virtually all of the increases in activity rates in 2000 were forecast to be amongst women (Hall and Torrington 1998). The alternative is to see the competitiveness of the industry decline in the face of foreign competition in an increasingly global construction-industry market.

Performance benefits of workforce diversity

Throughout the 1990s a strong business case was developed for workforce diversification. This essentially rests on three premises:

- that industry is under-utilising the full range of skills and talents in the population because of continuing unequal opportunities for some groups within society (Bagilhole 1997);
- that it should be possible for organisations to increase their efficiency and effectiveness by projecting a more pluralistic self-image and thereby widening their pool of potential customers (CIB 1996);
- that it leads to a better-informed, more innovative and adaptable organisation which is closer to its customers (Greenhaus and Callanan 1994).

Many of these arguments have been supported by recent research exploring how the proportional representation of certain demographic groups influences the functioning and effectiveness of teams and work-groups. This has shown that there are tangible performance benefits attributable to balanced workgroups. Ely and Thomas (2001), for example, identify two benefits of workforce diversity, both of which are relevant to the construction project environment: first, that workforce balance encourages better performance from those in the minority; and, second, that diverse teams perform more effectively than those composed of the traditionally dominant group. These arguments are discussed in more detail below.

The influence of proportional representation on minority employees

A great deal of research within many industrial sectors has examined whether increasing the numbers of minority groups such as women and ethnic minorities has a positive or negative impact. Those arguing that proportional representation will have a positive benefit suggest that barriers such as stereotyped perceptions should be removed and hence people will be allowed to contribute to their maximum potential. Others counter this argument by suggesting that radically increasing the numbers of women and minorities may actually lead to a backlash against minority entrants (Yoder 1991). Clearly, simply increasing the number of currently underrepresented groups without addressing the power relations and culture of the workplace is unlikely to bring about a long-term improvement in the positions of these groups. Any changes must be made carefully, and the benefits of proportional representation are only likely to be felt if it is implemented with an effective communications strategy and alongside robust diversity and equal opportunities practices.

The influence of diverse workgroups on team performance

Many proponents of diversity argue that balancing work teams can have tangible benefits for team performance. One area in which increasing diversity might influence workgroup values is that of ethical decision-making, some research indicating that women adopt a stricter stance than men on ethical issues at work (Ray *et al.* 1999; Zarkada-Fraser and Skitmore 2000). It could also be argued that teams which reflect the end-user group or local community in which a project is being constructed are more likely to be responsive to their needs. For example, architects designing a facility for the disabled are more likely to meet their needs if they have a first-hand understanding of disabled access and building use themselves. Employing disabled people as part of the design team is an obvious way to achieve this.

In terms of cultural diversity, it has been known for some time that different nationalities perform differently in different contexts. For example, the pioneering work of Hofstede (1980) found that people from different cultures vary along four main dimensions: masculinity/femininity (attitudes towards women, assertiveness, aggression), individualism/collectivism (the value of group membership and personal relationships), power distance (acceptance of power inequalities and authority) and uncertainty avoidance (the degree to which people are threatened by ambiguity and uncertainty). Subsequent research has introduced further dimensions, but Hofstede's work highlighted the value of different cultures in different contexts. For example, people from highly individualistic cultures may operate less effectively in teams than people from collectivist cultures. People from high uncertainty-avoidance cultures may feel more uncomfortable working under extreme time pressures in high-risk situations than people from low uncertainty-avoidance cultures. Of course, Hofstede's work did generalise and people can be found in each culture who vary along each of his continuums. Nevertheless, his work was incredibly important in highlighting the value of and need for culturally sensitive management practices. Diversity increases the number of different perspectives, styles, knowledge and insights that the team can bring to complex problems, and the world's most innovative organisations, such as Microsoft, take advantage of this by deliberately creating multicultural teams (Jehn *et al.* 1999; Ely and Thomas 2001).

Unfortunately, in contrast to sectors like IT, the construction industry has not harnessed the cultural perspectives and influences of different groups. All the evidence points to an *assimilationist* attitude which largely ignores the needs of different groups, expecting them to adapt to the dominant industry, organisational or national culture (Loosemore and Al Muslmani 1999; Loosemore and Chau 2002). However, current thinking in equal opportunities seeks to value diversity explicitly, to adapt to it and use it generate improvements in work performance and team effectiveness. Still, it should also be noted that linking group diversity to project outcomes is controversial. Our understanding of the behavioural dynamics of diversity is still in its infancy, and, although women and ethnic minorities may well bring different perspectives and styles, the necessary conditions, likely consequences and overall performance implications have yet to be proven (Ely and Thomas 2001).

Barriers to diversity and equality in construction

Prejudice is the enemy of diversity and equality, and is rooted in pervasive and systematic assumptions about the inherent superiority of certain groups and the inferiority of others based on differences in race, culture, ethnicity, national origin and descent (Thompson 1997). It is grounded in two forms of human behaviour which are central to people's cognitive processing – ethnocentrism and stereotyping. *Ethnocentrism* refers to the natural human tendency to see one's own culture as superior to others, and *stereotyping* refers to people's tendency to categorise others into distinct social groups and arbitrarily to generalise about their distinguishing traits, normally in a negative way (Sawin 1995). This arbitrary social categorisation and stratification is usually based on salient and physically identifiable features such as age, colour, appearance, disability, gender, race, ethnicity and social status. Furthermore, in many countries it is based on learnt fact. For example, Thompson (1997) points out that in the UK the link between race and class became strongly established after the Second World War as an increasing population of black migrants became overrepresented at lower levels in Britain's socio-economic class system. Soon black people were seen as the cause of poverty rather than victims of it, and the social problems experienced by them were translated by their white 'hosts' into matters of personal failing, weakness or inadequacy. The final step that led to the emergence of racism was to blame black people for the problems experienced by white people.

Both ethnocentrism and stereotyping are coping mechanisms that serve an adaptive function for people in novel, complex and uncertain situations, by reducing the world into a more manageable number of categories. They are also valuable 'sense-making' mechanisms to explain the actions and behaviours of 'out-group' members, particularly when things go wrong or are threatening. In these circumstances people try to preserve a sense of self-esteem and personal power by blaming groups that are 'different' and in a 'minority' (Altmeyer 1988). In particular, they help people to orient themselves to the realities of inter-group life and legitimise actions that would otherwise be socially unacceptable. For example, Hunter *et al.* (1991) investigated how violent acts become justified by Protestant and Catholic students in Northern Ireland and found that each group made 'external attributions' (e.g. retaliation and fear of attack) for their own group's violence and 'dispositional attributions' (e.g. aggressiveness and psychopathy) for 'out-group' violence. Stereotypes and ethnocentrism produced social explanations of behaviour which were supportive and self-serving of 'in-groups' but derogatory to 'out-groups'. When reinforced by the power of the group, this generated a dangerous cycle of self-fulfilling prophecies which legitimised otherwise anti-social actions and encouraged members to act in a way that was consistent with 'in-group' expectations.

Thus individuals employ a variety of defence mechanisms to protect themselves from unacceptable internal impulses and threatening environmental circumstances. In these situations negative internal impulses, personal internal conflicts and inadequacies are projected in an arbitrary way on to minority groups perceived as different, threatening and inferior (Eagly and Chaiken 1993). Furthermore, such behaviour is often reinforced by the mass media, which plays an increasingly important role in moulding and reinforcing attitudes towards other nationalities (Eagly and Chaiken 1993). As Wood (1997) argues, in an increasingly busy world most people rely upon the media for their information and, to be newsworthy, social comparisons tend to be oversimplified, exaggerated and caricatured. In other words, the media shape, validate, legitimise and perpetuate value judgements and understandings of how different cultural groups vary, playing a powerful role in determining how these groups value themselves and others.

The scientific origins of prejudice

Prejudice has an inherent cognitive component that performs an important adaptive function for people and is a natural and inevitable response to the increasingly uncertain and culturally diverse world in which people find themselves. Indeed, research indicates that the behaviour is so natural that it can often be activated automatically without conscious awareness, even among people who embrace egalitarian beliefs (Devine 1989). Until relatively recently, racism (a form of prejudice based around one's race) has had scientific justification from biological determinists who argued that behavioural, social and economic differences between human groups (races, classes and sexes) arose from inborn biological distinctions. These theories were based on the natural-selection ideas of Herbert Spencer and Charles Darwin and, later, on the now discredited science of eugenics (the science of 'improving' population by controlled breeding). They developed during the eighteenth and nineteenth centuries and lent intellectual justification to the imperialist expansion of many European countries. Indeed, the emergence of psychology during this period also contributed to these theories through research which sought to use measurements of physical characteristics such as skull size as proof of the superiority of the European (male) brain (Hannaford 1997). These ideas have now been discredited by genetics, which has shown that the concept of race has no basis in biology and that racial categories are largely arbitrary. For example, it has been found that there is more genetic variation *within* most racial groups than between them (Phinney 1996). Genetics has shown that people are far less variable than was once thought and it would seem that externally visible physical features such as skin colour only have relevance to an individual's worth when a society arbitrarily loads it with differential social value. Nevertheless, the nineteenth-century notion of biological superiority has proved to be resilient in some sectors of society, where it remains a highly salient political and social construct. One such instance appears to be the construction industry, which has a particularly poor reputation for representing and protecting the interests of minority groups.

Expressions of prejudice

Together, the dual processes of ethnocentrism and stereotyping are the foundations of prejudicial behaviour. This manifests itself in discrimination or harassment, where one group is treated (physically, emotionally and/or economically) unfavourably, unjustly and unequally compared to another, and therefore feels oppressed, devalued and intimidated (Hollingsworth *et al.* 1988). Most developed countries now have laws which make such behaviour illegal and differentiate between three main types of discrimination, namely direct discrimination, indirect discrimination and harassment (RDA 1975).

Direct and indirect discrimination

Discrimination can be direct or indirect. *Direct* discrimination occurs when a person is treated less favourably than another person on grounds of sex, race, gender, age, etc. An example of direct discrimination is the practice of placing advertisements with a maximum age limit for applicants or which excludes women or certain ethnic groups without any justifiable reason. *Indirect* discrimination occurs when an employer applies to one group the same requirement or condition which was applied to another but:

- it is such that the proportion of employees in that group who can comply with the condition is considerably smaller than the proportion of people not of the employee's group who can comply with it; or
- it is to the employee's detriment because he/she cannot comply with it.

An example of indirect discrimination is advertising for people over a certain height, which indirectly discriminates against people of an ethnic background where the average height is shorter than in other groups and against women, who are also generally shorter than men. Care must be taken in discerning where a requirement is unjustified and where it forms a legitimate occupational requirement. If it is shown to be a genuine requirement for the job to be performed effectively, then this would not amount to indirect discrimination. However, with the development of new technology in the form of off-site fabrication and advanced mechanical plant, physical justification to discriminate against women is becoming increasingly hard to justify. For example, although women's lack of physical strength is cited by many as an explanation for their underrepresentation in the construction trades, there is practically little reason why this should be the case. Many general labourers in developing countries are women (Wells 1990), as are half of those employed on building sites in Russia and China (Yates 1992).

Care must also be taken to ensure that established and accepted work practices are not discriminatory. Companies should recognise that the changing profile of the workforce must be accompanied by appropriately modified work practices. The increasing number of working mothers and the expectation that domestic and family care responsibilities be shared between both partners is of particular concern. Increasing numbers of working parents are juggling work and family responsibilities. Given these changes, work practices which demand excessive hours or regular overtime could preclude the involvement of some employees. The unswerving imposition of male-oriented views of the 'nature of the job' irrespective of participants' family status or responsibility could be construed as discriminatory. The importance of providing flexible work options as part of an equal opportunities strategy is increasingly recognised, and construction companies should bear in mind the legal precedent established in a recent Australian court case. In *Hickie v. Hunt and Hunt* (solicitors) it was ruled that the termination of a legal partner's contract on the basis of her part-time work status constituted indirect sex discrimination (Bourke 2000).

In most developed countries attempts are being made to make up for past mistakes, and law is being developed to encourage and provide training for people of one sex or racial group which has been under-represented in the past. For example, Australia has introduced legislation to encourage Aborigines into the workforce and has introduced *affirmative action* policies to support women's career advancement. The purpose of affirmative action is not to discriminate positively in favour of particular groups, but to put in place policies which redress imbalances arising from historical discrimination. However, although advertisements can explicitly encourage applications from a particular group, all other applicants must also be treated fairly and no discrimination must be allowed to take place. In essence, most legal systems require all employees to be treated equally with regard to terms and conditions of employment when they are employed on the same or broadly similar work. Of course, there are different jobs within the workplace, some of which may justifiably be paid at a higher rate, and the lower-paid jobs may by coincidence be carried out by women. Thus, even when it is proved that the woman is doing similar work to a male, she will have no right to equal pay if an employer can show that there is a material difference, not based on sex, between the two employees which justifies the difference in payment. What constitutes a material difference depends on the facts of each case and may relate to issues such as diligence, quality of work or experience.

Project managers may operate with considerable autonomy in terms of recruitment, management and development of their staff within the site workplace. They therefore must take on responsibility for ensuring that direct and indirect discrimination are avoided. The 'grapevine' or the word-of-mouth way in which recruitment is often undertaken at project level renders it more susceptible to allegations of discrimination. For this reason, project managers should scrutinise work practices and recruitment policies, and also work with HR managers in order to monitor their workforce and ensure that their team is representative of the organisation as a whole and is offering appropriate opportunities for those within it.

Harassment

Harassment refers to repeated and irrelevant references or innuendo to a person's gender, sexuality, race, disability, etc. in the form of jokes, verbal abuse or written abuse which creates a hostile or humiliating work environment for the recipient. Other commonly used words for harassment are bullying, victimisation and intimidation. Harassment is more difficult to prove than discrimination because of its intangible nature, which means that the treatment has not necessarily resulted in some loss such as lower pay, poorer working conditions, disciplinary action, dismissal, transfer or failure to promote or train.

According to Ansari and Jackson, there are several common causes of harassment in the workplace which an effective equal opportunities policy must take account of:

- race, ethnic origin, nationality or skin colour;
- sex or sexual orientation;
- religious or political beliefs;
- willingness of the individual to challenge harassment leading to victimisation;
- membership or non-membership of a trade union;
- physical, mental or learning disabilities;
- ex-offender status;
- age;
- actual or suspected infection with HIV.

(Ansari and Jackson 1996: 46)

Any of these factors can lead to bullying or intimidation by other colleagues forming part of the dominant group. The role of supervisors in identifying instances of harassment and bullying is crucial. Employees who are being bullied or harassed may not inform managers through fear of further retribution. Hence, project managers should observe workers closely and speak to those appearing to be marginalised or excluded by the main group in order to establish why this is the case.

Tokenism

A particular problem for members of under-represented groups is that they can be seen as being representative of their gender – as 'token' employees rather than as individuals with equal rights. People in token positions are not treated 'normally' or seriously, are not seen to acquire them by merit, and are normally given positions simply to pacify external or internal pressure groups. Consequently, they come to be associated with negative stereotypes and become victims of ridicule and discrimination (Renzetti and Curran 1992). Tokenism also leads to what Kanter (1977) termed 'boundary heightening', where the workers of the dominant gender exaggerate the differences between them and the 'token' employees. For example, in construction, women and other minority employees can often find themselves sidelined into particular tasks and excluded from informal networks of which membership is essential for career enhancement. By organising social events that do not revolve around the dominant culture, such as drinking after work, the project manager can engender a greater level of inclusion of non-traditional entrants.

Case study

Racism towards Asian operatives in the Australian construction industry

This research investigated racial discrimination towards Asian operatives in the Australian construction industry. It explored the extent of this discrimination, the form it is taking, the impact it is having and how it is being managed. The research was motivated by increasing numbers of complaints to union officials by foreign workers of being called insulting names, being forced to eat in separate locations, suffering poorer career prospects, working longer hours, and being 'cheated' out of legal rights to employment benefits and compensation, fair rates of pay and safe working conditions (O'Rourke 1998). This problem is not unique to Australia. For example, a recent report in the UK found that 39 per cent of ethnic minorities in the construction industry had been subjected to racist comments in work and felt that their employment opportunities were less than their white colleagues' (Cavill 2000).

The method consisted of two parts, the first being structured interviews with a random sample of 141 Asian operatives from a cross-section of large commercial projects in Sydney's metropolitan area. The second part of the research involved the use of focus groups with 97 supervisors working on similar projects in the same geographical area.

Forms of discrimination

Forty per cent of our operative respondents felt that they had suffered discrimination, in that some aspect of their employment rights had been detrimentally affected by their race. Interestingly, the areas of lowest discrimination were the 'softest' and least easily monitored (involvement in decision-making and ability to voice concerns), and the highest areas were the 'hardest' and most easily monitored (earnings, welfare facilities and protective equipment). Although the differences in emphasis are not strong, this pattern of behaviour does not reflect the contemporary trend in society towards discrimination of the more symbolic and less tangible kind. Discrimination in the construction industry would still appear to be mainly of the 'blatant' kind. This was supported by 61 per cent of operative respondents indicating that they had experienced some form of racist intimidation (harassment) at work. This is a worryingly high level that would indicate potential for a considerable degree of corporate and individual criminal liability. While any form of harassment is of concern, particularly disturbing was the significant level of physical abuse that appears to be occurring on construction sites.

It is interesting that the perceptions of the managers who participated in this research were quite different from the operatives'. While managers identified racial discrimination and harassment as a problem, it was not perceived to be a major problem, receiving relatively low priority among the many problems they faced. It is also interesting that the problems of racism and discrimination were seen as occurring mainly between operatives themselves rather than between managers and operatives. 'Joke telling', 'offensive graffiti', 'fighting' and 'swearing' were the most common forms of harassment identified by our managers. This was despite a number of supervisors (albeit in the minority) coming close to discriminating in their own supervisory practices, arguing that 'it is only fun', 'they deserve it', 'why should we waste money on people who are different to us?', 'it's easier to use nicknames because you can remember them easier. They don't know what they mean anyway'.

Coping with racism

The results indicated that the main mechanism amongst operatives for coping with racism was to ignore it or to turn to friends, typically of a similar culture, race, etc. Racism appears to create a socially fragmented workplace, where different cultural groups collect into 'ghettos' of highly cohesive groups which are likely to have little communication with the outside world. Given the cultural diversity of construction sites in many countries, this would have a serious and detrimental impact upon workplace communications that would ultimately impact on other areas such as safety and productivity. The results indicated that the productivity impact of racism is less direct and obvious than people simply working less or sabotaging work. Racism appears to act on productivity by reducing levels of cooperation, goodwill and open communications in the workplace. There is also a motivational effect, in that respondents felt undervalued and indicated that racist behaviour adversely influences morale and creates a considerable degree of resentment and malevolence in the workplace. Indeed, in many cases this was also transferred to our respondents' home life. Complaining to managers was not a coping mechanism that our respondents tended to employ, which is not surprising given the history of inaction in response to any complaints that had been made. However, it also reflects a lack of knowledge of complaints procedures and legal rights, language barriers which made complaining difficult and a fear of victimisation or intimidation as a result of doing so. Overall, our results suggest that racism is seen as an inevitable consequence of working in the construction industry, and one which is largely ignored by managers and accepted and tolerated by workers.

This perception was confirmed by our managers, who saw racial discrimination and harassment as a problem that is largely ignored and 'brushed under the carpet'. Most found the range of different cultures on site overwhelming and were uncomfortable with their ability effectively to prevent and manage discrimination or harassment activities. In general, managers identified a severe lack of infrastructure to support efforts to break down discrimination, such as effective and positive training and opportunities to share different site experiences. Indeed, no role models were identified by any of the respondents and only 26 per cent of the respondents had undergone any form of training in this area (although 87 per cent expressed a desire for more). Interestingly, the only training that had been provided had been with the express aim of informing our respondents of their legal rights and obligations under the law. It is not surprising therefore that managers' perceptions of equal opportunities issues were negative and defensive. One comment which reflected the reactive manner in which racism is typically managed on site was 'We will always have this problem and we can't do much about it. The only way is to split the groups and programme the job to keep them apart.' One respondent said that he insisted that the site toilets were repainted every month to cover up the racist graffiti, while another had put signs up in Portuguese and translations of site instructions in Lebanese. While there was a widespread realisation of the potential benefits that a more harmonious workplace could bring, there was a sense of defeatism (less in the female respondents) about bringing about cultural change in the workplace due to other pressures such as costs and time.

Sources of racism

One of the most interesting results was the perception among operatives that levels of racism increased with managerial seniority. Assuming that perceptions reflect behaviour, this supports other research which has found that people with power are more likely to stereotype others. It also indicates that racism in the construction industry is perceived to be a result of company policies rather than cultural disharmony within a multicultural workforce. The perception was that managers were a significantly greater source of racial tension than colleagues of another race. This is worrying for the construction industry because research is unanimous in indicating that the solutions to reduce racism must be 'top down'. Interestingly, this is precisely the opposite of the perceptions of the managers in this research, who saw the problem as best resolved at operative level by the supervisors of each cultural subcontract gang. Given the operatives' perceptions reported above, it is not surprising that our managers 'found it difficult to get subcontractors to take the issue of diversity seriously'. Clearly, what is needed is a sense of collective responsibility to resolve this problem, but the research indicated that this is unlikely to occur if each interest group continues to see it as the others' problem.

Company policies

The research indicated a worrying level of discrimination on construction sites, a level for which employers could be criminally liable. However, liability does not extend to organisations which can demonstrate that they have taken all reasonable steps to prevent racist behaviour. While the research did not seek to inspect company policies relating to equal opportunities and to measure their effectiveness, our operatives were generally unaware of the existence of such policies, their rights under law and the procedures for making a complaint. While many of the companies who employed the respondents might have had equal opportunity policies in place, it is clear that they were not being communicated and implemented effectively.

The same scenario of ignorance existed among the managerial respondents, with only 3 per cent having seen the equal opportunity policies of their employers and 57 per cent unsure whether a policy existed or not. None of our managers understood their equal opportunities responsibilities and grievance procedures, and none was aware of how to get more information on the subject. One comment reflects the general lack of awareness of supervisors' legal responsibilities in this area and the tendency to use the lack of information as a defence for doing nothing: 'If I do not know about a problem because it's not visible, then how can I be accountable for it?' Indeed, there was a widespread fear that communicating operatives' legal rights would open the 'floodgate' to claims, which would detract from the task of getting the job built on time and on budget. In this defensive environment it is unlikely that any changes in managerial practices will occur voluntarily, and the challenge is to reverse the perception of equal opportunities issues as a barrier to success which needs suppressing, to make it a mechanism which could be used to competitive advantage.

Increasing diversity and reducing disadvantage

As has been discussed earlier in this book, the principles of HRM indicate that competitive advantage can only be achieved through the efficient utilisation of employees of all types and at all levels. The fundamental starting-point must be to ensure that balances of different types of people are recruited into an organisation. In recent years the construction industry has begun to attempt to redress the imbalance in its workforce with targeted action aimed at promoting workforce diversification and providing a workplace environment conducive to their long-term employment. Within the UK an industry-backed task force was set up to explore equality issues (CIB 1996), and individual organisations, professional bodies and national training organisations have all taken steps to improve women's representation and the level of their involvement. Most recently, a diversity 'toolkit' was released to provide a practical checklist and measurement tool for companies attempting to improve their equal opportunities practices (Respect for People 2000). Within the UK, initiatives such as these have contributed to women's employment increasing over the decade to 1994 by 14 per cent, whilst men's declined by 7 per cent over the same period (Court and Moralee 1995: 13). This implies that efforts to attract under-represented groups may be beginning to have a positive effect within the industry.

Despite this considerable success in increasing the representation of women within the industry, these initiatives have not been based on good empirical evidence that women will have the opportunity to progress their careers in parity with their male colleagues once they have entered it. Anecdotal evidence points towards the construction workplace continuing to present a problematic environment for women to develop their careers in (Hanson 1995), and suggests that only 25 per cent of women believe that they could reach the top of their profession (Finch 1994). Research in other sectors has shown that lack of career opportunities is the principal reason why women managers leave organisations (Brett and Stroh 1994), and so there are now real concerns that women's underachievement within construction may lead them to leave the sector. This demands that an employment climate is developed in which people are valued for their input and abilities, and not by their gender, ethnic background, age or able-bodied status.

In order to increase diversity and overcome the disadvantages that minority groups suffer, it is important to influence the attitudes and preconceptions held about them by our society, and to introduce and implement legislation to prevent discrimination. In most developed countries legislation has been introduced intermittently since the mid-1970s in an effort to reduce the disadvantages certain groups experience. Dedicated commissions have also been created to promote equal opportunities issues and safeguard the interests of under-represented groups.

The legal framework for equal opportunities

Although difficulties for minority groups in construction are unlikely to be legislated away, an important legal framework does exist that outlaws discrimination, and those in breach can face severe statutory penalties. For example, America has many state and local laws going back to the Civil Rights Act (1964). In the UK legislation was first introduced in 1965 and 1968 with the Race Relations Acts, later replaced by the Race Relations Act 1976. In 1970 the UK saw the introduction of the Equal Pay Act, followed by the Sex Discrimination Act 1975 and, more recently, the Disability Discrimination Act 1995. The signing of the European Community Social Chapter in the late 1990s also had some impact on its equal rights legislation. In Australia there are twelve pieces of legislation, which include the Racial Discrimination Act (1975), the Human Rights and Equal Opportunity Commission Act 1986, the Racial Hatred Act 1995 and the Workplace Relations Act (1996), supported by two pieces of state legislation in New South Wales and one in each of the other states.

Although the exact details of legislative requirements vary internationally, the current legal position in most countries prevents organisations from discriminating against individuals in selection for employment on grounds of race, sex, marital status or disability. In addition, the law states that employees must be treated equally in respect of rewards, including benefits, training and promotion, once they have been employed. Finally, if the situation should arise where redundancies or short-time working are under consideration by an employer, all employees, regardless of race, sex, marital status or disability, should be treated equally and not be subject to discrimination. Courts are becoming tougher on companies that have not dealt effectively with discrimination issues and they have increasing powers to taking action against organisations that flout the legislation.

Whilst legislation has undoubtedly played a valuable role in defining and outlawing discrimination against particular groups, statistical evidence over the past 20 years indicates that it is not a particularly effective vehicle for influencing people's attitudes. Influencing attitudes is a very slow process and cannot be done coercively, and it is important to note that in most cases the law merely sets out minimum standards of performance on equal opportunities. Furthermore, in most countries positions of power in organisations are still dominated largely by indigenous groups, which means that those who are in the best position to promote change have no personal experience of discrimination and no interest in eliminating it (Cooper and White 1995). For example, in the UK many contractors do not even keep records of how many women and ethnic minorities they employ (Chevin 1995). A recent survey showed that although construction companies claim to be 'equal opportunities employers' their interpretation of this statement differed greatly (CIB 1996). Typically, construction employers exhibit a 'minimalist' approach to equal opportunities, defined as a base-level commitment to its principles with little effort towards their practical implementation.

Equal opportunities commissions

Equal opportunities commissions exist in most developed countries, their job being to work towards the elimination of discrimination, to promote equality of opportunity and to keep under review legislation in this area. They may also represent individuals in discrimination cases at an industrial tribunal, and in some countries they have the power to instigate formal investigations, issue a non-discrimination notice or, in extreme cases, obtain an injunction from a court to force changes in behaviour.

Equal opportunities commissions also promote research and educational initiatives, and may produce a code of practice which gives guidance on how to promote equality of opportunity in employment. This may require employers regularly to monitor the effects of selection decisions and personnel practices and procedures in order to assess whether an equal opportunities policy is being achieved. There may also be a requirement for employers to keep records of the breakdown of job applications, of successful appointees, as well as of staff seniority within their organisations. This data is useful for investigators if a candidate complaining of discrimination challenges an employer. For example, the UK's Equal Opportunities Commission and Commission for Racial Equality recommend that ethnic and sex monitoring information be kept separate from application forms or curricula vitae when processing candidates, and that the names of candidates should not be known to the people responsible for shortlisting since names could imply their sex or racial origin. Recent disability legislation in Europe requires that the same procedures are followed by organisations for the monitoring of disability in employment. The equal opportunities commissions provide valuable support for companies and project managers attempting to manage diversity and ensure a parity of opportunities for their employees and potential employees.

Best practice approaches to managing diversity

Recently a considerable amount of advice on how to manage diversity effectively has been published by both government and special interest groups. For example, the UK government's Department of Trade and Industry has produced a workforce diversity 'toolkit' through their 'Respect for People' initiative (Rethinking Construction 2000). Similarly, an initiative called 'Change the face of construction' has published tools to support good equal opportunities and diversity practices (CFC 2001). The Equal Opportunities Commission and Commission for Racial Equality also publish their own tools and checklists (EOC/CRE/DRC 1999). These offer construction companies guidelines to follow in order to create a fair and equitable workplace and are equally applicable to other countries.

Recommendations from most equal opportunities commissions begin by suggesting that organisations should develop written diversity/equal opportunities policies, which should be clearly linked to the aims and objectives of the organisation. They also stress the importance of communicating the policy to employees and their representatives, to job applicants, customers and clients, shareholders, suppliers of goods and services, as well as to external bodies such as pressure groups and the general public. This can be embodied within an equal opportunities policy and implementation strategy, for which a step-by-step approach should be taken:

1 Develop an outline policy containing clearly defined aims and objectives for the policy.
2 Arrange consultation with all employees and their representatives and any external specialist advisory bodies, and amend the policy accordingly.
3 Prepare an action plan, including targets for senior management, detailing responsibilities for policy implementation, resources for any changes necessary to fulfil the policy, timetables for action and implementation, and intended methods for measuring effectiveness.

4 Provide training for all employees, including all senior managers, to ensure consistency of approach and an overall understanding of the importance of equal opportunities. Arrange specific training for those people responsible for recruitment, selection, appraisal interviewing and training.

5 Carry out an audit to review current procedures for recruitment, selection, appraisal or performance review, disciplinary and grievance procedures, promotion, training, health and safety, and selection criteria for redundancy, transfer and redeployment.

6 Write clear and justifiable job criteria for each job in the organisation to ensure that they are objective and job related.

7 Consider pre-employment training, where appropriate, to prepare job applicants for selection tests and interviews.

8 Examine the feasibility of flexible working schemes such as career breaks, job sharing, flexitime, shift work, childcare, prayer breaks or areas for prayer, etc.

9 Set up monitoring systems, adjusting documentation for recruitment accordingly, to collect appropriate information as necessary for monitoring.

10 Consider the introduction of an equal opportunities committee to maintain a positive approach to the issues surrounding equality of opportunity and to review equal opportunity in practice.

11 Disseminate the equal opportunities policy throughout the organisation and set a regular review date to ensure that it is kept up to date.

As with many aspects of good business practice, construction can learn from the exemplary practices of other sectors. A good example of an effective equal opportunities policy is British Petroleum's (BP), whose equal opportunities policy handbook highlights the purpose of the policy thus:

> An equal opportunity policy is important for the individual. But it is also important for the company. Such a policy helps to identify, attract and make the best use of the skills and talents we need to conduct our business efficiently, wherever these may be available. In essence, race, religion, colour, nationality, ethnic or national origins, sex or marital status must not influence, directly or indirectly, the way in which a person is treated.
>
> (BP's equal opportunities policy handbook)

The policy goes on to detail each employee's responsibilities in relation to equal opportunities, as well as mechanisms for dealing with situations of perceived inequality. It also documents specific actions BP has taken to redress the balance of race and sex in its workforce. For example, the policy relating to women's issues revolves around 'family-friendly' approaches, which include childcare vouchers issued to female employees, career breaks with a guaranteed job after a two-year break, and preference over external applicants given to returners after a five-year break. Flexible working arrangements such as variable hours, flexitime and job sharing have also been introduced. Similar policies exist for other equal opportunities issues.

In most countries over 25 years have now passed since the first legislative steps were taken to eliminate unfair discrimination in the workplace, and organisations have employed a vast range of approaches to deal with this issue. The public sector was the first to respond and some rather bureaucratic systems were implemented, which sometimes served to make the situation more complex than it needed to be. Many organisations now have separate policies for racial strategy, equal opportunity on grounds of sex, disability equality and diversity policies, and there appears to be no right or wrong approach. It is up to each organisation to design policies and procedures to fit with their individual approach to business operations and culture while adhering to the spirit and intention of legislation.

According to the diversity 'toolkit' developed by the Respect for People initiative (Respect for People 2000), the procedures that need to be covered by the diversity/equal opportunities policy should include the following:

- *Recruitment/job presentation*: advertising jobs is one of the most crucial aspects of an effective diversity policy. Care should be taken to ensure that jobs are advertised in a place that will reach the widest possible audience and that the wording is as inclusive as possible. A reference to the firm's diversity/equal opportunities policy will help to convey a positive message about the company. Application forms should be offered in a way that allows disabled people to use them (e.g. in large print or Braille). There should be a robust monitoring procedure to record the types of applications being made for jobs, but this should be kept separate from the main applications and should be kept anonymously.
- *Selection/promotion*: competence-based criteria provide a fair and equitable way of assessing a candidate's suitability for a post. Decisions made against these competency criteria should be written in a transparent manner and made available to candidates. The gender, ethnicity and/or disability of those not invited to interview should be recorded and monitored. The success rate of those being interviewed should also be recorded in a similar manner in order that any inequalities in the process can be established and dealt with.
- *Retention/exits*: staff turnover should be broken down by gender, ethnicity and disability, and monitored carefully. Exit interviews can be used to establish the reasons why employees leave the firm. Any issues related to discrimination or harassment should be dealt with in a way which accords with the diversity/equal opportunities policy statement.
- *Training*: training in diversity and equal opportunities should be provided for those responsible for interviewing/recruitment and site

induction programmes so that the company perspective on equal opportunities issues can be explained to all new employees and those working for the company on projects. At a more general level, access to training should be made available to all, and the numbers of days recorded and monitored accordingly.

- *Management procedures and practices*: guidance must be provided for the operations managers implementing the diversity/equal opportunities policy. This should clearly explain how they should implement the policy in practical terms. The allocation of responsibilities should also be clearly explained and should include responsibilities from board level down to operations staff. Clearly defined complaints procedures must be put in place, and any complaints made should be monitored and action taken to deal with any issues arising, particularly multiple examples within certain projects or environments. Arrangements such as helplines, mentoring and informal networks should be put in place to provide employees with advice and support where necessary.
- *Monitoring and review*: Regular monitoring of all aspects of the diversity/equal opportunities policy is essential. Progress reports on its implementation and effect should be produced regularly and reported to senior management. This should include reports on employee satisfaction surveys; the make-up of the workforce at each hierarchical level in terms of the numbers and proportions of women, ethnic minorities and disabled people; and the age of employees. Finally, an overall assessment of the effectiveness of the strategy should be made in order to help determine future changes to the diversity/equal opportunities policy.

These interrelated steps can help to provide a coherent and robust policy which is closely aligned with the organisation's objectives. Regular updating of the policy to reflect the changing circumstances of the business will further help to ensure its relevance and effectiveness.

Case study

Working towards the diversification of Gleeson Construction

Gleeson Construction (Northern Division) is a large contractor operating in the north of Great Britain. At the beginning of 2001 Gleeson undertook a review of its employment practices as part of a trial programme initiated as part of the 'Rethinking Construction – Respect for People' initiative. It measured the performance of its people-management practices using the toolkits developed by the Respect for People group in order to identify any shortcomings in the company's existing approaches and to develop plans for improving practices in the future. An important aspect of this evaluation exercise was to measure Gleeson's diversity performance in terms of how representative the workforce was of the local population in the areas in which it worked.

Following an extensive evaluation of its working practices, Gleeson identified that the workforce did not reflect the working population generally, or the local communities in which the company operated. Accordingly, it has formulated a set of HRM policies to address the shortcomings that have been identified. Gleeson is now working towards the implementation of these strategies in order to ensure that the company will employ a more representative workforce in the future. Some of the areas currently being addressed are:

- *Better communication of the diversity policy*: Gleeson is aiming to ensure that employees understand the company's policy on diversity and equal opportunities, and that they understand the value in employing a balanced workforce that reflects the breadth of skills and talents available in the labour market.
- *Setting up links with local ethnic groups*: Gleeson is taking steps to form and develop links with local communities in the areas in which the company operates in order to help it to secure access to the local labour market.
- *Advertising in more appropriate places:* by advertising in publications where women and ethnic groups are likely to see adverts for positions, the company is hoping to increase the number of applications for vacancies from non-traditional entrants and hence improve the likelihood of attracting the best person for the job.
- *Improving disabled access in the head office*: ensuring that head-office facilities are fully accessible to disabled people increases the likelihood that people with disabilities will be attracted to the positions advertised, as well as ensuring that disabled visitors and clients can visit the premises in order to conduct meetings.
- *Measuring the turnover of women and minorities*: monitoring the workforce composition is an important first step in improving the workforce profile. However, an equally important task is to measure the numbers leaving the organisation and establish reasons for this wastage of employees. Accordingly, Gleeson is developing employee exit surveys so that it can identify weaknesses in its equal opportunities and HRM policies, which it can address in the future.
- *Equal Opportunities training*: training operations managers in good equal opportunities practices will help them to appreciate the importance of diversity and the importance of maintaining a fair and equitable workplace environment for all employees. It is hoped that this training will eventually extend to all those with a

managerial or supervisory role to ensure the commitment of the entire project team to good diversity practices.

- *Complaints procedures*: those with a legitimate complaint against colleagues who have harassed, bullied or generally discriminated against them should have a formal but discreet route by which to voice their complaints and concerns. Gleeson is currently working towards developing such procedures in consultation with its workforce.
- *Employee satisfaction surveys*: building equality-related questions into regular employee satisfaction surveys is a good way of allowing employees to raise issues of concern before they are allowed to escalate. Gleeson is exploring the use of such surveys as a potential route for allowing employees to raise concerns anonymously, rather than as a formal complaint.

Gleeson is hoping that by working towards the implementation of this integrated package of measures it will eventually have a positive influence on both the composition of its workforce and the performance of its projects.

Work–life balance

The interface between the work and non-work experiences of people in paid employment has become an area of increasing concern in recent years as a result of increasing work pressures and dramatic changes to traditional family roles and structures in most industrialised countries (Lobel *et al.* 1999). In the nomadic and increasingly 'lean' construction industry this conflict is becoming particularly acute because of the growing need to work longer, non-standard work schedules accompanied by lower security of employment from increased outsourcing, cyclical workloads and greater competition. There is compelling evidence that all of these characteristics are not conducive to a well-balanced life. For example, research suggests that long work hours are negatively related to family participation and positively related to divorce rates (Aldous *et al.* 1979). Furthermore, people engaged in 'commuter marriages' have reported significantly less satisfaction with partner and family life than people in single-residence families or relationships (Bunker *et al.* 1992). Irregularity of work hours has also been identified as an important variable affecting low marital quality among shift workers (White and Keith 1990), and non-standard work schedules have been found to affect separation or divorce rates among married people with children (Presser 2000). Finally, research has found job insecurity to be negatively related to marital and family functioning (Larson *et al.* 1994) and to be associated with burnout, for which there is a direct crossover effect from husbands to their wives (Westman *et al.*, in press). This is all reflected in a recent survey of engineers which reported that 33 per cent of respondents were dissatisfied with the balance between their work and family life.

While the extent to which construction-industry employees experience work–life conflict is not well researched, preliminary work undertaken in Australia suggests that there is a problem. A recent survey of 182 civil engineers revealed that the number of hours worked, feelings of being overloaded and the level of responsibility were inversely related to the engineers' satisfaction in their relationship with their spouse or partner. Also, the greater the number of hours the engineers worked each week, the more conflict they reported in their relationship with their spouse or partner. This conflict was also related to lower levels of satisfaction with the engineers' pay. The results also suggest that relationship satisfaction varies according to parental demands. The higher the number of children the engineers had, the less satisfaction they reported in their relationship (Lingard and Sublet, in press). Furthermore, the research showed that family demands interact with work demands to produce a syndrome of chronic stress known as burnout. While this research is in its early days, these results suggest that work–life conflicts exist in the construction industry.

Managing work–life balance in the construction industry

Construction-industry employers place great importance on the flexibility of their employees, who are expected to balance their work and family commitments, often without organisational assistance (EOC 1990). For example, very few construction companies provide childcare facilities, flexible working hours or career-break programmes, and part-time work is almost non-existent (Kirk-Walker 1994). This means that many employees, particularly women, experience difficulties in balancing work and family life. Accordingly, women often find that they have to choose whether to have a career *or* a family (Toohey and Whittaker 1993), and women who do take time out of the industry to have children may be severely disadvantaged in terms of career progression. However, work–life balance issues are not necessarily exclusive to one gender. This is consistent with research by Mauno and Kinnunen (1999), who found that it is not only women who experience difficulties balancing commitments in their work and non-work life and argue that gender differences in work experiences are minor and diminishing. In some cases men have reported more acute difficulties than women. For example, Wallace (1999) found that male lawyers actually reported higher levels of frustration than their female counterparts. This is because, despite women's education levels and participation in paid employment increasing, married couples still assign more income-earning responsibility to the husband as the main breadwinner (Gorman 1999). Gorman suggests that this puts them under greater pressure to succeed in their careers, resulting in longer hours, higher expectations and greater conflict between home and working life. In recent years this has been compounded by a significant change in fathers' expectations regarding their role in parenting. This new tension between bearing the breadwinner role and the expectation of a fulfilling and nurturing role in parenthood imposes significant role-stress on working fathers.

There are compelling reasons why the construction industry should be concerned with the impact of its demands on the quality of participants' experiences outside work. For instance, both men and women have been found to experience home-to-work 'spillover' effects, whereby subjective experiences in one context impact on performance in the other arena (Barnett 1994). A recent study by Adams *et al.* (1996) found that when work interferes with family life it also reduces the satisfaction from job and from life as a whole. Furthermore, this spillover effect is bi-directional and the impact of home events on behaviour in the workplace has implications for organisational performance. It is clear that the implementation of family-friendly work policies and practices has the potential to enhance organisational efficiency, morale and productivity, and this has been supported by a number of research studies (Cass 1993; Butruille 1990). Moreover, research suggests that the quality of family and marital life moderates the impact of job-role quality on psychological distress (Barnett *et al.* 1992). Thus, workers with positive subjective experiences of family and marital life are less likely to suffer mental health problems. Indeed, there is considerable new evidence to suggest that work–life balance initiatives may be a more effective means of enhancing employees' morale and fostering a sense of loyalty to the company than providing traditional rewards such as increased remuneration. For example, there is evidence that the provision of work/life balance initiatives such as flexitime or childcare/eldercare services pays greater dividends in terms of increasing employees' commitment to the organisation and yields beneficial outcomes for both the individual and the organisation (Perry-Smith and Blum 2000).

Despite the growing interest in the work–life interface, traditional management theories and practices presuppose a lifestyle that segregates family and work spheres. Dual-income couples, whether as spouses or parents, participate in many roles simultaneously. Managers may live with family responsibilities themselves, yet they are taught that successful managers must remain detached and rational, not concerning themselves with the family concerns of employees. This approach was largely predicated on the view that the workforce is homogeneous, comprising males of European ancestry married to full-time homemakers. In most developed countries this presumption no longer holds true (Popenoe 1993), even in the traditionally male-dominated construction industry. Furthermore, increasing numbers of dual-income couples mean that men and women now share, to some degree, parenting and family responsibilities. Traditional management theory fails to recognise this diversity, basing theories of motivation on employees' individualistic needs for self-actualisation, achievement and power (Bruce and Reed 1994). New changes in workforce characteristics require a shift in management approach to re-examine the values, roles and stereotypes, and to meet the increasing expectation that a balance between work and family life be achieved. Modern management theory presents a boundary between work and family which is artificial and contrary to systems thinking. This boundary is in fact permeable and the two spheres are interactive, in that what is positive or negative in one affects the other. There is also an increasing body of evidence to suggest that policies that facilitate stronger families lead to stronger companies, which enjoy enhanced productivity and improvements in long-term competitiveness. For this reason, many developed countries have ratified the International Labour Organisation's (ILO) Convention 156, which obliges them to develop policies to enable people to work without conflict between work and family lives (Cass 1993). This commitment is reflected in a growing body of legislation imposing requirements on employers in respect of the family responsibilities of employees (Napoli 1994).

Case study

'Build a Life': the Construction, Forestry, Mining and Energy Union (CFMEU) of New South Wales, Australia

In 2002 the CFMEU began a campaign to secure more paid leisure days for its members, limits to the amount of overtime each week, a ban on weekend overtime, six compulsory days off next to public holidays and six fixed and non-negotiable paid weekend days off per year. This was initiated by increasing evidence that excessive working hours damage employees' health, destroy family life and cause accidents on site. Many years before, the CFMEU had won regular days off, but members were increasingly complaining that managers in the construction industry were pressurising members to work through them, denying people the opportunity to spend the occasional weekend with family and friends. The CFMEU was becoming increasingly concerned that an unwritten expectation was developing that members work 60 or more hours a week over six or often seven days a week. These sorts of hours are the norm rather than the exception on many large Sydney construction projects.

The economic argument being put by the CFMEU was that it would be cheaper for employers to avoid paying weekend overtime and that if they provided more leisure time, productivity would be higher for the rest of the week and accidents caused by fatigue would be reduced. On 1 October 2002, 1,000 enterprise bargaining agreements are due to expire and the CFMEU intends to present a formal claim as part of this process.

Practical employer-led actions to facilitate employees' work–life balance

There are many ways for companies to assist employees with family responsibilities. Some of the options are presented below, although the needs of individual employees will differ and change over time. It is therefore important that companies examine the needs of their employees and ensure that policies address them through regular consultation.

Childcare

In Australia companies' provision of childcare surpasses that available in most other member countries of the Organisation for Economic Cooperation and Development (OECD) (Cass 1993) but company-sponsored child care is still rarely available to construction-industry employees. While it may be difficult to provide on-site childcare centres due to the limited space and the temporary nature of construction work, there are other options for childcare provision which construction companies may be able to provide. These include:

- off-site single-employer childcare centres for company employees;
- joint-venture childcare centres;
- the purchase or lease of places in existing centres;
- the provision of land for a childcare centre;
- the addition of places in a government-funded family day-care scheme;
- the provision of out-of hours childcare;
- a childcare information and referral service;
- employer contributions towards employees' childcare fees.

(Napoli 1994)

Where there is insufficient demand within a company to warrant a dedicated single-employer childcare centre, companies may join together to use community resources, collaborate to support existing services or form partnerships to provide day-care facilities. The need for the care of children outside school hours, during school holidays and when children or parents are sick should also be considered. For example, one American company is reported to have provided an emergency care-giver programme which subsidises payment for in-home care of sick children by trained professionals, while another provides a summer day camp for older children (Butruille 1990).

Eldercare

In some countries the eldercare needs of an ageing population may come to eclipse childcare needs. If eldercare is needed, then support in the form of special family leave or an information and referral service may be helpful.

Flexible work practices

Flexible work arrangements are one of the most frequently used ways to assist employees with family responsibilities. Evidence suggests that increased flexibility lowers absenteeism and tardiness and can yield tangible productivity gains (Napoli 1994). Flexible work arrangements cover a range of practices, including the following:

- flexible work hours
- job sharing
- working from home or telecommuting

Flexible work hours may involve compressed work weeks, shorter weeks with less pay, adjusted starting or finishing times, fewer breaks to shorten the working day, the option of half-day vacations and informal flexibility to accommodate needs on specific days. The needs of employees should be carefully assessed to determine which of these options would be the most beneficial. While there will always be a need to have on-site management and supervision of construction work, the increasing availability and use of IT in the construction industry should increase the options for work from a remote or home site in certain tasks. However, the main problem is one of information supply, because on many projects information management is poor and numerous unexpected problems result in frequent requests for decisions and further information. In this environment flexible working on the part of consultants could play havoc with site progress and create further uncertainty and delays. The key to flexible working in the construction industry is improved planning.

Permanent part-time work

Part-time work can assist employees to maintain a balance between work and family. From a company's point of view, part-time work can improve the retention of employees, reduce absenteeism, increase productivity, reduce overtime, provide flexibility to cater for peak periods and make recruitment easier (Napoli 1994). Permanent part-time work differs from casual work in that employees have a 'permanent' contract of employment with the company and retain benefits such as annual leave, sick leave, maternity and long-service leave. In introducing permanent part-time work for employees with family responsibilities, it is important that eligibility criteria are clearly established, that part-time workers are not marginalised and that they enjoy access to identified career paths.

Parental leave

Parental leave allows employees with a new child, either natural or adopted, to care for their child at home on a full-time basis in the child's first year and still retain employment and accrue entitlements. Parental-leave arrangements should form an integral part of a company's work and family programme, and employees should be clearly informed about their parental-leave entitlements. Unpaid maternity leave for up to 52 weeks has been available to women employees in Australia since 1974, following the Maternity Leave Test Case. From 1990 the federal Industrial Relations Commission handed down the Parental Leave Test Case, which extended the maternity- and adoption-leave standards to paternity leave where a male employee has had 12 months' continuous service with the same employer. The decision also provided for either or both parents to work on a permanent part-time basis, with pro-rata remuneration and conditions up to the child's second birthday or for two years after a child is adopted. This entitlement is only available subject to the employer's consent. The standard for unpaid parental leave has been widely adopted through legislation and awards. While Australia's provisions for parental leave are improving, they still lag behind those of many OECD countries (Cass 1993). For example, in all OECD countries except Australia, New Zealand and the US, at least twelve weeks paid maternity/parental leave is provided. Indeed, since the provisions are voluntary most private-sector employees in Australia do not enjoy such a benefit. Also excluded are women employed on a casual or part-time basis in both the public and private sectors and women who do not have a sufficient period of continuous employment. This includes students, unemployed women or women caring for previous children. Construction firms serious about attracting and retaining female employees may consider the provision of paid parental leave or consenting to women returning to work on a part-time basis up until the child's second birthday.

Other initiatives

Companies that actively seek to be at the forefront of best practice in supporting staff with family responsibilities do not have to limit themselves to the above provisions. Other initiatives intended to elicit commitment and loyalty from employees include the following:

- salary packaging of childcare costs, school fees or eldercare costs to provide a tax benefit to employees;
- providing work experience for employees' children or job-seeking skills courses for children in the later years of secondary education;
- health and dental insurance;
- family-related phone calls to enable employees to check on children or elderly relatives;
- employee assistance programmes offering counselling for employees with personal or family difficulties;
- family-oriented Christmas parties.

Case study

The family support needs of employees in a large construction company

A recent survey of a large Australian construction firm sought to identify the most pressing needs of employees; 284 employees responded. Employees were asked to express what level of priority they placed on the provision of a list of potential 'benefits'. The list included both *transactional* benefits, such as the provision of childcare assistance or parental leave, and *relational* benefits, such as recognition from one's supervisor. Relational benefits were included to assess cultural issues associated with the degree to which employees felt they worked in a supportive environment. This is important since the existence of cultural barriers to the utilisation of work–life balance initiatives has been identified as an important determinant of their success or failure, particularly in traditional, male-dominated industries. The results indicated that cultural issues are universally considered important. In particular, treating people with respect, the creation of an environment in which people enjoy their work, personal contact with senior management and recognition from one's supervisor were deemed to be of the highest priority. However, the time-related transactional benefits were also deemed to be a high priority by a large proportion of employees. Items such as ensuring that annual leave is taken regularly, more flexible work hours, providing special family leave and allowing part-time work options in the event of family crises were deemed to be a high priority by between one-third and half of the respondents; 37 per cent of the employees also valued the provision of an employee assistance programme to help them deal with family problems.

Initiatives designed to assist employees in meeting parental demands, such as assistance with childcare costs and the provision of extended leave after the birth or adoption of a child, were deemed not to be applicable to a large proportion of respondents. This probably reflects the fact that 45 per cent of respondents had no children and highlights the importance of considering workforce demographics in formulating work–life balance initiatives. However, in the interests of promoting greater workforce diversity, initiatives designed on the basis of an existing workforce should not be allowed to impede the successful recruitment of under-represented groups in the future. Human resource managers should therefore undertake a proactive assessment of the work–life balance needs of potential employees, paying particular attention to groups targeted under affirmative-action policies, and ensure that the company is in a position to address these needs in the future.

The research also suggests that the needs of site-based employees may be more acute than those who work predominantly in an office environment. The data reveal that the average hours worked each week were significantly different for employees who worked in different locations. Employees who worked on site in direct construction activity reported an average of 62.47 hours each week, compared to 56.11 among office-based site staff and 48.95 among employees in the head or regional office. The frequency with which employees were required to work weekends, work unpredictable hours and unexpectedly give up leisure time also differed significantly by work location. In all cases the frequency was higher among site-based employees than among those in the site office, and higher among site office-based employees than among staff in the regional or head office. The ease with which employees can adjust their work schedule to suit the needs of family members, to work at home and to work flexible hours to pursue leisure interests was also significantly lower among site-based employees than among employees in the head or regional office. In all cases employees in direct construction activity reported lower flexibility than those who worked mostly in the site office. It is therefore important that the needs of site-based employees be given special attention in the formulation of work–life balance initiatives.

This very recent research indicates that construction organisations should pay closer attention to the types of work/life balance initiatives they provide. It suggests that employees who fulfil different roles in their non-work environments, for example parents compared to non-parents, will have different experiences at the work–life interface. This suggests that different employees will value different initiatives and that their needs will change over time. In the context of an increasingly diverse workforce and the changing nature of family structures, employees' non-work obligations will take many different forms. Helping employees to meet these obligations will require more flexible 'cafeteria-style' benefit programmes that can be suited to individual and changing needs. In deciding what to provide and how to provide it, firms should seek to meet the needs of key constituent employee groups. Construction firms should identify their employees' priorities and be creative in formulating ways to respond to employees' work and family situations. As the working population ages this is likely to help significantly in creating a committed, motivated and satisfied workforce.

However, it is important to note that the provision of such practices is not sufficient in itself. A workplace culture must exist within which employees feel comfortable taking advantage of family-friendly work options. A lack of take-up of such initiatives, particularly by men, has been widely reported (Haas and Hwang 1995), and this is related to traditional gendered assumptions about the separation of family and work which results in greater valuing of male employees without family commitments. Lewis (2001) argues that only in the context of dramatic cultural change will these negative effects be overcome. The need for cultural change is particularly acute in the traditionally male-dominated construction industry, and if a better balance between employees' work and family lives is to be achieved, construction organisations and their employees will need to deviate from the socially constructed norm of rigid, long work hours that predominates. Current business trends towards downsizing, outsourcing, leanness and re-engineering are more damaging than helpful in this respect. Furthermore, employees with family responsibilities must not be regarded as being low in commitment to their job. Sensitivity training for managers and supervisors to assist them in understanding the family needs of their workers may also be required. Employees at all levels must recognise that, ultimately, ensuring flexibility at work is about balancing the needs of the employee and the employer through employment practices which are sympathetic to the requirements of both parties.

Conclusions

This chapter has shown that workforce diversity must be part of the formula for success in the future if companies are to be competitive on an international basis. Many organisations now have data showing that they employ more people from minority groups than they did 20 years ago. However, most organisations are still failing to make the best use of these resources. It is clear that diversity is not just about meeting legal and social responsibilities, but also, and primarily, about attaining organisational effectiveness through cultural change. As has been discussed throughout this book, influencing attitudes is a slow process and not always an easy one, but sound policies for equal opportunities and supportive training programmes can do much to influence behaviour and alter attitudes. For success to be achieved, senior management and trade union leaders must support equal opportunities. As with many initiatives, diversity and equal opportunities policies often fail because they are not promoted enthusiastically enough or are not implemented in a comprehensive enough manner. There is little point in setting up workforce monitoring procedures if there is not a robust implementation mechanism for taking action to improve the performance of the organisation. However, rather than trying to follow a single formula for success, managers need to initiate policies and practices that best suit their unique environment. In this way they can provide the optimum climate to foster the talent, performance and creativity so vital to future organisational effectiveness and growth.

Groups currently under-represented are unlikely to be attracted in significant numbers until construction careers are perceived as a natural vocational choice for them to pursue. This is only likely to occur when women, ethnic minorities and other under-represented groups are visibly successful in the industry. Currently, however, the industry is a long way off the 'critical mass' required to promote such a change. It is important to note that in the interim the goal should be to ensure that minority entrants are not grouped at junior levels within project hierarchies or restricted to junior operational positions within the industry. These entrants can then provide positive role models to encourage new entrants to the industry from under-represented groups.

Construction firms must also appreciate that employees have a life outside work. There needs to be recognition of the needs of employees with family responsibilities and a genuine attempt to accommodate these needs. This is likely to require the restructuring of these employees' jobs to provide greater flexibility while retaining career paths so that people who opt to use work–life balance initiatives are not disadvantaged.

Discussion and review questions

1   Consider the specification, design and construction of a community centre within a deprived inner-city area. Identify how a diverse project team could bring benefits to the successful achievement of the project.

2   Discuss the practical actions that a project manager could take to ensure the integration of a female site engineer in an otherwise male construction management team.

3   What sort of work–life balance initiatives could a construction project manager implement to assist site-based employees in balancing their work and family lives?

9 Employees' health, safety and welfare

People are an organisation's most valuable asset and so safeguarding their health, safety and welfare should be central to an organisation's HRM strategy. It is an organisation's ethical responsibility to do so and it is also good for business. In construction, threats to health and safety come from a myriad of sources, for example the physical nature of tasks, the attitudes of the employees, the culture of the industry, cost and time pressures, the uncertain production environment, client and management priorities, onerous contracts and a fragmented system of organisation. It is these issues and their management that lie at the heart of construction accidents. This chapter will discuss the management of occupational health, safety and welfare within the construction industry. It will look at what an organisation can do to prevent injuries and ill health amongst its workforce and, just as importantly, how a culture of safety can be engendered which is transferred from project to project.

Introduction: the importance of health and safety in construction

In spite of an increased awareness and recognition of the inherent dangers of construction activity, the industry remains one of the most hazardous sectors in which to work. Virtually all of the literature and guidance on construction safety issues begins with some kind of statement about the industry's poor safety record and the need for improvement. Statistics reveal that in many countries construction is amongst the most dangerous industries in which to work. For example, in Europe construction accounts for over 15 per cent of workplace accidents despite representing less than 10 per cent of the working population. In the US construction accounts for 19 per cent of workplace fatalities, whilst accounting for only 6 per cent of the labour force (United States Department of Labor 1999). In Australia the rate of compensated injuries and disease in the construction industry is 37.4 per 1,000 workers, which is 63 per cent higher than the all-industry average of 22.9 (NOHSC 1999). The statistics are even more alarming when one realises that they neglect to consider the thousands of unrecorded injuries and latent illnesses that occur as a direct result of construction work, often later in people's lives.

It might be reasonable to assume that advances in technology and management techniques in the industry would have made it a safer place to work in recent decades, but this appears not to be the case. Within the UK death rates continue to rise despite the introduction of the Construction (Design and Management) Regulations (1994), and in Australia, construction-site deaths have remained at the same high level for the past three years, the industry killing more workers than any industry except agriculture. The reasons provided for this lamentable performance have been numerous, and they include the industry's 'macho' culture, time and cost pressures, the uncertain and technically complex nature of construction work, the fragmented organisational structure, and the relatively hostile and uncontrollable production environment, etc. (see Suraji and Duff 2000). Whatever the causes, the industry's occupational health and safety (OHS) performance is unacceptably poor and construction sites remain one of the most hazardous environments in which to work.

The legislative imperative for safer working

Since the industrial revolution, most developed countries have introduced a multitude of Acts and regulations to protect the health, safety and welfare of workers across all industries. Typically this early legislation was very detailed and prescriptive, and was introduced in *response* to new hazards as they became apparent. The ad-hoc way in which this was enacted led to a complex situation which, despite the existence of hundreds of regulations, did not account for many hazards. Furthermore, the legislation was rigid and unable to adapt to technological change and did not encourage employers to be innovative in improving their working methods. In the UK the complexity and failings of the legislation came under scrutiny and in the early 1970s a commission of inquiry was established under the leadership of Lord Robens. The Robens Commission issued a report in 1972, which became the cornerstone of preventive OHS legislation in the UK and in most Commonwealth countries. Robens rejected the traditional punitive approach, in which government enforcement agents were charged with inspecting workplaces and determining compliance with prescriptive standards. Instead, it was argued that employers and employees were in the best position to manage OHS and that a system should be developed which fostered employer–employee collaboration with regard to OHS. Therefore the OHS regulatory system should encourage workplaces proactively and voluntarily to manage health and safety for themselves. In line with this thinking, it was decided that there should be one all-encompassing legislative instrument establishing general responsibilities, and the Health and Safety at Work Act was introduced in 1974. This established a common-law 'duty of care' by employers to their employees, and it established a broad set of responsibilities for employers, employees and other relevant parties. Although the precise nature of health and safety legislation varies around the world, countries such as Australia, the US and much of Asia have mirrored this trend, with prescriptive requirements being replaced by a 'performance-based' approach in which managers can exercise more discretion as to how they meet required standards. Another common feature of modern legislation is the requirement for employee participation in OHS management, which clearly establishes the need for a more inclusive management style. Employer–employee consultation through mechanisms such as elected employee health and safety representatives and safety committees has brought OHS to the forefront of workplace relations.

Another important element of Robens-style legislation is the establishment of an 'enabling' Act. This means that under the provisions of an overarching Act, such as the UK's Health and Safety at Work Act (1974), sets of regulations specific to a particular industry, hazard or process could be established with relative ease. For example, in the UK the construction industry's poor performance in OHS has led to the enactment of many regulations specific to the construction industry, including the Health Protection Regulations 1992, Health, Safety and Welfare Regulations 1996, and Construction Design and Management Regulations 1994. Together these regulations provide an enforceable set of measures aimed at ensuring that people do not come to any harm in the course of their employment.

Legislative responsibilities

The last decade has seen a steady increase in the number of legal cases stemming from the long-term effects of undertaking particular operations in the construction industry. This has, in part, been the result of more stringent legislation aimed at ensuring compliance with safer work practices, coupled with an increasing awareness among workers of their legal rights. In the UK, for example, the Construction (Design and Management) Regulations (1994) obliged designers *and* clients to consider safety as well as the contractor. The Construction (Health, Safety and Welfare) Regulations 1996 also introduced the concept of *duty holders*, handing a degree of responsibility to every person present on a construction site. This has stimulated contractors, acknowledging the impact of safety on their profits and image, to take a more professional attitude towards the management of health and safety. Formal safety induction, toolbox talks and safety audits are now standard practice within the UK and Australian construction industries, and many larger construction sites now display their safety statistics for all to see as they enter the site.

Although specific laws differ from one country to the next, there is considerable similarity in the provisions that they make. For example, all

OHS law aims to provide job safety and health protection for workers through the promotion of safe and healthy working conditions. Furthermore, all OHS laws tend to place similar responsibilities on employers and employees and have provisions for inspection, complaints and penalties. These generic requirements can be summarised as follows:

- *Employers*: employers are usually required to provide a place of employment free from recognised hazards that are causing or are likely to cause death or serious harm to employees, and to comply with occupational safety and health standards issued under the law. Furthermore, it is normal for employers to be able to show that they have, as far as is *reasonably practicable*, put systems in place to detect and monitor OHS risks and that they have responded accordingly to those that have been found. Increasingly, laws are placing the responsibility for OHS with the directors of a company. Finally, a major requirement of most laws is that employers keep standardised records of illnesses and injuries, which include calculations of accident ratios (frequency and severity rates). Indeed, reports may have to be submitted to an OHS inspectorate body (discussed below), which may undertake an inspection if they are poor.
- *Employees*: normally each employee is expected to comply with all occupational safety and health standards, rules, regulations and orders issued under the law that apply to his or her own actions and conduct on the job. In the temporary and transitional environment of a construction project, the role of employers and employees and the duration of their responsibilities are often difficult to discern. For example, while the architect has an employer responsibility to his/her staff in the office, their status on a project is less clear. Recent laws in Europe have attempted to overcome this by placing functional responsibilities on project members. For example, designers must ensure that they consider safety issues when designing a building and so should not specify hazardous substances, heavy materials or produce designs which are dangerous to build. Indeed, all members of the project team contribute to OHS in some way and now have to consider this much more than they used to. This legislation has been a response to research which indicates that many accidents can be traced back to decisions made early in a project. Indeed, perhaps the most important contribution of this legislation has been to highlight the fact that OHS is an issue of collective responsibility rather than one of single-point responsibility.
- *Inspection*: in most countries OHS laws are policed by an OHS inspectorate. For example, in the US the Occupational Health and Safety Administration polices industrial facilities, while in the UK this is the job of the Health and Safety Executive (HSE). In Australia the situation is more complex, as each state or territory has its own OHS Act

and enforcement agency and employees of the federal government are covered by a separate Act, irrespective of the state in which they reside. Most laws provide these bodies with the right to inspect premises and OHS records with little or no notice, and most laws require that a representative of the employer and a representative authorised by the employees be given an opportunity to accompany the OHS inspector for the purpose of aiding the inspection.

- *Complaint*: employees or their representatives normally have the right to file a complaint with the nearest OHS inspector requesting an inspection if they believe unsafe or unhealthy conditions exist in their workplace. Employees may not be discharged or discriminated against in any way for filing safety and health complaints or otherwise exercising their rights under the law, and an employee who believes he or she has been discriminated against may file a complaint with the nearest OHS inspectorate.
- *Penalties*: commonly, if, upon inspection, the OHS inspector believes an employer has violated the law, a citation alleging such violations should be issued to the employer. A citation will normally specify a time period within which the alleged violation must be corrected. Furthermore, the citation may have to be prominently displayed at or near the place of alleged violation until it is corrected, to warn employees of dangers that may exist there. Penalties vary from country to country, but are generally on the increase. Most violations are dealt with by standard penalties, which vary depending on their seriousness. There may also be penalties for failure to correct violations within the proposed time period and any employer who wilfully or repeatedly violates the law may be subject to a magnifier for each violation. Finally, most laws provide for criminal penalties, which may include serious fines or imprisonment for corporate manslaughter (Craig 1996). These usually apply to wilful violations resulting in the death of an employee.

Although penalties are imposed widely for OHS violations, the aim of most modern legislation is to encourage self-regulation and prevention. The aim of most governments is to encourage employers and employees to reduce workplace hazards collectively and voluntarily, and to develop and improve safety and health programmes in all workplaces and industries. There are many public and private organisations that can provide information and assistance in this effort if requested. It is wise to consult them and to be able to demonstrate that you have taken their advice seriously.

Corporate criminal liability

It is important to recognise that an organisation's legal responsibilities for OHS are not limited to their statutory duties. In the event of a workplace death or serious injury it is also possible for companies and/or individual employees to be charged with serious criminal offences, including manslaughter. For example, Australia experienced its first case of corporate manslaughter in 1994, when a Melbourne construction company pleaded guilty to a charge following a fatal accident in which a plant operator was crushed to death when the excavator he was operating overturned.

The basis of corporate criminal liability was established in the case of *Tesco Supermarkets Ltd v. Nattrass*. Under *Tesco*, criminal conduct must be committed by the board of directors, the managing director or another person to whom a function of the board has been fully delegated in order for it to be attributable to the company. This principle has been widely criticised because it limits corporate liability to acts and omissions performed at the top of the corporate ladder (Field and Jorg 1991). The problem with this is that, although senior managers create a company culture, the entire organisational structure has a role in enforcing it. Furthermore, it is invariably very difficult to trace liability to this level, especially in large and complex corporations, because most accidents arise from a multiple chain of errors or omissions that stretch through an entire organisation. Indeed, this prevented a manslaughter conviction following the *Herald of Free Enterprise* ferry disaster, which killed almost 200 people. Despite the public inquiry into the disaster finding that 'from top to bottom the body corporate [of the ferry company] was infected with the disease of sloppiness' (cited in Sheen 1995), no fault could be attributed to any single controlling officer of the firm (McColgan 1994). The British Court of Appeal upheld the direction of the coroner of the inquiry into the *Herald of Free Enterprise* ferry disaster, ruling that 'although it is possible for several persons to be guilty individually of manslaughter, it is not permissible to aggregate several acts of negligence by different persons, so as to have gross negligence by a process of aggregation'. In the Court of Appeal's opinion, the view that corporations are clones of individuals and therefore that the law should seek to identify a corporate guilty mind at the head of the entity is a fallacy, since corporate decision-making is typically diffused.

The difficulties associated with proving corporate manslaughter in the case of work-related deaths has led to repeated calls for reform of the criminal law. For example, in the Australian State of Victoria the Crimes (Workplace Deaths and Serious Injuries) Bill is currently being considered by parliament. This creates new criminal offences of corporate manslaughter and imposes criminal liability on senior officers of a body corporate in certain circumstances. Section 14A of the Bill overcomes the problem of identifying a guilty 'controlling mind' by stating that the conduct of an employee, agent or senior officer of a body corporate acting within the actual scope of their employment or within their actual authority must be attributed to the body corporate. Furthermore, Section

14B allows the conduct of any number of employees, agents or senior officers of the body corporate to be aggregated in determining whether a body corporate has been negligent. In other words, company directors will be criminally responsible for the actions of any employee. Section 14C identifies the criteria required for senior officers to be guilty of manslaughter or negligently causing serious injury. First, it must be proved that the body corporate committed the offence. If this is established, it must be demonstrated that the senior officer:

- was organisationally responsible for the conduct, or part of the conduct, of the body corporate in relation to the commission of the offence by the body corporate;
- in performing or failing to perform his or her organisational responsibilities, contributed materially to the commission of the offence by the body corporate;
- knew that, as a consequence of his or her conduct, there was a substantial risk that the body corporate would engage in conduct that involved a high risk of death or serious injury to a person; and
- having regard to the circumstances known, was unjustified in allowing the substantial risk to exist.

These amendments to the criminal law would overcome the current legal difficulties posed by diffused decision-making and radically change the basis on which senior managers could be convicted of serious criminal offences in the event of workplace death or serious injury. The Bill has gone through its second reading in parliament and, if passed, will make the conviction of companies and their senior officers easier to establish. Not surprisingly, the Bill, welcomed by trade unions, has faced widespread opposition from employers' groups.

The use of the mainstream criminal justice system in the prevention of occupational injury and death remains a contentious issue. However, the trend towards holding senior managers personally liable for incidents affecting the health or safety of employees provides a strong incentive for them to prevent such occurrences, to remain abreast of such developments and to put OHS at the very top of their HRM strategy.

The economic case for safer working

There are economic as well as legislative reasons to pay greater attention to OHS performance. For example, in the US approximately 6.6 million non-fatal injuries and illnesses occur at work each year and about 17 workers are killed in American workplaces each day. This results in about 6.1 million lost days of work, a total cost of over US$100 million and incalculable pain and suffering. Indeed, in the UK a study undertaken by the government's HSE estimated that, on average, these costs amount to about 8.5 per cent of a construction project's tender price (HSE 1993). Therefore OHS efficiency has significant competitive implications. This estimate included 'near misses', which also have a financial impact through damage to equipment, process interruptions, diminished employee morale, industrial relations problems or a poor corporate image, etc. These types of costs are 'indirect' and cannot be insured against. They include reduced productivity, job-schedule delays, overtime (if necessary), added administrative time, replacement and retraining costs, and damage to both workers' morale and company reputation. This is in contrast to the direct and more tangible costs of accidents which can be insured against. These stem directly from an incident and include employees' compensation, medical costs and property damage costs, etc. Alarmingly, the HSE (1993) study found that the typical ratio of direct to indirect costs associated with any one accident is about 1:11. This highlights the severe resource implications for companies with poor safety records and the hidden nature of these costs, which is probably the reason why many companies fail to make safety a high priority.

So how much should an organisation invest in health and safety? From a purely rational economic point of view, this depends upon the level of risk they face, which in turn depends upon the type of work they undertake and the effectiveness of their safety systems and work culture. The more risky the situation, the greater the investment justified. This is because each dollar invested produces a corresponding reduction in incident costs, making it possible to identify an optimum point beyond which investments produce diminishing returns (see Figure 9.1).

However, such a rational approach to safety is both ethically and economically questionable. For example, ethically it is difficult to place an economic value on a person's health, and in safety terms there is no such thing as a tolerable risk. Economically, it may be questionable because the costs of work-related injury and disease are shared between employers, workers and the community at large (Hopkins 1995). This would mean that a purely economic model would produce insufficient investments to cover total OHS costs. For example, several studies in Australia and the UK have found that the average cost of an OHS incident is spread 30/30/40 per cent, respectively (Industry Commission 1995). Indeed, in the case of serious debilitating injuries or chronic illnesses, the proportion of costs borne by employers is even lower. The figures suggest that a socially responsible organisation will seek to reduce risk below the level of economic equilibrium and invest at higher levels. However, since most organisations have a limited social conscience, these figures provide compelling reasons for continued government intervention in the protection of employees' health and safety at work. Until the construction industry takes OHS seriously, the current legislative trend is likely to continue. Indeed, there are good economic reasons, as well as social reasons, for investing beyond the point of economic equilibrium, because

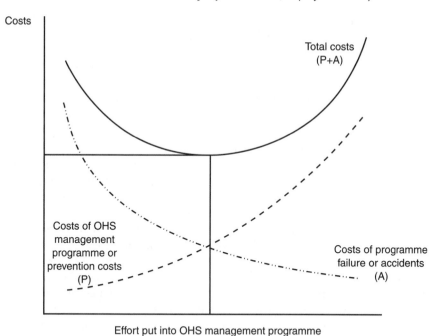

Figure 9.1 Economic analysis of OHS
Source: HSE 1997: 47 © Crown Copyright material is reproduced with the permission of the Controller of HMSO and the Queen's Printer for Scotland.

many *benefits* accrue to companies that invest in OHS. One problem with Figure 9.1 is that it only considers the costs of OHS. For example, Eisenberger *et al.* (1990) found that employees working in companies that invest in OHS are diligent, committed and productive. So, in considering investment levels in OHS it is important to consider both costs and benefits to organisations and to society as a whole. While state intervention is still important to effective OHS in the construction industry, it is not sufficient in itself. Indeed, Quinlan and Bohle (1991) found that safety legislation, even when well enforced, may have little impact on accident statistics.

Attitudinal and cultural requirements for safer working

In this age of unsurpassed technological development it is often tempting to rely upon machines, electronic or mechanical, to improve OHS performance. However, there are as many dangers associated with new technologies as there are potential benefits. Many of these dangers are psychological. For example, there is evidence to suggest that introducing safer technology can lead to more risky behaviour because most people have a natural level of risk and compensate accordingly by behaving in an unsafe manner (often called risk compensation). It is now widely accepted that the majority of accidents in industry are in some way attributable to human factors, in the sense that actions by people either initiated or contributed to the accident or acted better to avert them. Recent data indicates that approximately 80 per cent of industrial accidents, 50 per cent of pilot accidents and 60 per cent of nuclear-power accidents are attributable to human error (Jensen 1982; Rasmussen *et al.* 1994). Human error can occur for a whole host of reasons. For example, people may be unaware of or underestimate the hazards associated with their work; feel it is not 'macho' to follow safety procedures such as wearing protective equipment; rationalise risks away, believing 'it won't happen to me'; or deviate from safety procedures to gain some personal benefit such as getting home earlier or receiving a bonus. In the context of the intense time and economic pressures typical of construction work there is a tendency for people to cut corners in the belief that they are acting in the interests of their employer in finishing the job early or on time to avoid penalties. Also, quite simply, humans are fallible and make mistakes or behave unpredictably, perhaps due to tiredness or preoccupation with other issues.

In the light of the considerable pressure for the construction industry to become more technologically advanced, the key to improving the industry's safety record will be to examine human error mechanisms and take steps to prevent such errors from occurring. This does not mean blaming more people for accidents, but seeking to understand the more fundamental issues influencing their behaviour and addressing them effectively.

Table 9.1 lists some of the organisation, person and job characteristics that can influence workers' behaviour. The human factors model is helpful because it demonstrates the futility of blaming individual workers for OHS incidents and suggests that OHS behaviour, and therefore the incidence of occupational injury and illness, can and should be controlled through managerial actions. This involves fostering a positive safety culture by making OHS a top priority and communicating this to everyone involved. Having a safety culture means that safety becomes an automatic and unconscious consideration in all decisions at all levels of an organisation.

Some companies have attempted to engender an improved 'safety culture' on their sites by instituting awareness campaigns which publicly state accident rates and rewarding exemplary health and safety perfor-

Table 9.1 Organisation, person and job sources of human error

Organisation	Person	Job
Policy/strategy	Recruitment	Task complexity
Planning	Person–job fit	Team work
Standards	Task/safety training	Work pace
Operating procedures	Stress	Work environment
Communications	Motivation	Man–machine interface
Monitoring/control	Personality	Conflicting goals

mance. Others have implemented behaviour-modification techniques, including safety goal setting and contingent performance feedback (Lingard and Rowlinson 1997). However, the effectiveness of such motivational strategies appears to be moderated by employees' commitment to their organisation and the availability of resources to work safely. This is often problematic in the increasingly lean construction industry, where the prevalence of subcontracting reduces loyalty and spreads the limited resources thinly. What has become evident is that any strategy to improve safety culture must be underpinned by a well-designed, imaginative, well-resourced and continuous training system which makes people aware of the OHS risks inherent in their tasks. These risks are considered in more detail below.

Workplace health and safety hazards

Manufacturing industry has changed considerably since the industrial revolution, when the factories were filled with numerous risks like cramped conditions, dangerous machinery and poor lighting. Unfortunately the same strides have not been made in the construction industry, where operatives still regularly work in unacceptably cold, hot or wet conditions with dangerous machinery and with relatively unsanitary eating and toilet facilities. While we tend to associate poor working conditions with site operatives, service-oriented office-based employees in the construction service sector also face workplace hazards. For example, a recent study of Australian engineers reported that engineers work long hours, including significant amounts of regular unpaid overtime (APESMA 2000). Respondents also reported that the amount of work to be done had increased (63 per cent), the pace of work had increased (62 per cent), and the amount of stress had increased (52 per cent). Furthermore, more than a quarter of respondents believed there had been an increase in health problems as a result of their working lives. The most common ailments they identified were those related to excessive workloads, such as continual tiredness (66 per cent) and stress (70 per cent).

Before progressing, it is important to differentiate between occupational safety hazards and health hazards, and also to consider the related issue of employee welfare in the construction industry context:

- *Safety hazards* are aspects of the work environment that have the potential to cause immediate and sometimes violent harm or even death. Examples of safety hazards on a construction site include poorly maintained equipment, unsafe machinery, exposure to hazardous chemicals, and so on. Potential injuries include damaged eyesight or body parts; cuts, sprains, burns, bruises, broken bones and electric shock.

- *Health hazards* are aspects of the work environment that slowly and cumulatively (often irreversibly) lead to a deterioration of health. The person may develop a chronic or life-threatening illness or become permanently disabled. Typical causes are physical and biological hazards, toxic and carcinogenic dusts and chemicals, and stressful working conditions. These can cause cancer, heavy-metal and other poisoning, respiratory disease and psychological disorders like depression.
- *Employee welfare issues* are aspects of the work environment that impact on the general day-to-day well-being of employees and, in turn, their health within the workplace. The provision of adequate facilities for site-based workers, for example, is essential for ensuring their commitment, motivation and occupational health. At its broadest level, welfare could be said to include the provision of social activities and counselling for employees that improve their mental well-being and hence help them to contribute fully to the achievement of business objectives.

It is important to recognise that these aspects of work-related wellness are interrelated. For example, the safety implications of excessive alcohol consumption have long been acknowledged, but there is also a growing understanding that workers who are stressed, distracted by personal issues or overtired are more likely to create safety or health hazards by making mistakes on the job. It is therefore important that HRM policies take a holistic view of employees' *well-being* and do not regard these types of hazard in isolation. Furthermore, it is important to appreciate that each OHS system must be developed to take account of the particular factors that impact upon health, safety and welfare in the particular environment and organisation under consideration. For example, work in remote locations may impose additional stresses due to spending long periods away from home. The guiding principle should always be the same: to ensure a safe working environment in which employees are encouraged to think and act in a way which preserves their own health and welfare and that of their fellow employees and/or members of the public.

Public safety is a concern because, whilst the potential impact of failing to heed OHS needs can be catastrophic for workers, it can also have a major external impact. Perhaps the worst example of the public impact of a work-related incident occurred in December 1984, when poisonous methyl isocyanate gas leaked from a storage tank at a Union Carbide plant in Bhopal, India, killing 3,500 people living near the plant and injuring another 10,000 (Shrivastava 1992). The incident was the result of operating errors, design flaws, maintenance failures and training deficiencies. Union Carbide was sued for billions of dollars. In many of the lawsuits Union Carbide's own safety report on the plant was used as evidence of negligence, because it 'strongly' recommended, among other things, the installation of a larger system that would supplement or replace one of the plant's main safety devices – a water spray designed to contain a chemical leak. That change was never made, plant employees said, and when the leak occurred protective spray was not high enough to reach the escaping gas.

Although construction projects are unlikely to harm such large numbers of workers and members of the public, well-documented construction accidents have demonstrated the seriousness of potential hazards associated with construction activity. Consider, for example, the over-ambitiously designed Tacoma Narrows suspension bridge in Washington State, which collapsed in high winds in 1940; the Silver Bridge disaster in Ohio in 1967, which killed 46 motorists; and the West Gate Bridge in Melbourne, Australia, which collapsed during construction, killing over 30 workers. Indeed, in a study of 604 construction failures in the US between 1975 and 1986, which collectively killed and injured over 10,000 people, Eldukair and Ayyub (1991) found that the majority were caused by insufficient knowledge, ignorance, carelessness and negligence on the part of the engineer and contractor. Interestingly, while many of the contributory mistakes were technical in nature, 40 per cent were managerial, relating to errors in work responsibilities and communications. The disturbing finding emerging from the investigation of man-made catastrophes is that they are usually the result of a sequence of foreseeable errors and a failure to learn from the lessons of the past.

Influences on workplace health and safety

An important principle in OHS management is that of multi-causality. In most cases OHS incidents do not have a single cause but occur as a result of a complex interaction of many causes arising from individual, job and organisational characteristics. Only when a number of causal factors coincide does the incident occur. It is important to identify these causes and determine the sequence of events leading up to an incident if lessons are to be learnt from their occurrence. Thus it is important that managers appreciate multi-causality and that full information is recorded in a database so that lessons can be shared within and between projects – and, ideally, companies. Using an epidemiological approach, the database can be searched to identify common patterns in incident occurrence and this information can be used to target high-risk operations, trades or activities. Scientific risk-management methods are particularly useful. For example, fault tree analysis uses Boolean logic to trace the causal sequences of an OHS incident to identify immediate and latent causes. This information can then be used in the prevention of similar occurrences in the future. A similar method, applied in reverse, is a Hazard and Operability Study, used in the design of complex processes, in which the outcomes of every possible combination of occurrences are identified, with a view to designing, so far as is possible, fail-safe systems. Unfortunately, most construction organisations are ill equipped in terms of training and attitude to undertake such complex analysis – analysis that is commonplace in many other high-risk industries.

One of the reasons why construction organisations have failed to develop the intellectual capability to analyse their risks thoroughly may be related to the complexity of potential causes of OHS incidents, ranging from inappropriate work practices and site planning to the careless behaviour of individual employees. However, broadly speaking, most problems arise from the following principal factors, which means that a successful OHS strategy should also be built around them:

- working conditions;
- the tasks being carried out;
- employees' attitudes and human error;
- economic conditions;
- management goals;
- government and institutional policy;
- industry structure.

Working conditions

Working conditions include a congested, cluttered worksite, poorly designed, inappropriately used or inadequately maintained machinery and plant, a lack of personal protective equipment (PPE), the presence of dangerous chemicals or gases and inadequate welfare facilities. These can be exacerbated by aspects of the typical working conditions prevalent in the construction industry, including long work hours (leading to fatigue); noise; lack of proper lighting; exposure to inclement weather and exposure to the sun; boredom; and horseplay or fighting.

The tasks being carried out

Another crucial factor affecting the success of the health and safety objectives is the nature of the task, although nowadays the association of health and safety problems with manual tasks is not as clear cut as it used to be. For instance, while *safety* problems may be more likely for construction operatives who are required to work at height or handle hazardous equipment or materials, there is a growing recognition of the *health* hazards in an office environment, such as stress, sick-building syndrome, repetitive strain injuries, etc. Furthermore, the latent effect of some operations may not be known, which can make it difficult to ascertain an appropriate level of protection for employees. An example of this is the unresolved debate concerning the hazards associated with synthetic mineral fibres commonly used in construction. In such cases it is better to err on the side of caution rather than ignore the problem, carefully considering the potential danger and taking steps to manage the risk appropriately.

Employees' attitudes and human error

It is important to understand that, no matter how careful an organisation is in managing safety, some responsibility for safety must be borne by the individual worker. Despite advances in technology, in most instances if hazardous events are to be reduced, then the construction workers themselves must first identify them as a risk. Employees' attitudes toward health and safety can vary from enthusiasm for safety programmes to apathy. If employees are apathetic the most thoroughly designed and implemented safety programme will not improve conditions. Apathy is a major concern in the construction industry at all levels, from managers to operatives, largely because of its 'macho culture', poor training in OHS issues, and priorities which primarily revolve around issues of cost, time and quality. There is also considerable data to indicate that some employees have more accidents than the average. Such a person is said to be an accident repeater or to be accident prone. In particular, employees who are under 30 years of age, who lack psychomotor and perceptual skills, who are impulsive and who are easily bored are more likely than others to have accidents (Quick *et al.* 1990). Whilst in recent years we have seen an increase in the management of safety risk, *human error* remains an increasingly frequent factor in many accidents (Rasmussen 1990; Smith 2000). As Groenweg (1994) has noted, as the number of accidents decreases, so the proportion caused by human error increases, largely because they are the most difficult to prevent.

Economic conditions

A fourth factor affecting health and safety programmes is economic conditions, because, as with any management decision, trade-offs have to be made between the costs and the benefits of safety programmes. The dilemma for safety decisions is that ethically it is extremely difficult to justify placing a value on someone's health, but this is what managers must do. Indeed, the legislation recognises that a cost/benefit trade-off exists. For example, in Australia Victoria's Occupational Health and Safety Act 1985 requires employers to provide a working environment that is safe and without risks to health 'so far as is practicable'. This is the case in many other countries in Asia and Europe.

Practicability is defined as having regard to:

- the severity of the hazard or risk in question;
- the state of knowledge about that hazard or risk and ways of removing or mitigating that hazard or risk;

- the availability and suitability of ways to remove or mitigate the hazard or risk;
- the cost of removing or mitigating that hazard or risk.

Clearly, few employers would intentionally provide dangerous working conditions or refuse to provide reasonable safeguards for employees. However, economic conditions can legitimately prevent employers from doing all they might wish, the ultimate responsibility on a project lying with the client who funds it. This is a particular problem in the construction industry, where time and costs pressures have increased dramatically in the last decade or so. Arguably, this problem is made worse by government initiatives (such as Egan 1998 in the UK) calling for measures to make the industry more efficient. Consequently, the costs of prevention programmes are still seen as prohibitive in terms of the relative importance of other project objectives.

Management goals

Another important causal factor is management goals, which vary from organisation to organisation depending upon its culture and priorities. For example, some socially responsible organisations had active safety programmes long before the law required them to do so and invest well beyond what is economically rational. They have made safety and health an important strategic goal, and implemented it with safety programmes which include rewards for good safety performance backed up by rigorous training. On the other hand, other organisations have not been so safety conscious and have done little more than fulfil minimum legal requirements. In these organisations safety is seen as a barrier to the attainment of corporate objectives and a necessary cost burden which provides little return. There is evidence to suggest that most small to medium-sized construction firms fall into this latter category (McVittie *et al.* 1997).

Government and institutional policy

A sixth factor affecting an organisation's health and safety environment is governmental and institutional policy. It was during the 1970s and 1980s that most countries saw the development of the extensive legislation which sought to hold an organisation responsible for the prevention of accidents, disabilities, occupational illnesses and deaths relating to hazards in the workplace. All of this legislation has impacted on the way in which organisations have approached and managed health and safety risks, although the impact on effectiveness has been questionable. Compliance can only be demonstrated by the introduction of comprehensive OHS management systems, the elements of which will be discussed later in this chapter. In addition, other industry institutions, especially the unions, have pressured employers in collective bargaining for better OHS programmes. Unions have also used their political power to get legislation passed to improve the safety and health of workers.

The structure of the construction industry

The structure of the construction industry has an impact on OHS performance. For example, the traditional separation of the design and build functions has prevented the consideration of OHS in design decision-making, where many safety risks are created. Indeed, the Commission of the European Communities (1993) claimed that over 60 per cent of all fatal construction accidents can be attributed to decisions made before construction work commenced on site. The adoption of non-traditional contracting strategies and management approaches, such as design and build, private finance initiatives and partnering, might overcome some of these difficulties, though their impact on OHS has yet to be evaluated. All of these approaches provide a more cohesive and integrated system where there is greater collective responsibility for the management of safety risks. Clients, in particular, have a considerable impact on project OHS performance, and the inclusion of OHS criteria in qualification processes is one useful area of client involvement. However, American research suggests that, while public-sector clients pay attention to contractors' safety performance, private-sector clients still focus primarily on the lowest bid. There are attempts to change this in the UK, although changing the culture of accountability in the public sector will be no easy task. Finally, the relationship between contractors and subcontractors poses a challenge for OHS. In most OHS legislation head contractors are responsible for the OHS of the subcontractors they employ. However, recent research in Australia found that often the head contractor does not provide basic safety infrastructure, such as suitable access equipment (Lingard 2002). This leaves the provision of such equipment to trades who are only on site for a short period of time, for whom the investment of appropriate resources is not economically practicable. In many situations the OHS risk is passed down the contractual chain, to be borne by the party in the worst position to manage it effectively.

In 1994 the UK attempted to overcome some of these structural problems by implementing a legislative model for enhancing OHS risk communication and integration in construction projects. The Construction (Design and Management) Regulations 1994 identify key parties to a construction project, including the client, professional advisors, designers, the principal contractor and subcontractors or self-employed persons. Each of these parties has a defined set of statutory duties for ensuring that OHS risks are managed during the life of the project. Generally, industry participants perceive that the regulations have acted as a positive driving force for improved OHS in construction. Safety has undoubtedly been elevated in importance as a project goal and there has been a perceived increase in cooperation between designers and constructors (Preece *et al.* 1999). However, the introduction of the regulations has not been an unqualified success, and a number of problems associated with their implementation have arisen. In particular, compliance has generated a massive amount of additional paperwork and excessive bureaucracy, which has often become the end not the means (Anderson 1998), and the costs of compliance have proved to be higher than expected (Munro 1996). Most importantly, in recent years there have been dramatic increases in death and serious accident rates. In February 2001 the UK's environment minister, Michael Meacher, branded the industry's safety record 'appalling', with deaths at rates not seen since the early 1990s. Despite six years of the Construction (Design and Management) Regulations, there were 92 deaths between April and December 2000, which accounts for one in three of all work-related deaths in the UK. Furthermore, of the 1.5 million employed by the industry, 96,000 have suffered back injuries and 5,000 have suffered noise-induced hearing loss (Mason 2001).

Case study

Attitudinal change in small Australian construction firms: the importance of training

Twenty-five Australian construction workers were recruited to take part in an experiment. At the outset of the experiment the workers' attitudes towards OHS were explored by in-depth interviews and their OHS behaviour on site was directly observed. The workers then attended a complete first-aid training course. Once they had completed the course they were interviewed and observed once more. The results of the study showed that the first-aid training changed the workers' attitudes towards OHS and they behaved more safely at the end of the training course than before it.

Before training, participants attributed their own experience of occupational injury or illness to factors beyond their control. Their own behaviour was not identified as an important determinant of the likelihood of them suffering a work-related injury or illness. For example, in the pre-training interviews participants tended to believe accidents that happened to others were attributable to a lack of care or complacency about OHS, but that accidents to themselves were attributable to factors beyond their own control, such as the negligence of others. For example, one participant said, 'There is always the risk of stepping into a puddle and finding out that someone has been negligent and dropped a power cord in there and there is a fault in the leakage switch.' Another commented: 'Well hopefully I won't, but things can happen where it is not your fault either. I mean, someone could drop a hammer and it could hit you in the head ... there is nothing you can do.' Before training, participants also expressed the fatalistic view that their own personal experience of occupational injury or illness was related more to luck or chance than to their own behaviour. For example, one participant said, 'Put it this way, in ten years I've had one injury that has taken me to hospital, so that is not too bad I think.' Another commented: 'I think it is hard to say. It's your own fate.' Following first-aid training, participants still largely attributed OHS risks to individual factors such as complacency or carelessness. However, the biased view that accidents 'happen to others' was reduced. Following first-aid training, most of the participants expressed an understanding of the importance of taking care and concentrating to avoid occupational injury or illness to themselves. The themes of chance or fate and the negligence of others were not significant in the post-training interviews.

In the pre-training interviews the majority of participants expressed the unrealistically optimistic belief that 'it won't happen to me'. In comparison to others in their workplace, most participants indicated that they were less likely to suffer a work-related injury or illness. For example, one participant expressed this by saying, 'You make scaffolds that aren't up to scratch – I would be the only one to walk on them because I know it's safe for me but I wouldn't want any one else doing it.' Many participants also believed they had a high degree of personal control over OHS risks and therefore that by exercising this control they could avoid injury. One participant expressed this by saying, 'I think if you have got your wits about yourself, you can deal with anything.' Before first-aid training another group of participants attributed their comparatively low probability of suffering a work-related injury to their experience in their job. For example, one participant said, 'I'm probably less likely [to have an accident] because I've been doing it a long time. Not like the young guys running around madly ... [they] run into things, fall off the roof and try to carry heavy weights too quickly.' Following first-aid training, the majority of participants indicated that they had a medium to high probability of personally suffering from a work-related injury or illness, and only three participants said the chance of them suffering a work-related injury was low. One of these was an office-based site manager, while another had just returned to work on 'light duties' having suffered a work-related back injury.

Before participants underwent first-aid training, when asked whether they ever knowingly took unnecessary OHS risks at work, the majority indicated that they did. There was a strong acceptance of risk-taking behaviour as 'part of the job', and only two participants suggested that risks should not be taken or that they were concerned about taking risks. When asked why they took such risks, the most commonly cited response was 'to get the job done' – reflecting a strong production orientation among the construction workers. Following first-aid training, participants did not express such a ready acceptance of risk-taking behaviour. Less than one-third of participants expressed an unreserved willingness to take OHS risks to 'get the job done'. Half of the participants indicated that they would still take OHS risks but qualified this by saying that they would only do so under 'certain circumstances'. Several participants indicated that they had taken such risks in the past but that they were less likely to do so now, or that they sometimes took risks that they now recognised that they should not take. Another group of participants indicated that they would consider the costs and benefits before taking an OHS risk and base their behaviour on a 'calculated risk', only taking risks where the benefits outweighed the costs and where they considered the risk to be 'worth it'.

The results of the observation of participants' behaviour also showed that the attitudinal change was accompanied by behaviour change. The workers' safety performance improved in the use of tools, PPE and access to heights.

The results show that before first-aid training participants expressed strong attitudinal barriers to the formulation of favourable safety intentions. However, it seems that first-aid training changed the workers' safety attitudes, making them more safety conscious and less willing to take risks. As such, it is likely that first-aid training has a positive preventive effect in addition to providing lay persons with the skills they need to manage injuries or the onset of sudden illness. To maximise this preventive effect, first-aid training could be provided to all employees rather than restricting its provision to a handful of designated 'first aiders'.

OHS management systems

Having discussed the legal and moral arguments for the effective management of OHS, we will now discuss the policy and implementation mechanisms necessary for ensuring a safe workplace environment. However, before progressing it is important to reiterate the main point to emerge from the previous sections – that no matter how sophisticated the system, the impact on safety performance will be negligible and, at the very most, short-term if a safety culture does not exist. It is the easiest thing to change people's behaviour but the most difficult thing to change it permanently.

Nevertheless, the key to developing an effective OHS system is to recognise that employees' health, safety and welfare are an important aspect of corporate strategy, which is linked to financial loss-control strategies, broader HRM strategies, marketing strategies, information-management systems and other operating policies, including quality and environmental-performance management. Given the potential impact of OHS on other areas of business thinking, safety, health and welfare policies should be aligned with company strategy and be systematically managed to ensure that strategic goals remain relevant and are attained. The elements of an OHS management system are depicted in Figure 9.2. These will be discussed in turn.

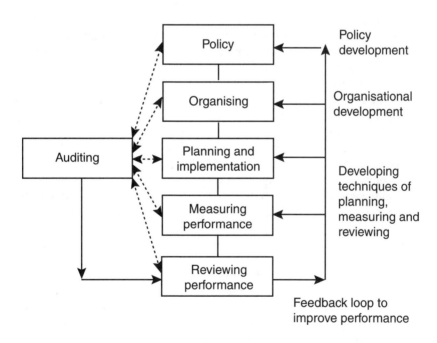

Figure 9.2 Key elements of successful health and safety management

Source: HSE (2000: 4) © Crown Copyright material is reproduced with the permission of the Controller of HMSO and the Queen's Printer for Scotland.

Health and safety policy

The formulation of a health and safety policy is the first step in the successful management of OHS. This normally takes the form of a simple one-page statement, signed by the company directors and hung on the office wall for all to see. This statement is then underpinned by a more detailed policy statement.

The purpose of an effective policy is to communicate clearly the core values of an organisation. In terms of OHS, this should emphasise the value of people to an organisation and the link between employees' well-being and corporate prosperity. Effective OHS policies should also recognise that most occupational injuries and incidents of ill health are preventable and occur as a result of a logical causal sequence in which managers can intervene. The policy should also be underpinned by a 'no-blame' philosophy, in which OHS issues can be discussed openly and honestly without fear of reprisals or finger pointing. The 'no-blame' culture requires an understanding of the multi-causal nature of accidents, that there is collective responsibility for the management of safety risks and an appreciation that most accidents are not caused by 'carelessness' on the part of employees, but by human errors attributable to failures in the management or control of risk. Health, safety and welfare policies should also be aligned to reflect that the company is not concerned just with preventing injury and ill health, but with positively promoting employees' health, fitness and satisfaction.

The responsibility for the development of health and safety policy varies from company to company. In large companies this may fall within the remit of the industrial relations function or a specialist OHS department, whereas in smaller companies a health and safety committee may be used which brings together experts from around the organisation, who regularly meet to discuss health and safety issues. In each case a close dialogue must be maintained with HRM staff in order that the health, safety and welfare policy is integrated with the mainstream HRM policy of the organisation.

OHS policies are often expressed in the form of written policy documents, often signed by senior managers. An example of a policy statement is given opposite, and this would be supported by a more detailed OHS policy document.

The contents of an OHS policy document should express the importance of preventing occupational injury and illness in company strategy and commit the organisation to continuous improvement in the preservation of employees' health, safety and well-being. It should also emphasise the need for all employees, at all levels, to cooperate in the pursuit of organisational health, safety and welfare goals, and for leaders to develop structures and a culture in which the health, safety and welfare of employees are paramount. The policy statement should also commit the organisation, and its members, to taking steps to protect the health, safety and well-being of the public and other people external to the organisation who may be affected by its operations. In a construction company this

Occupational health and safety

Baulderstone Hornibrook is committed to ensuring a safe and healthy working environment for all people at its project sites and offices. The Group believes that all work-related injuries, diseases and property losses are preventable and that safety is good business.

We will:

- be an industry leader in occupational health and safety;
- comply with all legislative and contract requirements related to occupational health and safety;
- establish challenging yet achievable occupational health and safety performance targets at all levels of our business; we will analyse and respond to our performance against these targets to ensure we continually improve our practices and our performance;
- ensure that occupational health and safety is an essential and integral part of management accountability; occupational health and safety considerations have equal status with other primary business objectives;
- ensure that systems are in place for the effective management of safety, including the development and implementation of safe work practices; if safety is compromised, operations are suspended;
- ensure that all work activities are carried out by competent, suitably trained people;
- ensure that all employees, subcontractors and third parties understand that they have an individual responsibility to conduct their work in a safe manner, adhere to Baulderstone Hornibrook's policy and procedures, and identify, eliminate and report any workplace hazards;
- systematically audit our occupational health and safety processes; we will analyse and respond to the results of these audits to seek continual improvements to our safety processes and the effectiveness of their implementation.

Signed
Chief Executive
Baulderstone Hornibrook

would include suppliers, subcontractors and the future users of the facility under construction.

In terms of delineating responsibility for health and safety, it helps to separate the responsibility for developing policy decisions from the operational issues of health and safety management at the project level. Thus there are usually two levels of safety committees, one at policy level and one at project level:

- At the *policy level* the committee should comprise major division heads and senior management and have responsibility for setting policies and rules, investigating major hazards and administering the health and safety budget. They should also collate accident statistics and investigate where problems have occurred, including 'near misses', where accidents have narrowly been avoided. Finally, they should keep up to date with the latest developments in terms of legislation and innovative methods for identifying potential hazards and mitigating health and safety risks. The development of policy measures relating to health, safety and welfare is discussed later in this chapter.
- At the *project level* OHS is normally managed by the establishment of an OHS committee, which is usually made up of managers, supervisors, nominated employees and union representatives. However, such committees are usually restricted to the construction phase of projects, although there is a strong argument for an OHS committee throughout a project's entire life-cycle. Nevertheless, the task of this committee is to establish and implement a site policy, within the overall organisational policy, and ensure compliance via communication, inspection, record keeping, training and motivation programmes. Typically such groups would meet regularly and in response to site incidents, reviewing and identifying OHS risks and establishing preventative strategies to mitigate them. The main challenge is to balance the competing demands of time and cost pressures with health and safety requirements.

Organising and communicating

Having established a company policy, the next step is to develop a system to implement it effectively. While the OHS policy document is an important statement of intent, practical measures must be taken to ensure that the organisation conforms to both their legal and moral obligations towards their employees' welfare. This involves establishing responsibilities and relationships within the organisation that enable OHS performance to be managed and controlled. Organising for the implementation of OHS policy also involves establishing achievable targets and practical measures to encourage project-based staff to meet the required standards of performance. In order to do this effectively, project-based staff must be aware of their responsibilities for OHS and accountable for their performance with regard to OHS. This can be done by communicating the OHS responsibilities of individual employees in job descriptions or organisation charts and implementing OHS performance reviews to be discussed during performance appraisals and other evaluative exercises.

Companies which have a specialist health and safety department are able to respond to legislative demands through the regular review of their management systems and by encouraging effective approaches to managing safety through face-to-face contact with site-based staff (Druker and White

1996a: 216). Moreover, these departments can offer a support service to projects by regularly visiting projects and providing external verification of the efficacy of their implementation policies. These can be more formally supplemented by health and safety audits, where documentation and record keeping are checked, as well as the more practical aspects of site safety. In this way, the specialist health and safety department can act as the link between the policy-making senior management and the project-level staff.

Key aspects of the organising effort are outlined below.

Communicating policy effectively

No matter how well written and thought through a health and safety policy, its value will be severely limited if it is not communicated effectively to the workforce. The success of the communication depends as much on its tone as on its content, on who communicates it and on how it is supported. The aim is to convey its importance, and this is the responsibility of managers at all organisational levels. The key to communicating health and safety policy effectively is to be seen to be taking the issues seriously. If employees are convinced that OHS has priority over profit and speed of completion, then they are likely to reflect this in their own priorities. Druker and White (1996a) suggest a number of useful mechanisms to achieve this:

- *site induction training*: an explanation of the rules, policies and procedures to be used on the project and a commitment to good practice on site;
- *written communications*: handbooks with concise and regularly updated information rather than lengthy guidance notes;
- *site notices*: both statutory notices and additional good practice notes displayed in prominent areas around the site, for example in canteens;
- *toolbox talks*: focused time set aside for project and senior managers to discuss health and safety issues with subcontractors and their employees, as well as directly employed workers, in order to nurture a health and safety culture on the site;
- *health and safety representatives*: trade union or workforce appointed representatives who make inspections, report health and safety problems and risks and, importantly, liase with the workforce to ensure that the issues are taken seriously;
- *safety committees*: these allow issues to be discussed by representatives of the various groups working on the site, and provide a communication channel for informing workers of particular risks and problems, policy and regulation changes, and for monitoring health and safety performance – involving employees in the management of the health and safety risk by incorporating their opinions, expertise and experience is a wonderful way to convince employees of the emphasis on construction safety;

- *training courses*: to keep people up to date with OHS regulations and policies, and to communicate expected standards of behaviour in relation to OHS.

While these mechanisms are useful, the most important element of an effective communication policy is consistency. If senior managers make safety a priority, then their actions, and those of middle and lower managers, should reflect this by adherence to safety protocols, rewarding those who perform well and commenting on OHS issues during site visits.

Training and motivation

Safety training is usually part of an orientation programme, but it should also continue through a project and an employee's career. The aim of traditional OHS training has been to provide workers with the knowledge, skills and abilities to do their job safely. Studies of the effectiveness of such training are mixed. For example, some studies indicate that methods such as *job instruction training* and *accident simulations* are more effective than others. Others contend that any form of training is effective if it creates a perception that management believes in safety as a top priority (Personick and Taylor-Shirley 1989). Some studies have found that, while safety training makes employees more aware of safety, it does not necessarily result in permanent changes in behaviour. For example, Goldstein (1993) has observed a low correlation between learning an ability to do something and actual job behaviour. With regard to OHS, this low correlation has been explained by the moderating effect of motivational factors to work safely (Lindell 1994). Training by itself is clearly not enough, and companies should not assume that their OHS training programmes are having the desired effect just because they are there. Consequently, Vojtecky and Schmitz (1986) recommend systematic evaluation to determine the effectiveness of a training programme in bringing about the desired outcome. In the absence of such evaluation, evidence as to the effectiveness of traditional OHS training remains inconclusive. Nevertheless, it would seem that effectively developed safety training programmes can make some contribution to providing a safer environment for employees, particularly when combined with other approaches.

Research suggests that the most effective training is based on behaviour modification, an approach which involves the identification and monitoring of specific behaviours representing safe and unsafe practices in an employee's job. The behaviours are prompted and then systematically reinforced. Behavioural prompts are defined by formal OHS goal setting and reinforced by team competition, the provision of rewards, the provision of feedback on individual performance and the provision of feedback on group performance. Applied behaviour analysis has been implemented successfully in the construction industry by Lingard and Rowlinson (1998).

However, the effectiveness of this approach may be moderated by other organisational issues, such as the degree of subcontracting, low levels of organisational commitment and piece-rate pay systems. Behaviour-based safety programmes should be carefully implemented to ensure their success. In particular, it is important that the objectives of performance-based remuneration systems are not in conflict with organisational safety goals.

Supervision

Whilst senior management and human resource managers have a very important role to play in developing a health and safety policy and deciding upon the most effective methods for its implementation, the most important group in ensuring safe working on construction sites are those in frontline supervisory positions. These are general and trades foremen, who have the responsibility to direct operations on site and who represent the link between operatives carrying out the work and the project management team. Their hands-on control of employees and their practical knowledge of site processes provide a uniquely powerful influence on health and safety performance.

Encouraging frontline supervisors to use their authority and control effectively depends on two key factors. First, they must understand that health and safety is the primary criterion defining project success. If a supervisor feels pressured by programme and cost-reduction targets this will detract from their role in encouraging safe working practices. Thus explicit policy and reward mechanisms must be put in place to ensure that safety issues always come first. Second, supervisors must receive detailed and regularly updated training and guidance on safe working and on their responsibilities in ensuring that this occurs. Written rules and procedures should provide a support infrastructure to enable supervisors to obtain advice from health and safety experts where necessary.

Planning and implementing

The process of OHS planning involves hazard identification, analysis and control, and is important for the effective management of OHS (Ridley and Channing 1999). This process of managing OHS risk enables decision-makers to make more informed decisions about safe methods of work and appropriate OHS control strategies, and to allocate scarce resources more effectively. A safety hazard is a situation with potential for human injury or ill health, and hazard identification is the process of recognising that a hazard exists, how it will arise and when it will arise.

Hazard identification is the first step in the risk-management process. Work activities should be carefully considered in every stage of the procurement process, particularly early on, to ensure that as many hazards as possible are identified. Methods to do this include checklists, brainstorming, cause and effect diagrams, soft systems analysis, etc. Once hazards have been identified, the frequency or probability of occurrence and the consequence of a specified hazardous event should it arise can be calculated. On the basis of risk assessment, risks can be prioritised according to their seriousness and decisions made about how urgently control actions are required. Quantitative methods for risk assessment exist, but in terms of safety it is often difficult to place a consequence value on damaging someone's health. Furthermore, few companies keep sufficiently detailed records of OHS incidents to make accurate quantitative assessments of probabilities, and, even when they do, the multi-causal nature of most accidents makes isolation of single causal events meaningless. Therefore, in the domain of safety, risk assessment remains a subjective process which relies on the assessors' knowledge and experience.

Once the magnitude of OHS risks has been determined, appropriate controls can be selected and documented in formal plans, and responsibilities for their implementation assigned. A hierarchy exists to guide the selection of controls for OHS risks which is based on the principle that preventative control measures which target hazards at source are more effective than measures which mitigate the impact of risks after they arise.

Although the early identification and assessment of OHS risks is important, it is not always possible to anticipate OHS risks at the commencement of a construction project, especially where design work is not complete when construction commences. The use of numerous risk-planning horizons (long term, medium term and short term) is useful to deal with such uncertainty, the degree of detail contained in each risk plan being inversely proportional to the planning horizon. Typically, as the planning horizon expands into the longer term the list of activities should become smaller and the specification of each activity more focused on ideas than precise facts and numbers. The role of upper management should be to prepare long-term plans with low levels of detail that are infrequently updated, while lower management levels should prepare detailed, short-term plans more frequently (Laufer *et al.* 1994). For example, a construction company might develop a strategic OHS risk plan spanning a timeframe of three to five years. This plan would state company objectives on broad terms and be formulated by senior managers and OHS specialists. Individual projects would then formulate project-specific OHS risk plans, identifying risks inherent in the major elements of the project and the general methods to be used for controlling them. The lifespan of the project OHS risk plan is the duration of the project, and a work-method statement should be developed to document the risks inherent in every activity (planning, design, construction, etc.), specifying risk control measures to be adopted. The lifespan of work-method statements is the duration of the activity or operation covered by them, and work-method statements would typically be developed under the supervision of the project manager and in consultation with OHS specialists. The lowest level of OHS planning is a job safety analysis. This is developed with the people who will perform a task, immediately before it commences. Once again, these tasks do not have to be physical site activities and can be design-based activities. Nevertheless, job safety analysis is commonly applied at site level and is normally undertaken by a foreman. It is very detailed and identifies the specific OHS responsibilities of individual crew members during the forthcoming operation.

Monitoring, measuring and reviewing performance

Monitoring, measuring and reviewing OHS performance is an essential part of successful OHS management. Indeed, it is the only way to determine whether OHS plans are being implemented and policy objectives achieved. This information should be fed back into the OHS management system and amendments made where necessary. The monitoring of OHS performance can be carried out using either proactive or reactive methods. Proactive monitoring systems provide important information before incidents, injuries or ill health occur. The main focus is on assessing compliance with specified standards or procedures, identifying deviations and correcting them before incidents occur. A safety specialist should undertake proactive monitoring during routine workplace inspections and seek to answer the following questions:

- Are safety rules being observed?
- Are safety guards, protective equipment and so on being used?
- Are there potential hazards in the workplace which safety redesign could improve?
- Are there potential occupational health hazards?

Inspection programmes need not be restricted to physical site activities and should be applied to documents such as designs, although they are rarely considered here. Nevertheless, the general rule in preparing an inspection programme is that levels of risk should be considered, high-risk activities demanding more frequent and detailed inspections. Laws and manufacturers' recommendations for plant and machinery should also be considered in developing an effective strategy.

Proactive monitoring systems are useful because they are positive and can act as a basis for rewarding successful attainment of OHS plans and good performance. Wherever possible, positive performance indicators should be identified for organisational activities rather than negative indicators. Positive indicators measure successes rather than failures in OHS management. For example, a positive performance indicator for purchasing could be the percentage of purchase orders issued in which OHS requirements were specified. A positive performance indicator for reviewing OHS management effectiveness could be the duration between reviews. Positive performance indicators can then be used as benchmarks on which to base improved performance. This is in contrast to the traditional emphasis on negative indicators, which result in a penal and negative environment focusing on failures rather than successes. The result is inevitably sub-optimal performance.

OHS performance monitoring can, and should, also be undertaken using reactive methods. Reactive monitoring systems are needed because it is impossible to create an accident-free environment and involve retrospective reporting of incidents, injuries and ill health. In doing this, it is important to collect data relating to minor injuries, since such occurrences are much more frequent and can yield important information to inform OHS decision-making. Furthermore, minor incidents, if left undetected, can often grow into major incidents.

In addition to regular inspections, organisations should instigate and disseminate accident research, which involves the systematic evaluation of evidence concerning accidents and health hazards. Data for this research should be gathered from both external and internal sources, and should include accident reports, inspection reports by the government and the organisation's safety specialists, and the recommendations of the safety committees. Accident research often involves the computation of organisational accident rates, which can be compared with industry and national figures to determine the organisation's relative safety performance. Several statistics are normally computed. For example, a common statistic is the *accident frequency rate*, which is computed per million hours of work, as follows:

accident frequency rate = number of accidents × (1,000,000 ÷ number of work hours in the period)

Another common statistic is the *accident severity rate*, which is computed as follows:

accident severity rate = number of workdays lost × (1,000,000 ÷ number of work hours in the period)

Accident frequency and severity rates should be used with caution as they are based on statistically rare events and are known to be unreliable, particularly when the number of work hours is low. Accident rates can also be deceptive and a low accident rate does not necessarily mean that risk is being controlled effectively. It is therefore important that both reactive and active monitoring of OHS performance be undertaken.

Auditing

Audits are an invaluable tool for identifying problems before accidents occur and should form a key component of every company's active monitoring activities. Health and safety auditing is similar to financial or quality auditing in that it aims to achieve a rigorous, independent evaluation of the validity, effectiveness and implementation of a company's OHS management system. All elements of the system are evaluated, including policy, OHS organisation, the planning and implementation of OHS policy, performance-measurement activities, review systems and the company's ability to improve when necessary. Audits can either adopt a vertical-slice or a horizontal-slice approach. The *vertical-slice* approach involves identifying one specific aspect of OHS activity such as PPE or fire safety and evaluating the effectiveness of the company's policy, organisation, planning, implementation, monitoring and review activities in this area. In contrast, the *horizontal-slice* approach involves one function of the OHS management system being evaluated, such as the relevance of OHS performance measurement. Audits can also be conducted at strategic, systems and operational levels. Strategic auditing examines the company's values and policies, looking closely at the attitudes of senior managers and the way in which OHS is integrated into the company's strategic planning activities. At a systems level, local procedures are compared with company standards and the degree to which company standards comply with legislation, codes of practice or national standards is also considered. Finally, at the operational level the way in which activities are performed is examined in terms of the degree to which employees understand and conform to good working practices and specified procedures. To obtain a complete picture of the effectiveness of a company's OHS management system, both horizontal- and vertical-slice approaches, as well as auditing at the three different levels, are necessary.

Occupational health programmes

Occupational health programmes deal primarily with the identification and control of substances that are hazardous to health. Construction work inevitably exposes project-based workers to health hazards, and these may broadly be classified as:

- chemical health hazards, such as gases, solvents, metal fumes and pigment dusts, and airborne particles and fibres;
- natural hazards, such as the sun, cold, heat, humidity and radiation;
- man-made physical health hazards, such as noise and vibration.

Health monitoring should be undertaken to identify changes in employees' health status due to occupational exposure. The aims of health monitoring are:

- to monitor exposure to a hazard, generally through biological monitoring;
- to measure biological effects, which may be reversible, which would require reduction or cessation of exposure;
- to collect data to evaluate the effects of exposure.

Part of the health-monitoring process should be a pre-employment medical examination, particular attention being paid to employees exposed to particularly high risks and on whose fitness the safety of others depends. Examples of such people would be crane drivers and demolition workers. Pre-employment medical examinations should also consider the requirements of the job, because certain jobs may require a certain level of physical strength or other mental or physical attributes to be conducted safely. Steel fixers and bricklayers might be among this category of worker.

Another component of health-monitoring systems is routine examinations to monitor the effects of workplace hazards on employees. For example, where there is a pneumoconiosis risk regular chest x-rays and lung function tests should be conducted. Initial pre-employment examinations provide a useful baseline against which to compare the results of subsequent medical examinations.

Employee welfare policies

More and more companies are switching from the traditional approach of focusing on safety to one that also focuses on the general well-being and health of a workforce. Managers are increasingly realising that they must be concerned about the general health of employees, including their psychological well-being, because an otherwise competent employee who is depressed and has low self-esteem is as non-productive as one who is injured and hospitalised. As a result, 'wellness' programmes are becoming more popular and widely accepted as a legitimate aspect of a company's health and safety strategy.

Essentially, the wellness approach is a preventative measure which encourages employees to make lifestyle changes through nutritional counselling, regular exercise programmes, orthopaedic rehabilitation, abstinence from smoking and alcohol, stress counselling, and annual physical examinations such as free mammography and blood pressure screenings (Thompson 1990). For example, one successful programme was based on a six-step model of behavioural change: *awareness, education, incentives, programmes, self-action,* and *follow-up and support*. Each employee was first made aware of their current health through a health-hazard appraisal (HHA), a statistical evaluation of that person's individual health risks. The HHA included suggestions for lowering risk and changing behaviour to live a longer, healthier and more contented life. Incentives included refunding the cost of weight-loss programmes if the loss was maintained for a full 12 months. Various programmes like the gym and nutritional counselling were also offered as support.

Research indicates that the 12 key elements of a successful wellness programme are:

- support and direction by the CEO;
- wellness as a stated priority in the company's policy statement;
- inclusion of family members as well as the employee;
- accessibility of the programme to the whole family;
- employees' input into programmes offered, times and so on;
- needs assessment before each phase of the programme was instituted;
- periodic in-house evaluation to be sure objectives were being met;
- ongoing communication of the programme's goals and components;
- HRM monitoring of related issues like AIDS, cancer and so on;
- community involvement;
- staffing with qualified healthcare specialists;
- establishment of a separate budget for the wellness programme.

Many companies employing thousands of employees have reported exceptional paybacks from adopting the wellness approach. Indeed, the best wellness programmes are producing a payback in savings (in reduced insurance costs, productivity benefits, and reduced turnover and absenteeism) of six times the investment (Caudron and Rozek 1990). For example, the Prudential Insurance Company reported that one of its offices reduced disability days by 20 per cent and obtained $1.93 savings for every $1 invested in its in-house exercise programmes. Its major medical costs dropped from $574 to $302 per employee in just two years. Similarly, a fitness programme including regular exercise and various health-education and lifestyle-improvement classes was started at the Canadian Life Assurance Company, resulting in a 22 per cent drop in absenteeism. Johnson & Johnson have also developed a wellness programme which includes nutritional and stress management, smoking education and cessation programmes and fitness classes. It compared 5,000 employees who enrolled in the programme with 3,000 who did not and found that hospital costs per person were 34 per cent lower for participants than for non-participants! While non-participants averaged 76 hours of absence each year, participants averaged 56 hours. While wellness programmes might seem like the preserve of larger companies with substantial resources, there is no reason why mid-sized and small employers cannot offer scaled-down versions and still reap significant benefits. However, it is important that the same 12 steps detailed above are followed, regardless of the company's size.

Despite the above evidence of great success, it is important to appreciate that the wellness or preventive approach is not foolproof, particularly if it is adopted without fully understanding the necessity of commitment by management and effective manager–employee communication. Almost 80 per cent of corporations offering a wellness programme continue without any quantifiable proof that the programme is saving them money or increasing productivity (Caudron and Rozek 1990). Furthermore, promised reductions in healthcare premiums from insurers are not always delivered and are minimal at best. Finally, the employees who need the help most may not participate, and some may resent the intrusion in their personal lives.

Nevertheless, it remains true that health-promotion programmes that can help employees develop healthier lifestyles, reduce their risk factors and use health services more appropriately are effective strategies to help management control the cost of health and disability benefits. However, it is important to note that the immediate benefits of a preventive approach to healthcare are minimal and that the ultimate payback is in the long term. In Australia there is a national wellness network for organisations interested in this approach.

OHS in the future

There appear to be as many safety and health hazards in the modern workplace as there were during the industrial revolution. The difference lies in the nature of these hazards, which has changed as a result of economic, social, and lifestyle changes. It would be impossible to address all the safety and health issues facing managers in the future, so the remainder of this section addresses some of the most interesting and pervasive, namely: job stress and burnout, workplace violence and intimidation, HIV/AIDS, and cumulative trauma disorders and repetitive strain injuries. As always, dealing with each issue requires project managers to balance the rights of the individual against the rights of the rest of the workforce and the economic imperatives of the project.

Job stress and burnout

Stress

Stress manifests itself in numerous ways, such as emotional discomfort, irrational behaviour and physiological or body reactions, such as higher blood pressure, increased heart rate and altered hormone levels. There is little doubt that job-related stress is associated with a vast array of diseases, such as coronary heart disease, hypertension, peptic ulcers, colitis, and various psychological problems, including anxiety and depression (Allen 1990). Workplace stress is becoming an increasing problem, particularly among white-collar workers, and it can cause serious economic and social problems for organisations. For example, in the US the costs of stress in the workplace – from increased accidents, absenteeism, lower productivity and reduced morale – have been estimated to be at least US$150 billion per year (Waxler and Higginson 1993). While the number of claims for stress-related illness represents only a small proportion of the total number of employees' compensation claims, the cost of these claims represents a major proportion of all compensation costs (Cotton 1995). It is clear that occupational stress-related illnesses result in substantial losses to both companies and the economy as a whole. Companies cannot afford to ignore the issue and need to develop intervention strategies to mitigate the problem.

It is widely accepted that there are two main dimensions to a comprehensive workplace stress intervention strategy. The first dimension is preventative and involves avoiding the problem by eliminating the stressor. Common stressors include dangerous environments, poor person–job fit, unhealthy workloads, lack of recognition and rewards, and conflicts with other employees. The second dimension is reactive, and involves developing coping mechanisms – such as meditation, exercise and diet – which help people relax, feel better about themselves and regenerate their energy. For example, some companies now provide seminars on stress management, which give practical advice to employees, helping them to examine their life goals, identify their vulnerabilities and modify their behaviour accordingly. Positive intervention programmes may also provide employees with individual well-being and fitness programmes, diet counselling, and fitness and exercise facilities.

Burnout

Since Freudenberger (1974) coined the term 'burnout' to describe a state of chronic emotional fatigue this phenomenon has been the focus of much research interest. The most widely accepted definition of burnout conceptualises the phenomenon as a syndrome of emotional exhaustion, cynicism and reduced personal accomplishment (Maslach *et al.* 1996). Emotional exhaustion describes feelings of depleted emotional resources and a lack of energy. In this state employees feel unable to 'give of themselves' at a psychological level. The second component of the syndrome is characterised by a cynical attitude and an exaggerated distancing from one's work, and the last component of burnout, diminished personal accomplishment, refers to a situation in which employees tend to evaluate themselves negatively and become dissatisfied with their accomplishments at work.

Case study

Are Australian construction professionals burnt out?

A survey was undertaken to explore the experience of 'burnout' among engineers working in the Australian construction industry. The data for this study were obtained from civil engineers engaged in professional practice in consulting and contracting organisations in

New South Wales and Victoria in Australia. Of 500 questionnaires distributed, the response rate was 36 per cent.

Results suggest that Australian engineers experience a strong sense of the social worth of their professional activity, independently of believing in their own individual competence as engineers. However, there was also a widespread belief that the rewards they enjoy as a result of their professional endeavours are not commensurate with their level of skill and responsibility.

The study found that two dimensions of burnout, cynicism and emotional exhaustion, were strong predictors of engineers' intention to leave their jobs. These findings indicate that when civil engineers experience emotional exhaustion or cynicism they are more likely to develop an intention to leave their job. Employees' disenchantment emerged as an important theme in the analysis of comments made by respondents at the end of the questionnaire and was often linked to turnover intention. For example, one participant commented: 'The issue of how our job affects personal leisure time, such as sport or socialising with friends, and how this makes us feel about our work has not been addressed. I am actively seeking a job that gives me my life back!' The significant association between burnout and turnover intention suggests that measures to prevent or alleviate burnout might assist employers in the construction industry to improve their employee retention rates. While we were not able to measure actual turnover behaviour, research suggests that intention to leave is a good predictor of actual turnover behaviour (Parasuraman 1982). However, as Hughes (2001) notes, intention to leave can be constrained by the availability of acceptable alternatives. Where such alternatives do not exist, employees may involuntarily remain in their job, which poses the problem of a further change in attitude and effort, and a possible decline in performance. Thematic analysis of the comments provided at the end of the questionnaire suggests that participants' disenchantment with their work was not restricted to the company for which they work but related to the engineering profession as a whole. For example, one participant expressed this as follows: 'My reluctance to look for a new job has to do with my dissatisfaction with my profession rather than my job'; while another posed the question: 'I often feel that life would be the same at other companies so why bother moving?' Given the perception that things would be no better elsewhere in the profession, it is possible that burnt-out and dissatisfied employees are remaining in their jobs because of a perceived lack of more satisfactory alternatives. This situation is one of serious concern to the construction industry because these disenchanted employees are likely to be underperforming. Ultimately, employees may become so disenchanted that they decide to change their careers and leave the engineering profession altogether. Several of the participants who provided written comments alluded to this. One participant commented: 'Bus driving looks like a good career move at times!', while another participant commented: 'I feel strongly that, as an engineer, I am undervalued by the community. While not totally dissatisfied with my remuneration, I get frustrated when I see other professions with similar levels of training and less responsibility getting paid far more than my peers and myself. This is leading to a level of disenchantment in the profession and resulting in many engineers leaving and going to areas where they are "properly paid for what they do" – even though they might enjoy it less.'

Significant predictors of emotional exhaustion included hours worked each week, overload and role conflict at work. Consistent with this, long work hours and the feeling of having too much to do in the time available were common themes emerging from the comments provided by some participants at the end of the questionnaire. For example, one participant wrote: 'The consulting engineering industry has undercut itself to the extent that projects rarely can be completed for the fee submitted – a profit can and still is made, but at the expense of working long hours and not booking the time.' Another criticised unrealistic client programmes, saying they resulted in 'insufficient time to do the work, hence long hours'. Emotional exhaustion was also predicted by individual personality and family-related variables, including neuroticism, the age of the youngest child and conflict in the relationship with spouse or partner.

Satisfaction with promotion prospects, responsibility, role clarity and satisfaction with pay predicted cynicism. Consistent with these findings is the fact that dissatisfaction with the level of job-related rewards in comparison with the responsibility borne by engineers emerged as an important theme in comments written by participants at the end of the questionnaire. The responsibility for the health and safety of workers was of particular concern, and there was a strong sense that engineers are undervalued by society as a whole. For example, one participant expressed this by saying, 'Tough OHS laws and heavy fines are now even more of a concern to engineers than the already busy workload. All this responsibility is not worth the pay.' Individual and family-related characteristics also predicted cynicism. These included the age of respondent, the number of children and satisfaction in the relationship with spouse or partner. Younger employees are more cynical about their work than older employees. Parents are less cynical than childless employees, and the more satisfied employees are in their relationship with their spouse or partner, the more cynical they tend to be.

The results of the study show that burnout is a multidimensional phenomenon and that the different dimensions of burnout – emotional exhaustion, cynicism and diminished personal efficacy – are associated with different job and individual characteristics. Burnout among construction industry professionals cannot be attributed to a single cause but occurs as a result of a complex inter-action of individual characteristics and issues in the work and non-work environment. As such, there is no single 'cure' for burnout and multiple intervention strategies are probably needed. However, the relative importance of job characteristics compared to personality characteristics in predicting burnout suggests that job redesign may be a particularly effective preventive strategy.

Demographic variables such as parental status and the employ-ment status of the spouse or partner also had a bearing on the engineers' experience of burnout. Different predictors of burnout emerged for the different demographic groups. Family variables were more salient among participants in dual-income compared to single-income couples, and among parents compared to childless participants. This suggests that organisations need to look outside the immediate work environment when devising preventive strategies for burnout. Work–life balance initiatives may help to alleviate the problem. Finally, the demographic profile of the workforce should be considered when developing preventive measures. The needs of different demographic groups should be considered, and flexible human resource policies may be required.

Violence and intimidation in the workplace

Unfortunately, the macho culture of the construction industry makes intimidation and violence more likely aspects of the workplace culture than in other industries. Workplace violence includes verbal and emotional abuse or threats, as well as physical attacks, to an individual by another individual or group. It also includes bullying, sexual assault, harassment and excessive peer pressure, and it may not always be obvious. Workplace violence can arise for a wide range of reasons, such as stress caused from overwork or personal problems, cultural differences and intolerance, financial disputes, personal conflicts, drug and alcohol abuse, psychiatric or psychological disturbance or intellectual impairment, dissatisfaction with a service or payment, or opportunism. The impact on a person depends on the severity of the violence, his or her experiences, skills and personality, and can vary from frustration, through depression to suicide.

Managing workplace violence demands a proactive and reactive strategy, and employee screening, security and counselling measures are all part of a comprehensive approach. The first step that needs to be taken in the prevention of workplace violence is to develop pre-employment screening procedures, such as taking up references and using psychological employment tests. This is quite difficult to implement because former employers are usually happy to let a violent employee go, psychological tests are notoriously unreliable and, furthermore, employers must be careful not to violate any of the applicant's civil rights. Thus post-employment measures are the most common and these include:

- reducing the causes of violence such as long working hours, late payment and unfair work practices;
- increasing workplace security, providing mobile phones and personal alarms;
- creating a culture of open communication, empowerment and recognition, which prevents frustration and encourages whistle blowing;
- delivering training classes in violence management, negotiation, communication, listening, team building and conflict resolution;
- referring troubled employees to employee assistance programmes to help deal with interpersonal, psychological, work, family, marital, financial and other personal problems;
- providing emotional support and outplacement programmes for laid-off employees;
- conducting exit interviews that will identify potentially violent responses to termination;
- implementing a clear, well-communicated, easily accessible grievance procedure and encouraging employees to use it;
- developing a confidential reporting system that allows employees to report threats or inappropriate behaviour which may indicate a potential for violence;
- strictly controlling access to the workplace with an up-to-date security system consistently enforced;
- training supervisors to recognise the signs of drug and alcohol abuse, depression and other emotional disorders;
- developing and implementing a crisis plan to deal with violent incidents, including escape routes, how to report the incident and how to avoid further trouble.

Managed together as part of an effective equal opportunities policy, these steps can lessen the likelihood and/or impact of intimidation and violence targeted at particular employees.

HIV/AIDS

The human immunodeficiency virus (HIV), which leads to acquired immune deficiency syndrome (AIDS), was first reported in the late 1970s and remains an incurable but preventable fatal disease. Each of the letters in the AIDS acronym stands for a word:

- *Acquired*: the disease is passed from one person to another.
- *Immune*: it attacks the body's immune system, the system that protects the body from disease.
- *Deficiency*: the defence system is not working.
- *Syndrome*: the status of AIDS is determined by the existence of any one of a group of AIDS-defining symptoms or illnesses.

HIV is transmitted by infected blood and bodily fluid transfer and is a potential health risk in the construction industry, where physical injuries are a problem. Here, employees could reasonably be expected to come into contact with blood at some time in their working lives. Globally, the problem is huge: in 1999 AIDS was the world's second leading cause of death (behind cancer) in the 25–40 age group. The problem is particularly acute in African countries such as South Africa, where one-third of the population is infected, and very special measures to protect employees need to be taken there. However, other countries are also experiencing major problems with HIV/AIDS. For example, in China 10 million people are now infected with HIV, and in the US the number is 1.5 million and rising. Approximately 97 per cent of those are of working age, and 75 per cent are between the ages of 25 and 44, the prime working cohort.

The problem with HIV/AIDS in the workplace is not just the productivity lost from illness, but the fear of the disease by co-workers, which can result in violence, discrimination and harassment from other employees and customers. A case of HIV/AIDS in a workforce is a serious issue for everyone concerned, and the complexity of the potential management problems which can arise requires that companies develop multidimensional policies to deal with it effectively. This should cover issues such as education and communication programmes to inform staff of risks and potential hazards; universal precautions to avoid contamination, such as cleaning protocols, personal protective equipment, etc.; control plans to minimise exposure and the spread of infection; and record-keeping systems of all possible incidents. The co-workers' fears can be reduced by an understanding that it is generally difficult to contract HIV outside unprotected sexual contact, since it is difficult for the virus to survive outside the bloodstream and difficult for it to get into the bloodstream. In addition, it is important that employers must understand that they cannot discriminate on the basis of HIV infection and that few jobs exist where the disease prohibits an employee from performing a task effectively. Should such a situation arise, the employer is expected reasonably to accommodate the worker. This may include changes in equipment, work assignments, job flexibility, part-time work, flexible hours and working at home. In particular, it is important that any manager who knows that an employee has HIV/AIDS preserves the individual's right to personal privacy.

Cumulative trauma disorders and repetitive strain injuries

Repetitive strain injuries (RSIs) and cumulative trauma disorders (CTDs) have always been a problem in the construction industry, where afflictions such as 'white finger' have been known about for some time. However, the problem has now also reached epidemic proportions in the professional world as people spend more time in front of computers. For example, in the US, CTDs have increased by more than 1,000 per cent in a little over 10 years, and workers' compensation payments now exceed US$2 billion annually. It is clearly a problem that cannot be taken lightly in the construction industry, where many tools and work processes may cause such damage.

The most effective way of preventing RSIs and CTDs is to understand the causes and symptoms. CTD usually refers to conditions that arise from obvious trauma or injury that occurs more than once. For example, if a man injures his back lifting a bag of cement at work and then injures it later, he could claim to be suffering from CTD. In contrast, RSI refers to a repetitive activity which is not in itself harmful or injurious but which is alleged to become harmful owing to the sheer number of repetitions. Common examples in the construction industry would involve 'white finger', associated with the use of compressed-air tools, or carpal tunnel syndrome, associated with typing activities and all jobs involving cutting, hammering, assembling small parts, finishing and cleaning. Carpenters, plasterers and bricklayers are obviously those at most risk in the construction industry.

In both cases the best way to tackle the problem is, first, to determine which employees and jobs are at the most risk, to identify and assess the risks and, then, to put in place an ergonomics programme to consider how jobs, facilities, technologies, methods of working, etc. can be redesigned in consultation with those affected to eliminate the problem. Training may be required to implement such a programme, and periodic review is required to ensure that people can cope with changes in staff, management, technology and the facility.

Conclusions

Safety risks are inherent in all human activity, have numerous dimensions and represent a complex problem. The construction industry has had a dreadful safety record and, despite increasingly penal legislation, continues to do so. The systems developed by construction companies are as good as those developed in any industry, which means that the problem is structural and cultural. It also means that it will be slow and difficult to change. The basis of safety management is to encourage people to identify, respond to and control the risks inherent in their work activities – those risks they create for themselves and those that they create for others. To be fair, the consideration of health and safety risks in the construction industry has become more widespread in recent years, albeit forcibly. Safety method statements are now common and operation plans almost always include assessments of the risks present in the work. Furthermore, specialist health and safety managers and professionals often have an input into both the design and the construction phases of construction projects, ensuring that health and safety is defined as a core project priority rather than an afterthought to be considered during the construction phase. The more advanced companies have even established indicator systems (for example accident statistics), and developed effective reporting systems and rules and reward strategies for those who manage safety effectively. However, the industry has a long way to go to improve its record to the standard of other industries, which in many cases face far higher levels of risk.

Discussion and review questions

1 Identify the legal, moral and economic reasons why construction organisations should manage OHS.
2 Identify one occupational health risk and one occupational safety risk pertinent to your work. Identify control strategies for these risks in accordance with the 'hierarchy of controls'.
3 Develop five positive performance measures for OHS in your workplace.
4 Work-related stress and burnout have negative consequences for both individuals and organisations. Discuss the extent to which organisations should intervene to address non-work sources of stress/burnout.

10 Strategic human resource development

Strategic human resource development (SHRD) ensures that people continue to add value to an organisation in a changing business environment, maintain their motivation and enthusiasm towards their work, and work in a way that supports the strategic objectives of the organisation. However, construction companies often treat HRD activities as someone else's responsibility or as an expensive activity which risks making employees more attractive to competitors. In this chapter we explore the role of SHRD within construction organisations and how it can be managed in line with the objectives of SHRM.

Introduction

In Chapter 5 we briefly discussed the role and importance of the SHRD function – the developmental side of the SHRM function that is used to improve the performance of the individual in line with a business's planned strategic direction. Such is the increasing importance of SHRD in today's ever-changing business environment that we have dedicated a chapter to exploring this important issue.

The necessity of investing in SHRD is driven by people's personal need for growth and the increasingly dynamic business environment in which organisations operate. It is based on the assumption that, no matter how effective an organisation is in attracting, recruiting and selecting people, this will not result in longer-term development and performance improvement without a sustained effort to train and develop them.

Beardwell and Holden identify ten principal motivations behind the need to invest in SHRD as follows:

- New employees are like 'raw materials' that need to be 'processed' in order to perform their tasks and fit into their workgroups and organisation. However, this must be managed in a way that respects their human qualities.

- Jobs change over time and so employees' knowledge, skills and abilities need to be updated so that they maintain their performance in the face of changing demands and requirements.
- New jobs will be created which will need to be filled by existing employees, who will need support and redirection.
- People need to be trained in order to perform more effectively in their existing jobs.
- People change their own interests, skills, confidence and aspirations with time, and the organisation must take account of this.
- Employees may move jobs either to be promoted or to broaden their experience and so will require further training in order to perform in their new roles.
- The organisation itself may change over time, and so employees' knowledge, skills and abilities must be updated regarding new ways of working together more effectively.
- The organisation may wish to ready itself for predicted future change by equipping employees with transferable skills.
- The organisation may wish to respond flexibly to its environment and therefore may require some employees to develop flexible, transferable skills.
- Managers require further training and development to allow for performance improvement and management succession via the development of new and potential managers.

(Beardwell and Holden 1997: 279)

Beardwell and Holden's reasons for investing in training and development point to a close conceptual relationship between employee training and wider development activities. In effect, training programmes offer a vehicle for individual and organisational learning. Managed effectively, they should create and maintain a healthy, motivated and adaptable core workforce that can respond to the changing demands of the modern business environment. Training is critical to ensuring that people have the skills that they need to perform their job and advance along their favoured career path, but ultimately should be viewed in the broader context of the need to align people-development activities with business objectives.

The concept of learning organisations

Essentially, the purpose of training and development in the modern dynamic business environment is to bring about a *learning culture* within an organisation – a systemic learning environment where knowledge is captured and transferred for the benefit of an organisation, its employees, its customers and other stakeholders. A learning organisation is one that treats learning as a continuous process of improvement which is fundamental to business success (Armstrong 1998). Such organisations are more proficient at problem solving, developing new ideas, learning from their experiences and transferring new ideas into practice. Dufficy (2001) highlights the importance of employee training and development in enabling organisations to survive in the face of a 'new industrial world' created by forces such as globalisation, the increasing power of the customer, dramatic social and environmental change, and the growth of e-business. Dufficy also argues that in the UK the automotive, electrical and electronics industries are much more effective at dealing with these forces than is the construction industry.

Unfortunately, evidence indicates that many construction companies do not exhibit the attitudes and qualities which underpin learning organisations. For example, Kululanga *et al.* (1999) reported low usage by construction companies of tools used widely in the general business community to enable corporate learning. D. N. Ford *et al.* (2000) suggest that one possible reason for this is the predominance of an engineering culture which focuses on technology instead of people and lacks an organisational development emphasis. Furthermore, the small subcontractors which employ the vast majority of the construction workforce confuse training responsibilities and are so highly geared that long-term investments in training have been difficult. Even in major organisations, training and development activities are often squeezed in the face of programme pressures and small profit margins, and there is little sense of paternalism towards the subcontractors they employ. For example, Scott *et al.* (1997) investigated the attitudes of civil engineers and their companies towards management training in the construction industry. They found that, although many participants stated that management training was necessary in the business world of today, many older engineers remain sceptical about management or the value of management training. Nevertheless, training is still a fundamental requirement for improving organisational performance, and so innovative delivery mechanisms must be considered which provide employees with the opportunity to develop the learning skills and attitudes that will allow them to function in a more efficient and effective manner.

Generalising about the provision and effectiveness of skills training and development in construction is problematic for two reasons. First, different countries have very different structures and training delivery mechanisms. For example, Beardwell and Holden (1997) point out that in many Asian countries, such as Singapore and China, HRD is often underpinned by strong government initiatives to improve knowledge and skills and to enhance the economic growth of the nation. Other countries adopt a much more laissez-faire attitude towards HRD. Thus it is important to recognise the potential socio-economic role of HRD in enhancing the well-being and performance of the industry as a whole. This does not enable easy comparisons to be drawn. Also, Burba *et al.* (2001) argue that instructional behaviours that are appropriate in one cultural context may be inappropriate in another. For example, they suggest that learners from societies with a strong norm of uncertainty avoidance, such as Germany, expect the instructor to be an expert and conduct him or herself in an authoritative manner. On the other hand, Saudi Arabian learners prefer a friendly relationship with their instructors, and Japanese learners expect instructors to adopt a humble demeanour and show respect for the learners by doing something to honour them. Burba *et al.* assessed students' expectations regarding appropriate instructional behaviour and found significant cultural differences, such as the use of gestures, eye contact or paralanguage. They suggest that when a trainer is not of the same culture as the population receiving the training the programme and its delivery need to be adapted to suit the target audience.

For the above reason, this chapter will explore, at a generic level, the objectives and characteristics of an effective training and development programme. We then consider the management-development function within construction, considering how it can engender improved organisational performance. Within both sections we discuss the barriers to training and development and explore the reasons for historical under-investment in construction. It is important to note that in this chapter we have taken an organisational-level view of training and development since the emphasis of this book is on SHRM practices that can lead to improved construction project performance.

SHRD in construction

Training and development are the two basic components of SHRD. In this section we shall look at training. The object of training is to alter the behaviour of employees in a way that will create improvements in the achievement of organisational goals. It should provide opportunities for an employee to learn job-related skills (such as thinking), change attitudes and help people to acquire knowledge. The type of training, the method of delivery and the objective of its provision will decide which of these three determinants of behaviour is changed. Since training is a form of learning, to be successful it is essential that the employee is motivated to learn, is able to learn, is able to transfer their learning to the job, and has their learnt behaviour encouraged and reinforced in the workplace. Thus, arguably the most important purpose of training in the modern business environment is to bring about a *learning culture*, which in turn provides an organisation with a skilled and self-motivated workforce that can adapt to the changing needs of the modern business world. The benefits of training and some obstacles to its provision in a construction context are discussed below.

The importance of training

Whereas in the past training may have been regarded by some organisations as a luxury, there is now widespread acknowledgement by both academics and commercial organisations of its importance to the success of the modern business. This recognition largely stems from globalisation and the related intensification of overseas competition in recent years. It has not gone unnoticed that those countries whose economies have performed well in recent years, such as Germany and Japan, have been those that emphasise the importance of training (Holden 1997). This is also reflected at industry level, where those industries which invest in training tend to perform better and have a more positive public image than those which do not. This subsequently influences the quality of new recruits to an industry, which in turn perpetuates the problem, leading to skills shortages and deficient performance in terms of product quality and delivery efficiency. Indeed, concerns over the quality of recruits to the construction industry have been a major problem in recent years and this has led to a range of training initiatives, which are reviewed later in this chapter. In essence, training represents the mechanism by which organisations invest in the intellectual capital of their workforce, and it lies at the very heart of achieving a vibrant, healthy, motivated, happy and efficient organisational culture.

A reason for the acceptance of training as a core component of SHRM relates to the incontrovertible link between key SHRM-related concepts and the need to impart the requisite skills, knowledge, attitudes and abilities to employees. For example, encouraging employees to work in a way which leads to better-quality products and therefore a more positive organisational image demands that they are trained and developed in a way which strives to achieve quality improvements. Also, ensuring loyalty, motivation and commitment requires that people receive support and encouragement through training and development mechanisms. However, perhaps the most important driver for greater training provision relates to the growing need for companies to develop adaptive capabilities which enable them to change in accordance with the increasingly dynamic environment in which they operate. For example, in construction the past two decades have seen countless changes and advancements in procurement practices, IT, construction technology, legislative demands and, perhaps most importantly, client demands in the industry. Each change has brought with it a need for construction companies to adapt to new demands and ways of working that cannot be achieved through external recruitment of skills alone.

Barriers to training provision in construction

Despite the undoubted importance of training, as discussed above, most construction companies do not engage in effective corporate learning.

Indeed, in an industry such as construction there exist many barriers, both real and perceived, to even the most basic training and development activities. These include:

- *The cost of training delivery*: training activities are assumed to be expensive in terms of both the cost and time. Therefore training programmes are often amongst the first expenditure items to be dropped in times of recession.
- *Clashes with production objectives*: there is a widely held view that the majority of formal training activities require key project-based staff to be removed temporarily from their operational responsibilities. In an increasingly lean construction industry this can cause additional pressure for already overstretched teams.
- *Existing legislative training requirements*: minimum training standards already exist that are protected by statute in most counties. This means that companies must provide minimum standards of training on issues such as health and safety. Additional training can be seen as an unnecessary add-on or a luxury within many construction organisations.
- *Staff turnover concerns*: providing employees with training and development support makes them more attractive to other companies. Construction is a highly predatory and transitory industry with a strong culture of nomadism. It is highly likely, in the common absence of retention strategies, that trained employees will take their skills elsewhere. Conversely, it is possible to attract trained employees from other companies through the use of remunerative incentives, negating the need for one's own training strategy. The overall effect is a training stalemate.
- *A macho environment*: the construction industry has a highly masculine culture, with a tradition of physical activities and an emphasis on production that cannot be learnt effectively in a classroom environment. Many employees have been failed by the traditional classroom-based educational system and perceive learning as a non-productive, feminine activity and associate it with failure. This is a major cultural barrier to training, and also permeates management positions in the construction industry.
- *A 'learn on the job' culture*: the historical attitude towards developing a career in the industry has been to value experience as the primary learning mechanism rather than formal training or education. The relative strengths and weaknesses of on-the-job and off-the-job learning are considered below.

The above discussion indicates a short-term and negative view of training which is not shared by many modern construction clients. They are increasingly demanding that construction firms demonstrate the competence of the team that will be working on a project through their training and development activities. To assist in satisfying and exceeding these growing market requirements, the following sections discuss the essential elements of an effective SHRD programme.

Planning and implementing effective SHRD activities

Essentially, training can be delivered either in a reactive, fragmented and unstructured way in response to immediate needs, or through a planned programme of interconnected activities. Whilst ad-hoc training provision can be valuable in terms of providing for immediate skills needs, getting value for money from a training programme requires more careful management of the function. Naoum (2001) put forward a simple framework when discussing training in the context of construction, which comprises the five steps listed below:

- *Define a training policy*, which involves defining clear links between organisational objectives and training provision.
- *Identify staff training needs*, expressed in terms of both organisational and individual needs.
- *Prepare a training programme* which is a carefully planned sequence of training activities.
- *Decide on methods for delivery*, which could include formal or informal approaches to training, including courses, training videos, job rotation or special assignments.
- *Evaluate* to review whether the training provision had the desired impact on the performance of the organisation.

In addition, we would argue that it is necessary to *engender a motivational climate* which ensures an appreciation of the importance of training and allows all of those involved to understand its value in terms of improving organisational performance. Taking each of these steps in turn, we have elaborated a detailed strategic approach to training which overcomes many of the barriers identified above. This comprises six steps, which are outlined below.

Step 1: defining a training policy

The primary role of an organisational training policy is to define clear links between the objectives of the organisation and the nature of training provision. This should be regularly updated and revised to take account of changing priorities brought about by new market opportunities, legislative changes, social expectations and fluctuations in the economy, all of which will place new demands on staff and their skill requirements. Essentially, an effective training policy should identify a hierarchy of skills and knowledge requirements for the organisation that allows it to prioritise its training resources. The training policy must not only define the training needs of individuals, teams and larger workgroups, but must also show how the provision of these skills will contribute towards meeting the organisation's objectives. Finally, an effective policy should reflect individual, group and organisational training priorities, linking them together in a complementary manner.

Step 2: identifying staff training needs

As was noted in Chapter 5, one of the first steps in managing a training programme is to identify the *skills gaps* in an organisation. These may be viewed as discrepancies between the skills needed for future business success and those which currently exist within the organisation. The HR manager undertaking a training needs analysis must explore the requirements of both the job roles required by the organisation and the needs of the individuals employed by the organisation. The matching of individuals' skills with jobs is a key HR management responsibility. As one senior business manager has said of his firm's philosophy, 'There is no such thing as a bad employee. There are only people who are not properly trained or who are in the wrong job and that is a management responsibility' (cited in O'Donoghue 2001: 257).

Assessing the skills required by a job role can be achieved through a job analysis – a formal assessment of required competencies for particular functions both now and in the future. The analysis of the skills needs of individual employees should be based on achieving planned organisational objectives, and should include interviews with staff to identify their aspirations and needs, inspections of company training records, and an examination of performance appraisals, which are then used to develop a profile of an employee's skills and abilities which can be systematically compared to the requirements of their role. Together, these analyses will identify the needs and objectives of the programme in the short, medium and long term, and will highlight the types of training that individual employees will need.

Step 3: creating a motivational climate

The next step is to create a motivational climate in which people want to learn and develop in order to better themselves and their contribution to the business. It is recognised that effective learning can only occur when individuals have both the ability and the desire to acquire new knowledge. The importance of this was highlighted in a recent study by Tracey *et al.* (2001), who found that supervisor, job and organisational supportiveness were significantly related to employees' level of pre-training motivation. Interestingly, employees' job involvement was an antecedent of their pre-training motivation, suggesting that the process by which employees become prepared for training is complex. However, it is known that the creation of a motivational learning climate demands the wholehearted support and commitment of top management for the training function. It also depends upon making people aware of the benefits of training, for themselves and for the organi·ation as a whole. For example, personal benefits may include better career prospects within and outside an organisation. Finally, a motivational environment requires that an organisation create the necessary time for individuals to undergo training, without it unduly affecting their job performance. This has to be done with the agreement of line managers, who will ultimately have authority and responsibility to release individuals from sites where progress must be managed, often within very tight programmes and limited resources. Some production techniques that have recently gained popularity in the construction industry, such as lean construction, threaten this aspect of creating a motivational learning climate because construction projects have little spare capacity in human resources to release them for training. Furthermore, these techniques increase the inevitability of unexpected problems and crises, which further prevent the release of staff for training (Loosemore 2000). Organisations choosing to implement these techniques should carefully determine their impact on human resource management and take steps to avoid unintended problems. Unless this occurs, project managers and site supervisors are unlikely to buy into the concept of the learning organisation and are unlikely to embrace training as a route to performance improvement.

Step 4: preparing a training programme

Having identified those skills required to meet the organisation's needs in both the present and the future and having motivated people to train, the next task is to define appropriate timescales for the development of these skills. This involves creating a *training schedule*, which includes individual *training programmes* for delivery, and making a realistic assessment as to when the benefits of the training are likely to be observable. It is important to note that every individual's training programme needs to be individually designed and developed according to the knowledge, skills and abilities required by that person. The HR manager should also evaluate the time necessary to impart the required skills and abilities and reconcile these with the urgency of requirement. This timeframe will depend on the nature of the skills to be learnt and the level of attainment required. For example, basic first aid skills, such as those required in many occupational environments, can be learnt in a 24-hour course. However, the development of management skills such as interpersonal intelligence or people skills cannot be achieved in a day, but requires considerable time, practice and reinforcement (Silberman 2001). Once established, the training schedule should be incorporated into a *training plan*, which will indicate a timescale for the acquisition of new skills and knowledge. This will include the strategic combination of different delivery methods in a way that ensures the effective and efficient understanding and take-up of required learning outcomes (Armstrong 1996: 543).

Step 5: deciding on methods for delivery

Having created a motivational environment for training and a programme for its implementation, the next step is to decide on an approach to training. This should be developed with the training programme in mind, in order to relate particular training delivery mechanisms to available timescales. Decisions about training delivery methods must take into account the important fact that the subjects of work-related training are almost invariably adults. Training delivery methods must therefore be informed by theories of adult learning, as distinct from those of learning in general. Cheetham and Chivers (2001) identify three broad approaches to adult learning, all of which view the learner as an active participant in the learning process and of broadly equal status with the teacher/instructor. These approaches are discussed briefly below.

Andragogy

Cheetham and Chivers (2001) summarise the principles underpinning the andragogy theory of adult learning as follows:

- mature adults are self-directed and autonomous in their approach to learning;
- they learn best through experiential methods;
- they are aware of their individual learning needs;
- they have a need to apply newly acquired knowledge or skills to their circumstances;
- learning should be seen as a partnership between teachers and learners;
- learners' experiences should be used as resources in the learning process.

These are widely accepted principles but their universal applicability is questionable. For example, self-directed behaviour and an ability to recognise shortcomings or knowledge gaps may not be characteristics possessed by everyone by virtue of their having reached adulthood. For example, technical training is likely to be best performed by an experienced practitioner who can demonstrate and explain methods of working, which can then be practised by participants before their competence in performing the tasks is tested at the end of the training course (Williams

2001). On the other hand, management training may best be achieved with minimal formal instruction, using team problem-solving activities and group-based assessments and reviews throughout the training course. Furthermore, the willingness to engage in self-directed learning as compared to instruction may differ according to training recipients' cultural background. Despite these criticisms, the theory of andragogy is valuable in highlighting that adult learners should be treated differently to non-adult learners. In particular, the relationship between teacher and learner is based on equality and the learners' practical experiences are drawn upon as a learning resource.

Experiential learning

The notion of experiential learning is based on the view that an individual's ideas are continuously being formed and changed by their life experiences. Common propositions of experiential learning were summarised by Cheetham and Chivers (2001) as follows:

- learning is best conceived of as a process rather than in terms of outcomes;
- learning is a continuous process grounded in experience;
- the learning process requires that conflicts between different views of the world be resolved;
- learning involves adapting to the world;
- learning involves an interaction between the individual and his or her environment;
- learning is the process of knowledge creation.

Kolb (1984) developed a Learning Cycle model to represent the experiential learning process. This model is depicted in Figure 10.1.

A variation on the experiential learning theme is *action learning*, a technique commonly used in work-based training programmes which is based around learning 'by doing'. This involves a group tackling a real-life work-based problem in stages. Between stages, group members are provided with an opportunity to analyse and reflect upon what has been learnt. This occurs with minimal external intervention, relying on a group's ability to learn 'from within'. The use of multidisciplinary action-learning groups may yield important benefits in the construction industry because it encourages practitioners to appraise their performance critically and enables the cross-fertilisation of ideas and/or transfer of skills between group members.

Symbolic interactionism

The last category of adult learning theories presented by Cheetham and Chivers (2001) is symbolic interactionism, which they base on a loose

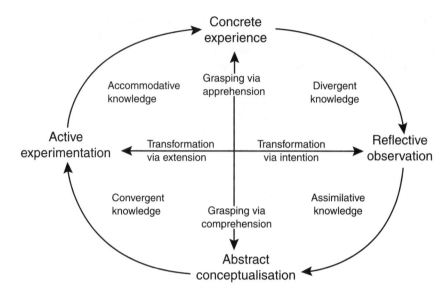

Figure 10.1   Kolb's Learning Cycle

Source: 'Experiential learning: Experience as the source of learning and development' by Kolb, D. (1984) © Reprinted by permission of Pearson Education, Inc, upper Saddle River, NJ, Martin

group of theories emphasising the importance of self-awareness and self-image in learning. These theories hold that adults are motivated not by objectives set by others but by their own desire to conform to an idealised self-image. This self-image is not fixed, but changes as new roles are taken on by the learner. These theories also hold that learning is more effective if it is self-directed and if adults view themselves as learners and have a high level of self-esteem. Interestingly, research by Tracey *et al.* (2001) supports the link between self-esteem and effective learning and has implications for supporting or enhancing employees' self-esteem in the training process.

Another important consideration is the location of training. In essence, the choice is between *on-the-job* or *off-the-job* training, or to adopt a *dual approach*. Each is discussed below in more detail.

On-the-job training

This form of training is the most widespread and can range from simple work-shadowing to fairly sophisticated experiential learning strategies. Work-shadowing simply involves an experienced worker monitoring and observing a learner as they assimilate the skills and knowledge necessary to carry out the new work activity, providing instruction and assistance when necessary. This approach is more effective if the person providing the training has undertaken training themselves in how to provide on-the-job instruction. An example of where this approach is used effectively in construction is the craft apprenticeship scheme, where employees are subject to guidance and assistance from a qualified and experienced colleague. However, definitions of apprenticeship reveal that they are actually a hybrid consisting of both on-the-job and off-the-job training. L. O'Connor and Harvey (2001) define an apprenticeship in the same way as the OECD, as a formal arrangement for the initial development of skills and including the following elements:

- training must provide systematic and long-term skill development to enable entry into a recognised occupation;
- training must be transferable;
- training is centred in an enterprise but has a component of instruction in an institution;
- the apprenticeship involves a contract of indenture between the trainee or his or her representative and a private or public employer, trade union or joint training committee, a public or quasi-public training organisation or recognised training body.

Thus, an apprenticeship combines practical and theoretical learning and comprises periods of off-the-job training and assessment. Traditionally, apprenticeships were based on the length of time an apprentice spent undergoing training. However, more recently apprenticeships have come to be based on the apprentice's demonstration of attainment of specified job competencies. However, research presented by L. O'Connor and Harvey (2001) suggests that there are difficulties in assessing competencies in on-the-job training. For example, site work may not suit the assessment of certain competencies and employers are reluctant to stage situations for assessment purposes. In the absence of suitable assessment circumstances, employers reported 'ticking the box' even if an apprentice had been unable to demonstrate competence in a specific area. This highlights the problems associated with on-the-job assessment and the need for employers to provide access to a range of situations for assessment purposes.

A similar approach can also be used for graduate entrants to an organisation, which involves assigning them to a senior and experienced colleague as a *mentor*. This individual need not necessarily be the line manager of the employee, but effectively takes responsibility for their development by ensuring that they receive appropriate on-the-job training and experience. In addition to the provision of formal training, the role of the mentor will also be to counsel and advise the employee on career-development issues and to ensure that s/he remains protected from being asked to undertake tasks for which s/he is ill equipped or from which s/he will derive little in the way of new skills, knowledge or experience. Research undertaken by Cheetham and Chivers (2001) reveals that mentoring is not very widely used in the construction industry and that the effectiveness of *formal* mentoring has often been limited. Employees who form informal relationships with mentors report greater satisfaction with the outcomes, and in all cases the effectiveness of mentoring schemes was related to compatibility. Unfortunately, informal relationships are difficult to manage in ensuring that all employees obtain the learning experience they need and are entitled to. However, it is possible to devise a two-stage mentoring strategy which incorporates formal and informal elements by allowing new employees a period of assimilation where they can build friendships before designating formal mentors.

On-the-job learning can also be linked to higher education, where students spend time in relevant work during the course of full-time or part-time studies. In full-time studies, on-the-job training may simply involve a brief encounter with industry on a short project or comprise a practicum period or sandwich placement of typically six to 15 months. Graduates may also be required to undertake on-the-job training to qualify fully in their professions, and to undertake formal continuing professional-development exercises to maintain their professional status thereafter. Employers must be aware of these professional requirements when they employ a member of a professional institution and provide the necessary training opportunities. The formality and extent of this on-the-job training will vary according to the requirements of the higher education institution, the employing organisation and the relevant professional bodies. However, if this type of work-based learning is to be used it is important that any tensions between the needs of an employer and those of a learner are resolved. For example, as government grants for full-time education become less common and generous it is becoming increasingly common for full-time students to have part-time jobs to finance their studies and for them to experience pressure from their employers to work longer and longer hours. This effectively destroys their educational experience by reducing the amount of time for rigorous study. For many full-time students further education is becoming a part-time experience, and this is becoming a serious problem for universities to deal with, particularly in larger cities, where the cost of living is relatively high. Similarly, full-time employees are often under so much work pressure that it is difficult to take time off for study. Once again, this devalues the educational experience, and it is important that employers allow employees necessary time for meaningful study.

Industrial-placement students and graduates entering construction organisations are often exposed to a job-rotation scheme. This involves them experiencing various departments and/or roles within the organisation for a short time in order to get a feel for where their skills could best be utilised to contribute to the business. In addition, it allows them to gain an appreciation of the role and function of each aspect of the business and how they interact. This can assist more effective integration and help in future teamworking activities. A development of this short-term job-placement approach is for employees to be deliberately moved into new roles after a specified period. This prevents employees from becoming bored with their roles and ensures that they remain functionally flexible (see Chapter 4). This in turn allows the construction organisation to cope with the changing demands of its business environment. An example would be allowing construction managers to work as assistant quantity surveyors in order to improve their understanding of the commercial aspects of works packages. As well as providing them with an insight into the work of their colleagues, this would allow them to develop multiple skills that could facilitate them managing both the production and commercial aspects of a package on a future project.

Off-the-job training

Off-the-job training takes place away from the workplace where the employee would normally be engaged. This has obvious advantages, particularly in relation to the construction environment, where the pressures of site-based activity can lead to training and development activities being considered of secondary importance in comparison to project requirements. On-site training may also be problematic due to high levels of noise and/or a lack of facilities. Formal training can take the form of formal exercises such as role-play, formal lectures or seminars, interactive workshops, films and videos or even computer-aided learning packages. A decision to use out-of-company training may be made if the available skills are not available in-house. Whilst outsourcing training will mean losing some degree of control, the different learning environment can enhance training outcomes in some cases. Indeed, there seems to be a dramatically increasing trend for managers to take this route by embarking on postgraduate study such as MSc and MBA programmes. For example, in Australia the number of enrolments in postgraduate MBA programmes has more than tripled over the last five years, from 2,600 to over 10,000.

It should be noted that training is often seen by employers as something which is the employee's responsibility and which does not need managing. Of course, nothing can be done about controlling people's voluntary training activities, but particular care must be taken when deciding on whom to sponsor for training and on which type of course is most appropriate to the job role of each individual and to the future needs of the business. Training can also play a strong motivational role in an organisation, and consultative approaches to training initiatives will go a long way to harnessing this potential effect.

Dual approaches to training delivery

In reality, a combination of the on-the-job (practice) and off-the-job (theory) approaches offers a potentially more varied and effective route to training delivery at both managerial and non-managerial levels. Bringing together the formal training and 'learning by doing' approaches also helps to reinforce and instil good practice within the recipient, and emphasises the relevance and importance of the material being learnt. This approach is also known as an *integrated managerial* approach when considered as a management-development activity. At a managerial level, the practice versus theory distinction can be used to define the difference between management education and management development. Management education is seen as more theoretical, emphasising a body of knowledge, while management development is concerned with the practical development of skills.

Developments in educational technology, particularly web-based technologies, are changing the way in which learning material is imparted and it is worth making special mention of the potential to use this technology in the delivery of workforce training. Computer technology offers the opportunity to increase the accessibility of learning and lower the costs of delivering effective training. This technology may be particularly helpful in an industry like construction, where projects are geographically dispersed and participants may find it difficult to attend training courses. K. Young (2001) suggests that web-based learning can be used to provide employees with:

- interactive self-paced multimedia instruction;
- assessment of knowledge and skills;
- performance-support materials, including databases, electronic indices, on-line journals, manuals, standards or guidance notes;
- on-line communication with instructors, expert advisors or colleagues.

However, to be effective, training programmes utilising this technology must be designed to meet the needs of both employers and the different user groups, whether they be local users accessing materials from computers connected to a network with high-speed connections, or remote learners whose access to the network may be limited and fluctuate in reliability. Standard systems are available for organisations to achieve this, allowing the monitoring of study time, on-line discussions, workshops and tutorials, assessments, etc. Ultimately, the usefulness of an e-learning system must be assessed on the basis of the extent to which it helps the learning process and the extent to which trainers or HR managers believe that learning objectives are being met. With this in mind, Sambrook (2001) developed a framework for assessing learners' perceptions of the quality of e-learning materials. She reports that the most influential factors are:

- *user-friendliness*: ease of use and instructions;
- *presentation*: clear and accurate with no mistakes;

- *graphics*: use of pictures and diagrams to aid understanding;
- *engagement*: interesting or boring materials;
- *information*: amount, for example too little or too much;
- *knowledge*: what knowledge was gained;
- *understanding*: ease of understanding the materials;
- *level*: too basic or too deep;
- *type of learning*: rote learning, memory testing or discussion;
- *language*: easy or difficult to understand, for example use of jargon, presentation of definitions, etc.;
- *text*: the amount and the balance with graphics.

It is recommended that in evaluating e-learning opportunities, training materials should be assessed against these criteria. Also, although e-learning can be used as a stand-alone delivery medium, it may be best to use internet technology to deliver training materials as part of a blended training process that also involves traditional training methods. For example, web-based learning could provide basic information to ensure consistency of knowledge among employees enrolled on a training course to be delivered using traditional methods, so that they can learn more effectively in the classroom.

Step 6: evaluating training effectiveness

There is little point in initiating expensive training programmes without evaluating their effectiveness. If an employee and organisation do not benefit from the training, then the investment will have been wasted. This evaluation should cover two interrelated outcomes of the training programme:

- How effective was the training approach in delivering the desired learning outcomes for the individual (skills, knowledge, attitude, etc.)?
- How has the training impacted on the performance of the organisation in terms of achieving its strategic objectives?

Establishing the effectiveness of the training under both of these measures and taking action to ensure its future improvement is complex, since opinions must be collected from many people involved in the training process. For example, line managers, personnel managers, training providers and the trainees themselves will have individual perspectives on the effectiveness of the training approach adopted. Indeed, when a dual approach to training delivery has been used, further difficulties might arise in ascertaining whether it was the on-the-job or off-the-job elements of the training that led to any observed effects. For this reason, the evaluation of training should involve a wide variety of different feedback mechanisms such as post-course questionnaires, tests or exams, structured exercises, tutor reports, interviews with trainees following the study, observation of training programmes and pre- and post-training appraisal trainees (Holden 1997: 394). These may have to be conducted over a lengthy period if the longer-term benefits are to be established.

An important element in the effectiveness of a training programme is the extent to which learning is transferred from training interventions into behaviour change in the workplace. This transfer has a key bearing on the ability of training to impact upon organisational performance (Holton 1996). Figure 10.2 depicts a model of the process by which *training* provides employees with new *knowledge, skills and abilities*, which are then translated into *individual behaviour change* and ultimately lead to enhanced *organisational effectiveness*. The model shows the central role played by employees' motivation in both the learning and the transfer stages of this process. Thus, employees' motivation to learn will intervene to moderate the extent to which employees acquire the knowledge, skills and abilities that the training aims to provide. Assuming that learning takes place, the extent to which this is transferred into positive behaviour change in the workplace will also depend upon whether employees are motivated to do this. As the model suggests, employees' acquisition of knowledge, skills and abilities is not sufficient to elicit behaviour change. Employees must also be motivated to apply this knowledge, these skills and abilities in the workplace and change their behaviour. Through this behaviour change, organisational effectiveness is enhanced. Motivation to learn may be influenced by the training course content or delivery. For example, perceptions of the trainer's ability or expertise may influence employees' motivation to learn. The training content and delivery can also impact upon employees' motivation to transfer learning into practice. For example, Holton (1996) suggests that intellectual learning may occur but trainees are not provided with the opportunity to practise the training in the work context or are not taught how to transfer this learning. One solution to this would be to ensure that the training programme closely reflects the work environment (Yamnill and McLean 2001). The model also suggests that organisational climate is an important determinant of employees' motivation to learn and apply new knowledge, skills and abilities. A positive learning culture, in which new knowledge and skills are valued and the ongoing development of employees is supported, is essential if employees are to be motivated to learn. Similarly, on completion of a training course, trainees return to the workplace and respond to cues in the environment. Cues that remind trainees of their training can facilitate transfer of that learning. Also, when the learning is put into practice it should be reinforced with praise or positive feedback to ensure that the desired behaviour is maintained. It is important to note that cues and feedback can emanate from supervisors or managers, and conflicting cues could prevent the successful transfer of training outcomes. The model is important because it illustrates the complexity of the training process and

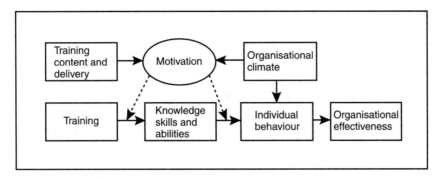

Figure 10.2 Training transfer model
Source: Adapted from Holton (1996).

suggests that the provision of training must be carefully evaluated. Unless training occurs in a supportive environment and is designed to engender in employees a motivation both to learn and to apply this learning, training will not be effective. Donovan *et al.* (2001) share this view, suggesting that a failure to address motivational issues in employee training programmes is one reason why some such programmes do not succeed.

Current issues influencing construction training provision

The demands on construction training and the associated mechanisms for its delivery are changing in response to increasing demands from clients, governments and those working within the construction industry. Some of the issues which might impact on training provision within the industry over the next few years are outlined below.

Equality in education

A major issue for the future of training in the construction industry is its continued reliance on the 'flexible firm' model of working (see Chapter 4), which is reflected in an increasing trend towards outsourcing. This is resulting in a workforce of core and peripheral workers, and in a tendency for organisations to focus on developing the central core of employees who they see as driving the organisation forward. Peripheral groups which provide the supporting functions are often afforded only minimal training opportunities, thereby increasing the knowledge gap between these groups of workers (Holden 1997: 381). It is a trend which was predicted two decades ago by Handy (1984/94), and which is resulting in ever-greater disparities between the educational opportunities afforded to different types of employees depending on whether they are classified as core or peripheral. Outsourcing is being driven by economic rationalism, and the long-term effect of outsourcing will be to reduce the sense of collective responsibility which is needed for an effective training regime and to fragment the construction industry further, into a collection of small organisations which are less able and willing to invest in long-term training initiatives.

It is also a disappointing fact that the lower down in the organisational hierarchy an individual employee is, the less training they are likely to be provided with (Holden 1997: 381). Whereas managers and professionals are proportionately likely to undertake a great deal of training, clerical and manual workers are far less likely to benefit from any provision beyond that which is required by statute. This raises an important question as to whether construction organisations are missing out on potentially talented people who could use their skills and abilities in a higher position within the organisation. For example, should a foreman or woman not be offered the opportunity to attend higher-level management courses which could potentially provide him or her with the skills needed for a construction management position in the future?

Globalisation

The need for employees to be able to work in overseas markets is becoming an increasing priority for many construction organisations. Contractors and consultants have begun to exploit lucrative overseas markets, often via joint-venture agreements between themselves and indigenous firms. Training employees to equip them to work in such environments now presents one of the key challenges for large construction companies. The skills needed extend well beyond language abilities, to include cultural assimilation in a business and general sense, and an understanding of a country's legal and contractual framework. Whilst some of this can be learnt on the job, there are established techniques to help managers and professionals to integrate themselves more easily into these environments.

The process of globalisation also raises equity issues, which were discussed in detail in Chapter 8. Essentially, they revolve around evidence which indicates that the construction industry in most developed countries is becoming increasingly culturally diverse and that non-indigenous groups are being afforded fewer opportunities for personal development than indigenous groups. It is a serious issue, which needs addressing if the industry is to ensure a harmonious workforce in the future and maximise the productive potential which cultural diversity affords.

Health and safety

An increasing awareness of the poor performance of the construction industry with regard to health, safety and welfare has also raised the importance of health and safety training in the construction sector. This is true of all developed countries. Notably, there is an increasing emphasis within the industry on people taking responsibility for their own safety and welfare. This demands that all employees have an awareness of the risks to health and safety, some insight into accident causality, and knowledge of the measures that they must take to avoid injury, ill health or death. Although basic site-induction programmes can help to explain the procedures, hazards and responsibilities, only thorough health and safety training can provide employees with the in-depth knowledge needed for safer working.

Externally audited training and development standards

The question of what level of investment in people is required in order to maximise the contribution of individuals towards the strategic objectives of the organisation has always proved a difficult question to answer. To facilitate this decision, within many countries voluntary standards have been developed against which companies' investments in training and development can be measured. Should firms achieve such a standard they can use accreditation under this standard as a lever to attract better employees in the future. In the UK an auditable standard for investing in people has been developed to provide a framework to improve the UK's competitiveness. It is known as the Investors in People standard (IiP), and organisations are assessed according to their commitment towards the training and development of their employees. Their performance is regularly reviewed against the IiP standard every few years to ensure that the organisation is making a sustained investment in its employees. Although the IiP standard is not solely about training and development, these activities inevitably form the cornerstones of an effective commitment to improving employee performance in line with organisational objectives. In addition, the standard also requires that the outcomes of training and development are evaluated in terms of the benefits to both the individual and the organisation. As the labour market becomes increasingly competitive, the need to demonstrate a commitment to people through an externally audited standard is bound to become increasingly important for construction companies. This is borne out by the number of construction companies within the UK that have now achieved or have committed themselves to achieving IiP status.

Responsibilities for delivering training in construction

The debate as to who should be responsible for training provision has been running for many years within the construction sector, particularly with regard to the provision of craft and technical training. In the UK construction has retained a central organisation which takes responsibility for overseeing training in the industry, known as the Construction Industry Training Board (CITB). The majority of the CITB's funding comes through a levy contribution from all construction firms with a turnover above a certain level. This money is paid back to firms which provide training deemed to be at an appropriate standard, as well as being used to subsidise new-entrant training for those without employed status. The advantage of having this national training board is that training levels can be maintained even in times of economic recession, when construction companies are less likely to train (Druker and White 1996a). This affords the industry some protection against skills shortfalls, although it does not necessarily ensure an adequate provision of new craft entrants to the industry to cope with demand. Despite having the CITB, serious skills shortages have persisted in recent economic booms, and the role and effectiveness of the CITB have come under close scrutiny.

The equivalent body in Australia is Construction Training Australia (CTA). CTA is the National Industry Training Advisory Body (ITAB) and is charged with developing the industry's competence standards and working with industry to package the standards into relevant industry-endorsed qualifications. CTA is also charged with the development of national strategic advice relating to issues affecting the implementation of structured training and with pursuing outcomes on behalf of the construction industry. It is an industry-owned body made up of employer and union representatives, and receives much of its funding from the federal government through the Australian National Training Authority, but it also receives important funding and internal support from industry member organisations and other industry sources. Over recent years CTA has operated as the industry's Competency Standards Body. Its job has been to develop, in consultation with employers and unions, the standards required for the various jobs in the industry. It has also developed a system of competency-based training (CBT) and assessment for the industry. The Australian building and construction industry is currently seeing some major changes – particularly in the areas of training and skills development. CTA is now engaged in the development of National Training Packages for the building and construction industry. These will facilitate the introduction of new apprenticeships and traineeships in the industry, and, in addition, will be available to assist the skilling of current employees. The employers and unions represented on CTA have worked together to bring these changes about. They should make the industry more efficient, more productive and more competitive. They will also improve the skills, training, work prospects and career opportunities of workers in the industry.

In addition, CTA undertakes strategic planning for the skills development of the industry. The last skills audit of the building and construction industry was carried out by CTA in 1990. However, since then there have been significant changes in the way industry operates, which in turn have influenced the structure of the industry. The changes in recent years are due, in part, to outsourcing of labour by large and medium-sized enterprises in both the public and private sectors. This has resulted in an increased reliance on contracting, subcontracting and specialisation. According to the 1996/7 Private Sector Construction Industry Survey carried out by the ABS, the construction industry experiences a higher degree of contracting and subcontracting arrangements than most other industries. It is likely that another comprehensive skills audit of the industry will be carried out in the next few years. Information gathered from skills audits helps the industry to make more informed judgements with regard to developing strategies to meet the skills requirements of the industry.

Qualification frameworks in the construction industry

It is important to note that, despite the historical division between craft and professional training within the UK industry, in recent years the establishment of the National Vocational Qualification (NVQ) structure has led to a uniform set of standards (Druker and White 1996a). These are competency-based standards, which come in levels from 1 to 5. Level 3 implies a level of competence compatible with that of a qualified tradesperson, whereby the holder would be expected to work relatively autonomously. This has recently been taken a stage further through the initiation of the Construction Skills Certification Scheme (CSCS), where site-based workers of all types will be required to carry a card which indicates their level of training and competence as measured against the NVQ competency standards. This approach provides a uniform benchmark level of performance for both craft and professional roles and could lead to a convergence in the provision of training in the future. This helps employers to select people with the necessary skills and competence, as well as setting benchmarks for future achievement. In addition, it ensures a minimum level of health and safety awareness from all cardholders, which could lead to fewer accidents in the future.

In Australia there is a similar national framework of competency-based qualifications called the Australian Qualifications Framework (AQF). The AQF was introduced Australia-wide on 1 January 1995 and was phased in over five years, with full implementation by the year 2000. It is a unified system of twelve national qualifications in schools, vocational education and training (technical and further education [TAFE] colleges and private providers) and the higher-education sector (mainly universities). The qualifications range from secondary/high-school level to post-school certificates, diplomas, degrees, masters and doctoral degrees. The AQF is designed to link all these qualifications together to form a highly visible, quality-assured national system of educational recognition which promotes lifelong learning and a continuous and diverse education and training system.

AQF qualifications certify the knowledge and skills a person has achieved through study, training, work and life experience. The ultimate aim of the AQF system is to make work-based qualifications and academic qualifications part of a single system, allowing maximum flexibility in career planning and continuous learning. Essentially, vocational qualifications are now *industry based*, with specified combinations of *units of competency* required by each industry for each qualification. These qualifications are designed in a *sequence*, allowing people to move steadily from one qualification to the next. Sometimes people will want to mix and match units of competency, and to facilitate this, units will accumulate on a person's record of achievement and help towards retaining their job, promotion, a change of career or further learning. To be assessed as competent for one of the vocational qualifications, people have to show that they can use their skills and knowledge under workplace conditions. This means that most of the competency training for AQFs will be in the workplace. Also, people can be assessed for the skills and knowledge they may already have gained informally in previous work. This assessment process is known as *recognition of prior learning* (RPL). Registered Training Organisations (RTOs) are accredited to provide training and issue qualifications according to the requirements of the AQF.

The CTA is responsible for developing the construction industry's competency standards and ensuring that the workforce meets them. CTA is also working on the development of a Skills Card to recognise the skills of the members of the building and construction industry workforce, and this continues to be an objective of CTA. From 1 July 2002 CTA will progressively introduce a system to issue nationally recognised Skills Cards to acknowledge skills achieved by the members of the building and construction industry workforce. The cards will be issued based on the outcomes of assessments conducted by RTOs. However, experience in the UK industry illustrates that industry and industry regulators have quickly recognised the attached benefit of such cards. The major constructors in the UK have decreed that by December 2003 all persons employed on their sites must have the relevant Skills Cards. The UK Contractors' Confederation requires compliance with the Skills Card system by December 2005.

Case study

Multiplex Construction's international benchmarking in training standards

The main objective of this project was to assess the applicability of competency standards developed by the CTA in Australia to determine if they could be used as benchmarks on Multiplex projects undertaken in the UK. If Australian competency standards were equally applicable outside Australia, then this would be a tremendous help to Multiplex in managing its offshore projects and in understanding how its culturally determined standards and expectations of competencies compare to those of other countries. It would also be valuable information in understanding the relative competencies of an increasingly transient workforce, where ever larger numbers of people are moving between the UK and Australian construction industries.

In November 2001 a delegation of CTA representatives met with representatives of Multiplex Construction Ltd (both from Australia and the UK) and of the CITB in the UK. It was agreed to implement this project using the new Wembley Stadium as a possible project for evaluating the practical applicability of the competency standards developed by the CTA in an international context. Forty competency standards were selected for this exercise and the conclusions were as follows.

First, the structures used to present industry standards in Australia and in the UK are different. In general, Australian competency standards have a broad focus with a broad outcome, whereas UK standards have a much narrower focus. Furthermore, the Australian standards are far more detailed and comprehensive than UK standards and can be used as stand-alone documents for training and assessment without any additional information. In general, the UK standards are much lighter on detail and provide basic essential information for trainers and assessors. In this sense, the UK standards are less standardised and more flexible than those in Australia. UK standards are meant to be used as a guide to help trainers develop their training programmes rather than to dictate exactly what the programme should contain.

A comparison of the assessment guides given to trainers indicated that the performance criteria used to rank people's competencies in Australia are comparable to the UK's. However, one important difference was the emphasis on OHS, which is integrated into every Australian standard and emphasised much more than in the UK, which has a stand-alone standard.

When the practical applicability of each set of standards was tested, it was found that the Australian standards are as applicable and assessable in the UK as they are in Australia. Some notable exceptions occur in relation to OHS regulations, where significant differences in practices make comparative assessment difficult. For example, in Australia any cables carrying electricity are required to be fixed overhead, whereas cables in the UK are commonly found on the ground.

In general, Australian competency standards developed by the CTA match those developed by the CITB in the UK. This means that Australian competency standards can be used outside Australia. However, although these competency standards are very similar, this project also showed that there could be difficulties in providing mutual recognition for qualifications based on competency standards. This relates to the different range of assessment variables used in each set of standards, and to the different requirements as indicated for the standards to be deemed competent. Another obstacle to mutual recognition is the different sources of evidence for proving competence. In the UK the main source of evidence for competence must come from the workplace, whereas Australia also gives recognition to simulated assessments from RTOs. A final problematical difference lies in the qualifications of assessors. In the UK each assessor must be not only competent in assessment but also technically competent in the area of assessment. In contrast, Australian assessors do not have to possess these two types of competence, although this is likely to change in the future.

Management development activities in construction

Most large construction companies have undergone a period of significant organisational change over the last 20 years and have effectively become managers of the construction process. Egan (1998) recognised this fundamental change when he identified the need for a commitment to people as one of the key drivers required to promote change within the industry. He called for a wider commitment to the training and development of managers and supervisors as part of his *Rethinking Construction* report (1998: 17). This reflects the return of training and development to the top of the strategic agenda within the UK, and an emphasis on employees' life-long learning and on national workforce development (Rainbird 1998). This is urgently needed since evidence indicates that the poor image of the construction industry is resulting in an increasingly short supply of construction professionals (Cavill 1999) and that salary levels are rising accordingly (Knutt 1997a).

Management development plays a vital role in ensuring that the managers and professionals responsible for the future performance of an organisation are provided with the skills, knowledge and opportunities to meet the organisation's strategic goals. This should be seen as a process of continual improvement; hence, the now ubiquitous term *continuing professional development* (CPD) is often used in construction to mean the development of managers in their professional function.

In Chapter 5 we identified management development as the organisation's primary tool for ensuring that is has the necessary employees to meet its present and future management needs. According to Armstrong, its primary aims are threefold:

- to ensure that managers understand what is expected of them and how their performance will be measured;
- to identify managers with potential and to encourage them to prepare and implement personal plans that allow them to take on more demanding managerial responsibilities;
- to provide for management succession within the organisation.

It is clear that to achieve these three objectives management development cannot comprise a single technique or programme, but should form a series of interrelated activities which together equip managers with the necessary skills to take the organisation forward (Druker and White 1996a). Langford *et al.* identify six key benefits of management development in a construction context, which can be summarised as follows:

- it helps managers to learn their jobs quickly, thereby minimising learning costs and lessening the chances of expensive mistakes;
- the present and future work requirements are likely to be met (especially because the organisation must be clear about its future work requirements);
- the organisation can grow on the back of management development and it is an essential ingredient of sustaining future growth;
- the morale of managers is improved and this reduces wastage linked to inadequate development opportunities;
- managers learn their jobs more rapidly, which leads to greater job satisfaction and a likelihood of long service in the organisation;
- it helps the organisation to cope with change in a proactive way, as newly required skills and knowledge can quickly replace those that have become obsolescent.

(Langford *et al.* 1995: 134)

These benefits are particularly salient to construction firms, which are becoming more management focused and have suffered in the past from high levels of staff turnover, a rapidly changing business environment and very tight programme, quality and cost pressures. The real advantage lies in the way in which a properly developed management-development strategy can prepare the organisation for change whilst simultaneously maximising the performance of its managers. However, this relies on the careful planning and implementation of a management-development system that responds to the needs of an organisation.

Developing a management development strategy

Achieving the objectives set out above requires an organisation to formulate a management-development strategy. Langford *et al.* (1995) indicate that such strategies vary considerably across the construction industry. Nevertheless, a common characteristic of all effective strategies is they are business led, in that the organisation has made a clear decision about what kind of managers it wants to develop and has put in place the mechanisms to achieve this goal. These management-development strategies are well resourced and put forward clear and practical solutions for providing the management resources necessary for future growth in terms of achieving core elements of the business plan. Flexibility is also important, because the goals and orientation of management development must be responsive to changes in market demands and opportunities. Finally, effective strategies tend to have overt support from senior executives, who attribute a significant level of importance to management-development activities and reward the achievement of development goals which support business priorities. A failure to define an explicit link between management development and organisational performance is one of the most common reasons why such initiatives fail. Without this, it is difficult to convince managers of the benefits of the inevitable commitment of time and effort that management development requires.

Responsibilities for management development

Once an effective strategy has been developed, the next stage is to decide who within an organisation should take on responsibility for management-development activities. There are a number of ways that managers can be developed within modern organisations, which have to be combined to ensure the effectiveness of the approach. These include self-development (acknowledging that the responsibility for development rests with the individual), organisation-derived development (the formal aspects managed by personnel specialists) and supervisor-derived development (actions undertaken by a senior manager) (Mumford 1993). The requirements of each approach are discussed in more detail below.

Self-development

Although an organisation can facilitate and encourage the individual manager to work towards his/her own development, a large part of the responsibility for development ultimately lies with the individual manager him/herself. It is therefore incumbent on all managers to develop their own development plan, which should contain goals which are *specific, measurable, appropriate, relevant* and *timely* (known as *SMART* goals). These goals should also be realistic in terms of resources available and personal work–life balance constraints. A personal development plan should detail the knowledge, skills and levels of competence that they intend to gain, a clear route to achieving these development targets and an approach for being able to demonstrate the achievement of development targets.

Organisation-derived development

The personnel specialist also has an important role to play in assisting employees to develop their personal development plans and in implementing general organisation management-development policies. Their role will be to advise on the types of management development that will support business activities, to provide guidance to managers on how they can develop within the organisation and to facilitate learning by setting up learning opportunities and courses, etc.

Supervisor-derived development

Supervisors must be committed to and involved in the management-development process of their subordinates if it is to lead to positive results. They should act as mentors, advise on appropriate development activities and generally provide encouragement and guidance to managers under their charge. The learning organisation concept discussed above emphasises the need for line management to take responsibility for the continuous learning of employees (Armstrong 1998).

Thus, management development must seen as involving the individual, line management and HRM specialists if it is to be fully effective.

Approaches to management development

There are several ways in which a management-development programme can be implemented, ranging from very informal to highly formal (Mumford 1993). Formal approaches rely on a structured process of coaching, learning programmes, training courses and development through work experience. Formalised development processes are planned processes, which are often kept distinct from normal managerial activities. There are clear objectives and structures which are aligned with organisational objectives and development strategies, but the processes is owned by the management developers rather than the managers themselves. Learning can be real or detached depending on whether it is carried out through a job or on a course.

In contrast, informal approaches use everyday experiences to develop new ways of dealing with problems and challenges. Informal managerial development is a series of natural development processes occurring within a manager's normal day-to-day activities. There are no clear development objectives and the delivery is unstructured. The result is that learning is direct, but unconscious and ultimately insufficient.

Another approach is to combine both formal and informal methods, which is known as *integrated* management development. In this approach the manager will be expected to undertake self-development activities through experiential learning, the organisation supporting this process by encouraging self-reflection, self-assessment and discussions of experiences with senior colleagues. The integrated managerial development process is essentially a series of opportunistic processes which occur within a manager's normal day-to-day work activities but which have clear developmental objectives linked to organisational objectives and strategies. The activities are structured and planned by the manager and his/her line manager and are subsequently reviewed as learning experiences. The result is that learning is conscious and more substantial.

Mumford's classification points to obvious advantages for integrated approaches, because the manager can learn on the job whilst obtaining support from their line manager and the organisation. Thus, instead of management development being viewed in isolation and as a rigid and mechanical activity, it becomes a flexible and responsive aspect of managerial work. Doyle equates this perspective to *open-systems* thinking in that it focuses attention and thinking on:

- management development as an adaptive system with identifiable parts and components acting together in an organised way to achieve improved organisational effectiveness;
- the environmental, social, technological and cultural variables that influence and change organisational processes;
- the mutual interdependence between management development and other organisational subsystems, activities and processes.

(Doyle 1997: 418)

Evaluating the effectiveness of management development

Just as training provision should be evaluated in terms of whether it has led to the desired performance improvements that it set out to achieve, management development should also be evaluated. As was mentioned earlier on in this chapter, evaluating the performance of management development is difficult, since not all elements of managerial performance are easily measurable (e.g. effects on staff attitudes and motivational levels) and it is often difficult to link measures of managerial performance at an individual level to overall organisational performance. Moreover, many of the benefits of management development will accrue over a long period of time as the skills and knowledge developed by managers are used to change and adapt an organisation to the changing circumstances which surround it. Thus, a range of evaluation mechanisms are usually used, including assessment by subordinates, self-reflection by the managers undertaking the development, observations by trainers and line managers, peer review by other managers and customer surveys, business performance analysis, etc. The approach adopted will depend on the nature and goals of the development activity undertaken.

Team perspectives on management development

As has been emphasised throughout this book, construction is a team-based activity. Thus, viewing management development as an isolated and individual activity would not engender the development of effective team-working, which is the hallmark of most successful construction projects. In a construction context, management-development programmes and policies must recognise the mutual interdependence of specialist managers and professionals operating across an organisation. The solution is to set objectives which relate to both team and individual performance, and to define a clear and unambiguous approach towards the development of teams where people are encouraged to understand their own role and contribution, but also the way in which this interacts with and supports the roles of others. Rewarding performance in terms of team performance outcomes as well as individual achievements will help to focus minds and ensure that managers communicate effectively for the achievement of core performance outputs.

Case study

Galliford-Try PLC: a supportive and proactive approach towards management development

Galliford-Try is one of the UK's leading contracting organisations. It operates across every major sector of the industry and across most of the country. The size and the diversity of its projects demands that the organisation recruits, develops and retains high-quality managers and professional staff who contribute to the development and continual improvement of the business. The company HRM strategy has been to invest in people and to promote good management-development practices throughout the organisation. The company brochure states:

> Our major resource is our staff and their talents, therefore we invest heavily in them to maximise their skills and experiences, and thereby achieve greater individual satisfaction as a team member to continuously deliver a better quality product to our customers. We are committed to the 'Investors in People' ideal.

In order to deliver this high level of commitment to its staff, Galliford-Try has focused its efforts on developing a culture where people are valued and have the best possible opportunities to develop their career in a way that accords with both their own needs and those of the organisation. The company has developed and implemented several mechanisms to engender this cultural environment:

- *Mentoring and support*: on a personal level, management development has been supported by coaching and mentoring schemes. This provides a point of contact for both newcomers and managers rising through the organisation, from which they can obtain informal careers advice, encouragement and support. This approach is also used to help instil the company values in all managers within the organisation.
- *Job-shadowing and induction*: whenever a manager is newly recruited or promoted, a higher-level or longer-serving member of staff will take the necessary actions to ensure that the manager is familiarised with company policies and practices. New senior managers are given the opportunity to 'job-shadow' an existing senior member of staff to ensure that they fully understand their new role, and also to provide the company with the opportunity to establish whether they are likely to comply with the desired work ethic and approach.
- *Developing potential and transferring knowledge*: clusters of managers and other personnel identified for succession management via the performance-appraisal system are invited to attend a management training course run in partnership with the Henley Management College. By bringing together clusters of people from different areas of the business, this encourages new practices and innovative approaches to be developed, which are then applied throughout the organisation. As part of the course, participants work collaboratively on projects linked to the strategic business plan, always based on real project and business scenarios. On completion of the course, participants then present the outcomes of their work to the main board of directors, who then disseminate the new and innovative practice around the group companies.
- *Encouraging innovation and shared good practice*: regular weekly meetings between senior managers and directors are used to discuss new ideas and practices emerging from individual employees and project teams in order to help transfer good practice throughout the organisation. These meetings also contribute

to efficient succession planning, as employees with particular abilities or potential can be identified and then groomed for senior management positions in the future.

The result of Galliford-Try's interrelated staff-development policies is that managers feel supported and empowered and are able to take advantage of the opportunities within the organisation. The succession-planning benefits that this provides in the longer term have meant that many of the organisation's key personnel are long-serving members of staff who have reached their positions through the promotion and development processes. They have learnt the approaches and practices of the company over many years and have adapted their own attitudes and behaviours accordingly. This cascades down the organisation to influence other operational staff. However, the organisation also benefits from encouraging newcomers to bring their fresh ideas and approaches, which helps to develop new ways of thinking throughout the organisation. Together, these management-development activities have ensured a culture of mutuality within a spirit of continuous improvement, both of which are essential to thriving in the competitive construction industry environment.

Conclusions

As we discussed in the introduction to this chapter, training should go hand-in-hand with wider HRD activities if it is to be effective in delivering better performance. It is important to note that training should be viewed as a means to an end and not an end in its own right. An organisation that invests in HRD will derive organisational performance benefits only to the extent that it facilitates learning and the development of teamwork and knowledge (Holton 1996).

Training and development are key aspects of the HRM function, but are often overlooked as a mechanistic activity within many organisations. This is dangerous, since it grossly oversimplifies one of an organisation's primary routes to competitive advantage. By defining the competency and development requirements of the individual employee in a team context and targeting the achievement of these competencies through training, construction organisations can ensure that their employees have the requisite skills to cope with a dynamic industrial environment. The ultimate goal for organisations should be to develop into learning organisations that have the capacity continually to develop in response to their changing business environment.

Discussion and review questions

1   Identify the key changes likely to affect the industry over the next five years in your country. Evaluate the probable managerial training requirements necessary to prepare a large construction company for these changes.

2   Devise an integrated management-development approach towards managing a particular construction operation with which you are familiar. Your solution should combine a framework for gaining both on-the-job and off-the-job training and an appropriate evaluation mechanism.

3   Identify specific behaviours or attitudes of employees and their managers that might hinder the transfer of learning to the workplace.

11 The HRM implications of management thinking, trends and fads

Cross-cutting HRM themes for the new millennium

While the construction industry is becoming more innovative in its management practices, there is a lack of critical debate about the implications from an SHRM perspective. The latest management techniques and ideas are often embraced without enough thought for those who will have to implement them and they are often imposed from the top down, with little consultation. It is important to recognise that any new idea needs people's commitment and acceptance if it is to be implemented effectively, without adverse effects on employees' morale, motivation and industrial relations, which could undermine any short-term improvements achieved. This chapter explores the potential impact of the latest management trends being adopted in the construction industry from an SHRM perspective. The aim is to encourage the reader to apply SHRM thinking to contemporary management ideas and techniques in order that the human dimension is embedded within any drive for performance improvement. This should lead to more balanced solutions in which organisations work towards performance improvements in a spirit of cooperation with employees, rather than merely imposing new ways of working upon them.

Introduction

A feature of management practices over the last forty years has been the increasing frequency and fervour with which new ideas come and go. This trend began during the late 1970 and 1980s, when there was unprecedented instability in the business environment. At this time, influential texts such as Toffler (1970), Kanter (1983), Peters and Waterman (1982), Hornstein (1986) and Peters (1989) convinced managers that their lives would never be the same again, and that they would be faced with the shock of accelerating and uncontrollable technological, economic and political change. Initially, the panacea was to be prepared for change. However, managers were soon told that success was dependent upon an organisation's ability to thrive on chaos and generate change rather than wait for it to happen. That is, the path to prosperity rested upon the courageous actions of managers in challenging the status quo and in embarking upon a constant process of revolution, self-examination, renewal, innovation and radical change. Importantly, Peters and Waterman (1982), in their best-selling book, which sold over five million copies, warned that the world's largest and most successful companies were most vulnerable to this new environment, a message reinforced by Pascale, who poignantly illustrated that 'nothing fails like success' (1991: 11).

These warnings struck fear into the heart of the business community and produced a new and lucrative market for managerial 'wonder drugs' to help managers successfully bring about organisational change (Huczynski 1993). There is considerable evidence that these fads are becoming increasingly popular in the construction industry, with many industry leaders and academics now suggesting that techniques such as BPR, partnering, strategic alliances, lean production and supply-chain management all offer better ways of managing the construction production function. Managing relationships is a key feature of these strategies, which recognise that vertical relationships in the supply chain are just as important to an organisation's success as worrying about horizontal relationships *vis-à-vis* competitors. This emphasis on relationships highlights the importance of people in the production process, since relationships are formed and maintained by people. Neglecting the role that people play in implementing new management approaches is therefore likely to threaten the effectiveness of the approaches.

Unfortunately, this is precisely what appears to be happening, the construction industry enthusiastically taking up these new ideas amid a notable lack of critical debate about their SRHM implications. For example, in recent years the notion of best practice has gained special popularity but, as Marchington and Grugulis (2000) point out, many examples of best-practice approaches emerging from the management field ignore the input of employees altogether, with the exception of where they contribute towards employer goals. In particular, Green (2001) has pointed to the contradiction between the cultural change being advocated by highly influential government reports such as Latham (1994) and Egan (1998) and the management techniques being advocated, which merely reinforce the industry's dominant ethos of command and control, which has been the cause of its past problems. In this chapter we explore the latest management trends being adopted in the construction industry and look in some detail at their SHRM impact. Our aim is to distil the core SHRM themes from within this book which must be addressed if they are to be implemented effectively.

The psychological origins of management fads

Pascale (1991) graphically illustrates that throughout the late 1980s the imagination of managers was captured by an explosion of new buzzwords such as 'management by objectives', 'decentralisation', 'delayering', 'supply-chain management', 'value-chains analysis', 'downsizing', 'TQM,' 'diversification', 'restructuring' and 'quality circles'. Indeed, the predictions of change were largely underestimated, and organisations have fallen behind in transferring new technologies into coherent managerial strategies and production processes (Crainer and Obeng 1995). The failure to keep pace with the relentless surge of technological change has encouraged a flow of increasingly radical managerial products to satisfy managers' seemingly insatiable appetite for change, despite them being surrounded by the skeletons of past fads which were once widely popular but which are now disregarded or discredited. As Huczynski (1993) argued, new management ideas have become a valuable commodity and are being marketed as such.

An example of one of the most recent buzzwords to catch the imagination of managers is BPR, which epitomises the increasingly radical approach to 'transforming' organisations. BPR is not about incremental change but about 'quantum leaps forward', 'starting from scratch', 'obliteration', 'big-bang solutions', 'fresh starts', 'wiping the slate clean' and radically rethinking the way a whole business operates (Crainer and Obeng 1995). As Oliver cynically points out, 'there is a new-look menu over at the consultant's café. Good old soupe du TQM and change management are off. Perhaps you would care to try some BPR instead?' (1993: 18). Maintaining the dietary analogy, it is possible to liken many modern managers to compulsive dieters, finding solace in the wonder programme for a few days and then moving relentlessly on to the next. However, there is little cynicism in the construction industry, where academics and managers appear to have been swept along by the euphoria which has accompanied the latest wave of management fads, showing an uncritical perspective towards them and a blind enthusiasm to experiment with new ideas that often have dubious records and offer questionable results (Green 1998, 2001). Green's arguments suggest that the 'groupthink' phenomenon may be at work and, indeed, trendy techniques such as benchmarking, value engineering, partnering, re-engineering, TQM and lean construction seem to saturate the professional and academic literature in construction, being hailed as the way forward in the face of declining profitability, increasing global competition and poor performance.

The concept of 'lean' thinking seems to have taken a special place in the heart of many managers in the construction industry. In the UK the momentum to accept lean thinking has been fuelled by influential government-sponsored reports such as Egan (1998), which use similarly emotive and rhetorical language to that which characterises much of the trendy management literature. The production of very similar reports in many other developed countries has meant that the idea has also taken hold elsewhere. Lean construction is a development of the lean production principle developed within the manufacturing sector. This states that production systems can be viewed in terms of the value-adding conversions and non-value-adding flows/activities. By eliminating non-value-adding activities and targeting value-adding conversions for improvement (using approaches such as just-in-time delivery), the theory holds that performance should be improved (see Koskela 1997). However, while there is undoubted waste in some areas of the industry's activities, some argue that construction is already operating too leanly and that the unconsidered application of lean thinking to all operating areas could be highly dangerous (Loosemore 1999a). For example, in Australia most operatives already regularly work a 60-hour week, which produces safety hazards associated with fatigue and resentment at spending too little time with their families (see Chapters 8 and 9). Indeed, Green (1999) points out that there are links between the use of lean production techniques and high levels of stress, the loss of labour autonomy and long working hours that have led to severe overwork and even deaths in Japan. Loosemore (2000) also argues that it is ironic that the construction industry is attempting to create inflexible construction organisations precisely when an increasingly uncertain and risky business environment is demanding more flexibility. Furthermore, there is considerable evidence that a key feature of all 'high-reliability' (low-crisis) organisations is redundancy and duplication. Sagan (1993) illustrated this by referring to a number of situations where organisations have learnt to deal effectively with high-risk environments. For example, US aircraft carrier operations stress the critical importance of having both technical redundancy (back-up computers, antennas, etc.) and personnel redundancy (spare people and overlap of responsibilities) in their systems. Overlapping responsibilities may seem inefficient in modern business terms, but on an aircraft carrier it can be the difference between life and death by ensuring that potential problems missed by one person are detected by another. The same principles are used to manage nuclear power stations, where independent outside power sources and several coolant loops are incorporated into system designs, should existing provisions fail. Finally, Sagan (1993) illustrates that redundancy is also a feature of our body's immune system, which is why we can survive if certain body parts are severely damaged. For example, if one kidney is removed the other can compensate. If our spleen is removed our bone marrow takes over the job of producing red blood cells. These examples illustrate that redundancy is essential to survival in a world full of potential risks.

The vulnerability of the construction industry to mainstream management trends is not only a consequence of pressures to improve performance. Many of the ideas which are being propounded and which seem to be gaining greatest acceptance fit very comfortably with the engineering mindset and culture which dominate the industry's management practices. Construction management's engineering foundations are represented by its early and important contributions to scientific management through the work of the American Society of Civil Engineers and the individual contributions of industrialists such as the Gilbreths (Sheldrake 1996). These foundations are still evident in, and reinforced by, the technically based nature of construction education, which continues to place relatively little emphasis on HRM skills (Russell *et al.* 1997). An analysis of concepts such as benchmarking, value engineering, BPR and lean construction reveals that they are all sympathetic to this mindset and represent an extension of engineering principles to the management of organisations. Indeed, Hammer (1990), who is often credited with popularising re-engineering, was a computer scientist at Massachusetts Institute of Technology (MIT) and in developing his ideas drew upon many of his experiences of systems-oriented change-management programmes. In each case the emphasis is very much upon structural and process change, and people with an engineering mentality are instinctively attracted by the neat and prescriptive solutions which many of these concepts appear to offer (Huczynski 1993). Furthermore, in the time-pressured environment of the construction industry it is not surprising that construction managers are seduced by the prospect of the quick fix and easy to follow solutions to their problems. This is particularly so when they are sold by so-called management gurus, using largely anecdotal, emotive and dramatic claims of improvement. Indeed, these techniques must often seem as if they were designed especially for the construction industry. For example, Heller reports on a manager of a manufacturing company who asserted that 'there is nothing that TQM's tools and techniques are not able to tackle' and who reported that service has improved by 30 per cent and inventories are down by 80 per cent (1993: 40). However, as with the lean construction example above, there is no guarantee that such concepts will translate unproblematically into a construction project context.

The problems with business fads

Whilst innovation and creativity in finding management solutions should be praised and encouraged, it is important to recognise the ways in which they can lead an organisation to make decisions without considering the implications for the people they impact upon.

Overuse, misuse and abuse

One of the problems with any business fad is overuse, abuse and misuse which eventually surround it with ambiguity and confusion. For example, Oliver (1993) questions the relationships between BPR and re-engineering, business process improvement, redesign and management, business transformation, incremental process improvement and core process redesign. In a construction context, Mohamed (1997) distinguishes between engineering and re-engineering, and McGeorge and Palmer (1997) even distinguish between the hyphenated and non-hyphenated versions of re-engineering. As Oliver (1993) complains, the range of offerings riding the re-engineering ticket is so vast and bewildering that it is not surprising that every company has its own understanding of what BPR is and that every management consultancy has its own approach to it. Indeed, in this context it is not surprising that only 5–10 per cent of re-engineered companies in America have seen any benefits from the process and that some researchers quote failure rates of 70 per cent (Lorenz 1993). Partly in response to these problems, but to add to the confusion, the latest variant to be offered to the construction industry is construction process re-engineering (CPR), which treats construction as a special case, offering a more pragmatic, incremental and flexible approach than BPR (Mohamed 1997). Yet CPR appears to be underpinned by very few convincing arguments to justify creating a watered-down, construction-specific version of BPR and, as Crainer and Obeng (1995) point out, re-engineering is about asking basic questions and seeks to recreate organisations designed around the needs of customers, owners, employees, suppliers and regulators. These are the core constituencies of any business, and their inextricable interdependency makes it difficult to see how they can be treated separately if the aims of re-engineering are to be achieved, unless of course the aims have also been diluted.

Complacency, invincibility and protection

Another problem with construction management fads relates to their origins in manufacturing, the argument being that the industry's product and production practices are unique and that it should be treated as a special case (McAlpine 1998). However, few have argued that the construction industry cannot learn from other industries, and a more serious problem is the almost obscene levels of hype and expectation which surround them, examples of which were provided in the previous section. Organisations with a vested interest in selling such techniques produce large amounts of vivid anecdotal information about its benefits, drawing attention away from statistical information about the probability of success, dissipating managerial responsibility for failure and making managers believe that they have to make little effort to seek alternatives. As Hilmer and Donaldson (1996) warn, the dogma, platitudes and grossly biased claims associated with many current management trends foster superficiality and put the reasoning powers of otherwise sceptical, thoughtful and pragmatic managers on hold. They also produce a sense of invincibility and immunity to crises, which paradoxically increases an organisation's vulnerability (Wildavsky 1988). Furthermore, they are misleading, dangerous and irresponsible, particularly for the small and medium-sized enterprises which employ the bulk of the industry's workforce, and which are particularly vulnerable to utilising new management ideas without a complete grasp of their underlying principles and knowledge of the necessary resources and commitment required to make them a success. The warning signs for such companies are not as clear as they ought to be, because, while examples of failures often appear in the re-engineering literature, they are not nearly as frequently and vividly reported as the successes.

The consequences are rushed and ill-conceived initiatives which do more harm than good. BPR, for example, involves reviewing all aspects of people, technology and process in one coordinated approach, and so costs a significant amount of time and money and has enormous implications for issues such as retraining, safety, industrial and public relations and staff morale (G. Hall *et al.* 1993). For BPR to be effective it must be carefully thought through, supported by a clear sense of strategic direction and, most importantly, senior management must commit resources to its implementation. Unfortunately, in addition to small and medium-sized enterprises, those who are most vulnerable to the evangelical hype surrounding management fads such as BPR are those who are desperate and who turn to new fads to avoid an impending crisis. Indeed, Huczynski (1993) notes that these people's motives may not be entirely altruistic, because management trends represent an important defence against possible blame for poor corporate performance since no manager could be blamed for attempting to implement the latest management techniques.

Stress, fear and crisis

Arguably the greatest problem with many scientifically based management trends is their ruthless emphasis on productivity improvement by producing sinewy, lean and toned organisations (Richardson 1996). While Huczynski (1993) recognises that some management fads are grounded in the human relations school of management thought, Richardson demonstrates how 'the scientific management movement, in its many guises, is widespread, growing and extolled as the way forward' (1996: 20). He goes on to argue that 'Scientific management is a "must" for modern day strategists. ... If it isn't practised then the organisation cannot expect to survive' (*ibid.*). Richardson argues that, while many of these new fads appear to be fresh ideas, many are out of touch with contemporary management thought. The contemporary view of management is that it is a social activity which involves coordinating purposeful individuals who are embedded in complex and constantly changing social networks (Rogers and Kincaid 1981; Tsoukas 1995; Furze and Gale 1996). This has transformed the traditional view of management as a regulatory activity designed to control some objective and static scene. In practical terms, contemporary management thought requires that managers attend to the spiritual needs of individuals, avoid the prescription of people's activities, encourage spontaneity, creativity and a sense of collective responsibility, foster and maintain trusting interpersonal relationships and drive fear out of the workplace (Ryan and Oestreich 1998). Richardson argues that these themes are conspicuous by their absence from many modern management trends and that, in many ways, these modern manifestations of scientific management are worse than Frederick Taylor's (1911) original conceptions. While Taylor's aim was to share some of the benefits of increased productivity, the new systems are leading to the long-term disadvantage of the workforce in terms of reduced job security, higher workloads and higher levels of work-related stress. The problem with many modern-day management trends is that the emphasis is very much on *processes* rather than *people* and on *efficiency* rather than *effectiveness*. Furthermore, the important distinction and balance between these interdependent issues is lost in the blaze of enthusiasm in which they are adopted. While many of these new trends do not necessarily mean delayering and downsizing, in practice these often happen, and indeed unscrupulous managers use these new buzzwords as managerial facades to keep their shareholders happy or legitimise their streamlining plans, making token gestures to the philosophies and processes involved and to the introduction of meaningful cultural and structural organisational change (Hall *et al.* 1993). Indeed, in labour-intensive industries such as construction, where the principal operating cost is manpower, it is very difficult to achieve any efficiency gains without some job reductions. At the very least, fundamental reorganisations of business processes will result in changes to more flexible employment practices such as part-time working, contract work, job sharing and outsourcing, with fewer permanent core employees. The impact of such practices on levels of motivation, work–life balance, safety and industrial relations is not well understood.

Ryan and Oestriech (1998) warn that workforce reduction programmes send powerful messages about how people are valued in an organisation and are a very effective way of instilling fear into the workforce. In direct reference to BPR programmes Richardson argues that '[t]he message is clear that people who operate other than at the very top of the management hierarchy are mere, expendable, aspects of the organisation's operating system' (Richardson 1996: 25). Gretton (1993) points out that the results of working within such an environment can be traumatic for those involved, and the damage to human resources permanent. The result is a workforce of 'corporate mercenaries' who lose the capacity to develop loyalty to their employers and who coldly drift from one organisation to the next, with an emphasis upon individual performance and securing an immediate and tangible return for their efforts.

Therefore the consequence of embarking on a process of continuous improvement by adopting a modern management technique such as BPR can result in people being overloaded, pressurised and stretched to the point where they have no spare capacity to deal with unexpected events. At the same time, this leanness deters the investment of resources in contingency plans or even the most basic risk-management practices, exacerbating organisations' vulnerability to crises (Berkeley *et al.* 1991). Chaos and catastrophe theorists such as Gleick (1987) argue that such environments are the spawning grounds for dangerous tensions, revolving around issues such as safety and industrial relations, which can trigger unpredictable and uncontrollable crises and rapidly destroy the viability of organisations. Indeed, one need look no further than the West Gate Bridge collapse, the Ronan Point tower collapse and the Challenger space shuttle disaster to have examples of catastrophes which were contributed to by an element of stress from overworking (Bignell *et al.* 1977; Jarman and Kouzmin 1990).

Flexibility and inequity

Despite the enormous impact that re-engineering programmes have on people and despite their enormous dependence on people for success, Crainer and Obeng (1995) point out that many find the human aspects too demanding and focus upon system changes. Consequently, as Handy (1984/94) predicted, management trends are producing an increasingly transient workforce where security of employment is a thing of the past, where fewer people do the same or more work, and where they will constantly have to learn and perform a greater variety of tasks in a variety of different situations or organisations. This is a recognisable trend in the construction industry and is consistent with the Flexible Firm model of operation discussed in Chapter 4, where managers at the core of the organisation offer their employer functional flexibility. However, this neo-managerial model of flexibility is ideologically grounded in the belief that companies facing new economic and technological challenges should respond pragmatically and simply expect employees unquestioningly to adapt to and accept new modes of working. One difficulty with the flexibility described in the writings of Atkinson is that the benefits to employees of implementing this model of flexibility are far from clear (Harley 1995). Thus it is unclear whether those employees at the core of the organisation do enjoy varied, skilled and rewarding jobs, or whether they are simply working harder and experiencing greater demands and stress. Similarly, even if core employees enjoy more favourable conditions they make up only half of the workforce and may come to be a much smaller minority. The danger is that employees in peripheral roles are likely to experience job insecurity and lower pay levels as these are driven down by competitive forces in the labour market, resulting in a 'ghettoisation' of peripheral workers. Even Atkinson wrote that '[t]he clear implication for employees is that one man's pay, security and career opportunities will increasingly be secured at the expense of the employment conditions of others, often women, more of whom will find themselves permanently relegated in dead end, insecure and low paid jobs' (1985: 25). The filling of peripheral roles by women, ethnic minorities, young workers, people suffering ill health or with a disability poses obvious problems for achieving equality and diversity in the workforce. Thus it is essential that companies which achieve flexibility by retaining a core base of employees and using subcontractors, self-employed and temporary employees ensure, so far as is possible, that these employees are not disadvantaged.

Cross-cutting themes and priorities for construction HRM

SHRM has a significant part to play in taking account of and coping with the industry's propensity to adopt new fads and approaches. Taking a strategic view of the impact of new management thinking on the people working for the business should in fact be the first consideration for senior managers looking at any reorganisation of their traditional working practices. By managing this carefully, this should help organisations prepare their workforce to cope with the significant attitudinal, process and technical changes which accompany many new management techniques.

Taking account of the SHRM dimension when implementing new management ideas is not simple. For example, practices such as BPR and lean production often require radically different SHRM support systems. This is because such techniques require a fundamental change in attitudes, working practices and relationships in response to the new ways of working, structures and cultures that they seek to induce. Sparrow and Hiltrop (1994) summarised the main HRM changes that can arise from the introduction of new working practices as follows:

- employees are exposed to new sources of information and new networks of relationships;
- the roles of employees change;
- managers think differently about the tasks that need to be done;
- decision-making processes are changed;
- the time span of discretion before the consequences of inappropriate actions are known is altered;
- the criteria for effectiveness are altered;
- there are shifts in the work content and business process flow;
- the choice of performance metrics and performance-management criteria is changed;
- there are changes to career aspirations, power, influence, credibility and potential problems in people not wanting to change their present roles and responsibilities.

These impacts present many challenges for managers introducing change, who have to respond with mechanisms which facilitate their smooth transition and acceptance by employees. To help managers achieve this the HRM literature has also evolved in response to these wider management changes. According to Sparrow and Marchington (1998), the main developments have been:

- new organisational forms and new psychological contracts;
- the need for partnership and involvement in the employment relationship;
- the drive for multiple and parallel flexibilities within organisations.

These themes have been explored in previous chapters and, in effect, represent the SHRM challenges facing construction if the industry is to innovate effectively in the future. The implications of each theme are discussed in more detail below, in the context of dealing with the challenges inherent in adopting new ways of working and adapting to new ways of management thinking.

New organisational forms and new psychological contracts

The shifting nature of job roles, organisational structures and the nature of the psychological contract have been recurring themes throughout this book. Management fads such as BPR and lean construction are bound to have even more fundamental effects, which will produce many SHRM problems for managers. For example, according to Sparrow and Marchington (1998), existing rewards systems are unable to cope with the changes induced by downsizing, technological change and new strategic priorities such as quality and teamworking. The reallocation of knowledge, information, responsibility and power that must occur to deal with this change creates the need to refocus towards person-related performance-management systems, away from traditional job-related systems (Sparrow and Marchington 1998: 14). Another potential problem for construction managers is the new organisational structures which are needed to cope with the re-engineering of business processes, rationalisation and downsizing. These place pressures on managers to get more out of their remaining staff. However, managers are finding that small, autonomous teams require radically different HRM techniques and solutions from managing people within large organisations.

The implications of these changes for managers in the construction industry are profound. Rather than look towards job functions, they must now shift their thinking towards people-based approaches which attribute value to the individuals employed rather than the jobs that they fill. When this approach is taken, psychological contracts become more important – the beliefs of the parties as to their mutual obligations within the employment relationship (Herriot 1998). However, while psychological contracts used to be based on long-term commitment and reciprocity, today's are more likely to be based on short-term arrangements which provide organisations with the necessary flexibility. This assertion is supported by trends in the length of managers' tenure with their employers. For example, while 80 per cent of managers had spent at least six years with their employers in 1955, only 10 per cent had done so by 1995 (Beardwell and Holden

1997: 5). This transitory environment in managers and operatives makes it more difficult to avoid violating psychological contracts and thereby increasing staff turnover. The solution is for the industry to begin to adapt its people-management practices through partnership and flexibility.

The need for partnership in the employment relationship

Throughout this book we have continually emphasised the importance of engaging with employees and encouraging them to exert their influence on decisions that impact on their own interests within organisations. Partnership has become a key term in the SHRM literature and, in a pluralist sense, refers to reciprocal arrangements made between unions and employers (Bacon 2001). However, we have also suggested that these new ways of working offer a different perspective on the individual employment relationship when viewed in the unitarist sense – one of *partnership* between the employee and the employer rather than adversarial and conflict-ridden relationships. The rationale behind partnerships is that, by increasing the amount of information provided to employees and encouraging their involvement with the decisions that affect their roles, organisations will increase their commitment and hence improve their performance. In Chapter 7 we described the evolution of the concepts of employee participation, involvement and, most recently, empowerment. The development of empowerment can be seen as reflecting a spirit of partnership, whereby the employee takes on more responsibility and works in a flexible way for the good of the organisation in return for greater autonomy and security of employment. The notion of partnership is therefore closely tied to concept of the psychological contract.

Construction companies would be well advised to seek partnerships with the workforce in both the unitarist *and* the pluralist sense. Forming participatory and flexible relationships with employees is a key strategy in securing longer-term commitment. Similarly, forming effective partnerships with trade unions and other employee representatives is an effective way of ensuring the cooperation of the workforce (especially with regard to flexible working) in return for informing and consulting them on employment decisions.

The drive for multiple and parallel flexibilities within organisations

The third cross-cutting theme identified by Sparrow and Marchington is that of pursuing multiple flexibilities within organisations. Throughout this book we have emphasised the dynamic and changing nature of the construction project environment and the resulting demands that this places on the organisations that operate within the sector and the people they employ. In Chapter 3 we introduced the concept of Atkinson's (1984a) Flexible Firm model. This shows how many construction companies adapt to the fluctuating demands of the business environment by employing different groups of employees who offer them two forms of flexibility: *functional flexibility*, provided by a central core of employees who can adjust and deploy their skills in different ways depending on the nature of the workload; and *numerical flexibility*, in which the organisation can adjust its levels of employment to cope with fluctuations in demand through the employment of temporary and outsourced workers and subcontractors. Although the Flexible Firm model was developed in the early 1980s, it remains a useful framework within which changes in employment and organisation can be analysed (Proctor and Ackroyd 2001). However, we must remain cognisant of the impact that flexible working can have on people working in construction organisations. Critics of the Flexible Firm model suggest that employers operating in a flexible manner are unlikely to manage this in a strategic way. Rather, flexible approaches tend to be emergent and to be developed in response to the business environment in which the firm operates. This raises a question as to whether people's needs are really likely to be taken into account through planned HRM policies that run alongside this highly flexible functional form.

Sparrow and Marchington (1998: 18) suggest that functional flexibility cannot be introduced without parallel flexibilities in structure, information systems and rewards, etc. In fact they identify seven discreet flexibilities which must be taken into account in parallel with the others. In addition to numerical and functional flexibility, discussed above, they add *financial flexibility*, in terms of the reward–effort agreement with the job holder; *temporal flexibility*, in terms of the working hours required by the job holder; *geographical flexibility*, in terms of where the job holder needs to be to carry out their tasks; *organisational flexibility*, in terms of how the organisation is structured and how it interacts with other members of the supply chain; and *cognitive flexibility*, in terms of the mental frames of reference and cognitive abilities required by the job holder to perform their job effectively.

It is arguable that an itinerant, project-based sector like construction demands more flexibility of its employees than most. For example, construction professionals are expected to work on a wide variety of different projects, undertake differing roles and responsibilities, work long hours in rapidly changing geographical locations, learn new ways of working that align with new legislation and organisational priorities, and be rewarded according to their performance as part of dynamic, multidisciplinary project teams. Thus the organisation must have in place HRM systems and policies which allow flexibility in the employment relationship.

Requirements for successfully taking into account the human dimension in re-engineering business processes

It is clear from the above that any move to re-engineer business processes requires commitment from potential stakeholders in the change. Oram (1998) suggests five barriers to the total commitment of people in this process and suggests solutions to these issues. These are valuable for any manager planning to introduce any new management idea. They are:

- *Inadequate leadership*: clear and decisive leadership is often lacking in the implementation of new management ideas. Good leadership is crucial to the development of trust, which is necessary to ensure the successful implementation of a re-engineering programme. According to Oram, new roles and behaviours are required, where managers must pass power and decision-making responsibility down to employees and where HR managers must disperse their traditional responsibilities amongst line managers. This places considerable challenges on leaders and employees alike, both of whom must readjust their views on giving/being given instructions. This requires support in the form of training and development activities for all parties.

- *Insufficient communication*: poor communication of the reasons for and expectations of new management ideas is often a problem. Communicating in an effective, appropriate and timely manner is a prerequisite to effectively managing a transition in business processes. According to Oram, the key to successful communication is to ensure that what is communicated is purposeful. This means that there should be communications for sharing the vision or mission of the change, integrating the efforts of those involved and securing their commitment.

- *Inappropriate structures*: put simply, if business processes change, then the structure of the host organisation must also change. This is often forgotten in the implementation of new management ideas. Oram points out that structures that have been designed with particular job roles in mind do not necessarily lend themselves to the need for flexible working. Job roles should emerge through process mapping the revised activities. However, the social dimensions of the new roles must also be taken account of through this process – human aspirations, interactions and behaviours are bound to change with different structural forms.

- *Inadequate preparation of new roles*: selecting people to fill the new roles created by undertaking and developing others so that they are equipped for them is one of the most problematic issues involved in introducing any new idea. Oram recommends clearly defined roles and adequate training to ensure the preparedness of people for their changed responsibilities.

- *Misaligned systems*: aligning the five HRM systems outlined in Chapter 3 (resourcing, reward, training and development, employee relations and management of the HR functions) is difficult in a stable environment. Re-engineering demands that all HRM systems be redesigned to ensure that none become redundant or even counterproductive in the context of new working practices.

All five of these potential barriers have been dealt with in previous chapters as areas where taking a more strategic approach towards the HRM function in construction offers a potential solution to the problems of managing people within the industry. We have emphasised that changing ingrained ways of working is problematic, time consuming and prone to meeting attitudinal barriers. Thus, coping with the industry's propensity for adopting new ways of working requires leaders who recognise the importance of the HRM dimension in achieving them. This in turn requires HRM representation at a board level in order that HRM becomes embedded within strategic decision-making. A failure to achieve this will result in not only failed systems and damaged organisational performance, but violated psychological contracts and a loss of the personnel necessary for driving forward organisational performance.

Conclusions

It was once widely thought that technological advances would transform our lives, producing a world of leisure where we would have more time to spend with our families and enjoy life to the full. However, technological advances have led to an unexpected gap between products and processes and the re-emergence of scientific management in many disguises such as BPR. As predicted, the 9–5 day has been consigned to the past, but it has been replaced by a longer working day rather than more leisure time. New entrants to the construction industry can expect to work longer hours with fewer resources than their predecessors and to suffer higher levels of work-related stress as a result. The potential long-term social implications of this trend are serious, as are the potential negative impacts on the employment relationship. As Panati (1991) argues, for a fad to succeed and survive it must provide pleasure or profit. While management fads may be found to provide short-term profits, these will not be sustained if the result is that they merely load more pressure and responsibility on to the shoulders of employees.

All too often the tendency of the industry is to jump on the latest bandwagon to resolve production efficiency problems without considering the implications for people or merely treating them as an afterthought. If the industry is to prepare better for new management trends and thinking, it must begin to prepare itself for the inevitable impact on SHRM that new ideas such as BPR will demand of the industry. In particular, they require new forms of employment relationship that build a level of trust and mutuality between the employer and their employees. Addressing the leadership, communication, structure, job-role and systems-alignment issues will help to facilitate this. This, in turn, requires new organisational forms and a redefinition of the psychological contract, partnerships in the employment relationship between employers, employees and their representatives, and a drive for multiple and parallel flexibilities within organisations.

Discussion and review questions

1 Outline the pressures that management trends and fads such as BPR place on employees and discuss the implications of these in the context of the modern construction business.
2 Devise a strategic HRM framework for how a large construction company could prepare its employees for the implementation of lean construction methods.
3 How does construction differ from manufacturing, and what implications might these differences have for the wholesale transfer of management practice from manufacturing to construction?

12 Conclusions

SHRM as a route to improved business performance

This book has explored many aspects of the SHRM function and has applied them to the construction industry. The aim of this concluding chapter is to bring them together in order to make recommendations to help managers improve the performance of their projects. It argues that construction firms should view SHRM as an important enabler of business performance and provides an approach to putting the concepts contained within the book into practice. We begin by identifying the tangible links between SHRM and organisational performance. Next, we explain the benefits of adopting an SHRM approach and outline the necessary long-term philosophy required for achieving this. Finally, a mechanism for evaluating SHRM performance is put forward. This provides a practical methodology for measuring SHRM performance and for ensuring that the organisation works towards continuously improving its people-management practices.

Introduction

We began this book by discussing the main challenges to SHRM in the construction industry. We explained that the industry represents one of the most complex and dynamic industrial sectors since it relies on skilled manual labour supported by an interconnected management and design input which are often highly fragmented right up to the point of delivery. Construction projects operate under extreme time and cost pressures, and involve many different personnel, from different occupational and national cultural backgrounds, working for many different organisations, whose sporadic involvement changes throughout the course of the project. These factors render it a difficult sector in which to apply good SHRM practice, and it is not surprising that construction companies have developed largely reactive approaches to this function. Whilst this may help them to cope with fluctuating demand cycles and project-based structures, there is a price to pay for relying on such approaches without first putting in place a strategic framework. This approach can result in the needs of employees being ignored, which could lead to a disillusioned workforce and hence to levels of staff turnover that damage the development of the business and its relationship with its customers.

Given increasing client demands for improved performance, the appalling record of the industry in areas such as OHS, the onset of globalisation and an increasingly competitive labour market, the role of SHRM in ensuring improved performance is one of the key questions which managers must address. The aim of this concluding chapter is to explore possible rejoinders to this complex question in order that a new agenda for SHRM in the industry can be identified.

The relationship between HRM and business performance

There has been increasing interest in examining the links between personnel and development activities and business performance ever since the introduction of SHRM ideas. This book has, for the first time, considered the link in a construction context and has argued that there is an absolute necessity to place SHRM issues at the centre of decision-making. We concur with the contemporary view of SHRM put forward by Marchington and Wilkinson (2000), who argue that while human resource considerations should reflect business strategies, business strategies should also reflect human resource considerations. However, the evidence presented in this book suggests that neither is widely achieved in construction, where HRM is too often treated as an afterthought. In many construction businesses the personnel or HRM function is regarded as an administrative overhead, or as a necessary burden for dealing with the mundane aspects of people management such as payroll, recruitment and the legal requirements of employment. This may explain why so few construction companies have HRM representation at a board level. However, despite the increasing tendency (and desirability) for companies to devolve many aspects of HRM responsibility to line managers, there remains a crucially important role for the HRM specialist in managing the *strategic* aspects of the function. According to Armstrong (1996), the personnel function should contribute to organisational effectiveness by adding value to its products and services and by contributing to competitive advantage. He argues that this is best achieved by:

- ensuring a positive quality and performance-oriented culture;
- recruiting the right people, motivating them, ensuring their commitment to organisational values and deploying them in such a way that they contribute effectively to the organisation;
- ensuring that SHRM initiatives are treated as investments on which a proper return will be obtained;
- delivering cost-effective personnel services.

Competitive advantage is crucial to the growth and prosperity of any business, and involves the achievement of an advantageous market position to enable the consolidation and expansion of market share in relation to competitors. In order to sustain a competitive advantage any organisation must commit itself to improvement, innovation and change (Porter 1990). Each of these relies on the quality of an organisation's human resources. Thus, a key strategic role for the SHRM function is to embed, within the psyche of its employees, the need to innovate, improve and adapt to change. However, this can be a difficult proposition within many construction companies, where there exists an ingrained opposition to changing working practices and cultures that have survived several generations. Changing this culture so that people are willing to adopt and accept new ideas will require the breaking down of occupational stereotypes and a sustained commitment to rewarding the attitudes, thinking and performance necessary to change and innovate. This may run contrary to many of the personnel-management practices upon which the industry has been founded.

Current industry issues and the role of HRM

In Chapters 6–11 of this book we identified some of the greatest challenges currently facing the construction industry for which SHRM approaches must be developed if they are to be successfully resolved. These included particular industry needs in areas such as employee relations, employee participation and involvement, equal opportunities and diversity, health and safety, and training and development. In this section we identify the HRM-driven issues that are likely to have an impact on the industry in the future. Trying to predict the factors most likely to impact on people-management practices is a difficult undertaking. Nevertheless, the analysis of the changing nature of the SHRM climate outlined within this book does raise some key recurring themes which are almost certain to permeate HRM thinking over the next decade. They are discussed in the following sections and all construction companies must be aware of these potential trends when formulating their future SHRM policies.

Changing demographics and workforce composition

Since the 1950s most industrialised countries have experienced unprecedented changes in workforce composition. In particular, there have been dramatic increases in the numbers of women in paid employment and, simultaneously, reductions in the employment participation rates of men. For example, in Australia between 1966 and 1998 the employment participation rate of women rose from 36 per cent to 54 per cent. During the same period men's employment participation rate dropped from 84 per cent to 73 per cent (ABS 1994, 1998b).

This convergence has substantially changed the profile of the Australian workforce, with many more employees now fulfilling family responsibilities in addition to working (Francis and Lingard 2001). No longer can employees be assumed to participate in 'traditional' family structures in which one partner, usually the male, takes on the role of breadwinner (Gorman 1999), supported by a full-time homemaker who takes on the responsibility for family care and home management. Instead, dual-career and lone-parent households are now commonplace. For example, in Australia at present, in 59 per cent of two-parent families both parents are in paid employment, and 47 per cent of all female lone parents and 63.1 per cent of all male lone parents are in the paid workforce (ABS 1998b, 2000). These changes have significant implications for HRM because they imply that traditional management practices based upon homogeneity of workforce are becoming less relevant. Practices based in an artificial separation of work and family life are no longer tenable. Instead, HRM policies and practices must be based on an understanding that employees occupy many domains and that experiences in one domain have an impact on satisfaction and performance in other arenas. Thus, emotional experiences at home can spill over into the workplace and vice versa. For example, concerns about the quality of childcare have been identified as a key influence on employed women's satisfaction at work. Human resource policies of the future must recognise the interconnected nature of employees' work and non-work lives and be designed in such a way as to allow employees to cross their work and non-work boundaries without difficulty. The concept of 'balance' is crucial to the effectiveness of such policies.

Another important demographic trend impacting upon SHRM practice is the declining birth rate and the ageing population. In Australia at present the average birth rate is 1.8 per family. However, professional women tend to have fewer children than non-professional women, with an average of 1.6, and also bear children at an older age (Francis and Lingard 2001). This perhaps reflects a perceived conflict between having children and the pursuit of a fulfilling career. Whatever the reasons, the declining birth rate represents a serious social problem in many industrialised countries. From an SHRM perspective it has two serious implications. First, the labour market is likely to become even more competitive and it may be increasingly difficult to attract young workers in the future. This will require that companies implement SHRM strategies attractive to younger employees and attempt to recruit employees or influence their choice of careers at an earlier age. This may require a deviation from traditional HRM practices, because research suggests that younger workers have a more cynical approach to work and are motivated by different things than are older workers. For example, Loughlin and Barling (2001) suggest that young children's understanding of the world of work is likely to be influenced by their parents' experience of work. Indeed, many of today's young workers observed their parents and others losing their jobs as a result of corporate 'downsizing' and other economic pressures during the 1980s and 1990s. This has led the new generation of workers to be sceptical about long-term employment and loyalty to an organisation. Consequently, young workers expect their work to deliver immediate pay-offs such as independence, flexibility and an enjoyable work environment. There is also evidence that the new generation of workers will have greater expectations of achieving a balance between their work and non-work lives. For example, a US study of older teenagers found that 80 per cent had mothers who work and 86 per cent had fathers who work, and 79 per cent said they want a job that allows for personal and family activities. How to attract, motivate and retain these young employees will be a major challenge facing organisations in the future. Loughlin and Barling (2001) suggest that this will require that companies accommodate the preferences of the new generation for 'non-standard' work, including part-time and temporary work, and ensuring that employees taking these work options are not treated less favourably than other workers.

A second implication of the ageing population for HRM practices is the burden this will place on employees who need to care for elderly relatives. Indeed, it has been suggested that in the future eldercare responsibilities may eclipse employees' childcare responsibilities. In many industrialised countries there has been a change in emphasis from institutional aged care to home and community-based care. This means that family members, more than ever before, now bear the responsibility for caring for elderly relatives. For example, in Australia at present 70 per cent of all providers of personal care and home help for the aged, terminally ill or disabled persons are also in the workforce (ABS 1994). Coupled with women's decision to delay childbirth, the ageing population is likely to result in a generation of workers who bear childcare and eldercare responsibilities simultaneously, the so-called 'sandwich' generation.

These demographic changes will undoubtedly have an impact on the construction industry. If construction companies are unable to devise HRM strategies that allow employees to balance work and non-work responsibilities and which are attractive to a new generation of young workers, it is likely that the industry's future human resource needs will not be met.

The competitive labour market and the need for diversity

Throughout the 1990s increasing recognition was given to the need for long-term solutions to cope with an impending skills crisis. In the past these concerns had been met with scepticism by some industrialists, who contended that predictions of skills shortages had not been supported by micro-economic data (Alexander 1991; Sheridon 1991). However, recent labour-market statistics in many countries have suggested that skills shortages are becoming a reality, at both craft and professional levels. Any significant increase in the industry's output is likely to cause the predicted skills shortages to become a reality for many organisations (Agapiou *et al.* 1995a). One implication of a highly competitive labour market is likely to be increased salary levels which have inflationary effects on the cost of construction work. Indeed, recent reports in the British trade press have indicated that skills shortages have already led to increased salary levels for professionals (Knutt 1997a).

A related consequence of increasingly competitive labour markets is the issue of increasing labour-force mobility (Sommerville 1996). Although the incidence of inter-company mobility has traditionally been high in construction in comparison with other industries, a recent UK survey indicates an almost mercenary culture, where 42 per cent of construction managers and professionals were actively seeking new positions (J. Ford 1997). Unfortunately, it is likely to be the better employees who will be lost as the internal labour market tightens and people move to other industries. Thus, construction companies must develop long-term approaches to labour-force planning if they are cope with future changes in labour-market competition.

One issue that could militate against the industry's ability to cope with increasing skills shortages is its poor image. Construction has one of the worst public images of all industries, being synonymous with high costs, low quality, low prestige, and unsafe and chaotic working practices (Ball 1988). Accompanying this poor image is a widely held perception that career opportunities within the industry are also poor, particularly for minority groups (Baldry 1997). This will further deepen its recruitment problems. For example, its image as a male-dominated industry in most developed countries is bound to work against the entry of women into the sector. Most women in the UK view the industry as a male-dominated, threatening environment with an ingrained masculine culture characterised by conflict and crisis (Gale 1992). Increased diversity is clearly a problem the industry must deal with. In addition to the moral and well-rehearsed demographic arguments for workforce diversification, there is also a strong business case for organisations to employ a more socially representative workforce. Diversity leads to a better-informed, more adaptable organisation which is closer to customers and more responsive to market changes (Coussey and Jackson 1991). Too much homogeneity in workforce profile is detrimental to long-term growth and to the ability of organisations to adapt to new markets, technologies, societal shifts and workforce expectations (Kossek and Lobel 1996).

Increasing performance expectations and the Respect for People agenda

The late 1990s arguably saw the dawn of a new era for the construction industry. Calls from influential clients, governments and other industry stakeholders have demanded that the industry works towards redeveloping its practices in order to find more productive and less adversarial ways of working. Within many developed countries government-backed reports have demanded lower costs, higher quality and better productivity from the industry. Furthermore, they have also demanded that the industry address its lamentable performance with regard to people-management practices and begin to respect the people working within the sector for the good of its overall performance.

The new imperative to respect people working within the industry has been provided with a focus in the UK through Rethinking Construction's Respect for People initiative. In November 2000 this group released a report entitled *A Commitment to People: 'Our Biggest Asset'* (Rethinking Construction 2000). This report set out to find practical ways for the industry radically to improve its performance on people issues. It developed a powerful business case for improving respect for people working within the industry, based on the premise that failure to improve the current situation will result in falling profits, an inability to achieve effective teamworking and partnering, and in it falling behind the leaders in process productivity and innovation. The report identified several themes which underpin their key performance measures for improving respect for people. Initiatives such as Respect for People demonstrate an increasing recognition and awareness of the link between good HRM practice and improved business performance. Many organisations are finding that by working towards providing better working conditions, improved career opportunities and more equitable workplace environments they are receiving greater loyalty, better productivity and more added value from their employees. This in turn is helping them to achieve improved levels of performance for clients and hence greater profitability.

New expectations of the employment relationship

The past decade has seen a sea change in terms of what employees expect from the employment relationship. Indeed, the current emphasis on the need to develop greater respect for people in the industry represents an implicit recognition of the need to meet employee expectations of the employment relationship. This may stem from increasing numbers of research projects showing that the nature of the construction workplace is having a negative effect on people's performance and motivation. For example, the construction-site environment has been shown to be a significant determinant of motivation in construction, with long hours, a chaotic work environment, lack of recognition and aggressive management styles all impacting on the demotivation of construction professionals (Smithers and Walker 2000). Lansley (1996) traced the changes in levels of morale amongst construction managers in the UK industry and noted that levels of intrinsic satisfaction derived from work have reduced dramatically since the 1970s. In particular, he found that a shift towards focusing on commercial concerns has eroded those aspects of working in construction which had previously been the most significant motivators. These developments will inevitably increase the pressure on HRM practitioners to ensure the continued commitment of employees through the management of the psychological contract. The implication of this is that, in addition to the 'hard' areas of the employment contract that have to be met, a 'soft' set of employee expectations also have to be organised and managed. Current practices fail to promote long-term organisational commitment from employees and lead to construction employers having to offer enhancements in the form of additional extrinsic rewards. This reactive approach is simplistic, outdated, inadequate and inefficient, and construction companies require a more sensitive understanding of employee expectations if they wish to engender commitment and loyalty in their employees in the future.

Globalisation

Another recurring theme in this book has been the issue of globalisation. Unavoidably, the world of business is changing from one grounded in national markets to one operating across international boundaries between nations, continents and cultures. ICT has revolutionised business practices and the capabilities of organisations in terms of being able to transfer information across the world. Furthermore, national boundaries no longer represent barriers to trading between nations as they once used to. For example, whereas in the past a construction company would have sourced its key resources locally, materials, plant and labour are now increasingly likely to be procured from overseas. Moreover, these overseas markets present construction organisations with excellent opportunities to expand and diversify their operations and to insulate themselves from vulnerabilities to economic downturn by only operating in a single market.

As we discussed earlier in this book, until now construction has remained relatively isolated from global competition. However, recently the industry has started to develop into a truly international industry. Even smaller firms are now exploiting international work opportunities and are beginning to branch out into foreign markets. Companies looking to exploit these new markets have to develop a clearly defined strategy in order that they can plan, resource and deliver projects successfully. For example, it will be necessary to decide where employees will be found to work on overseas projects. If the company wants to use its existing staff, then this has significant implications in terms of encouraging them to work in foreign countries and in preparing them for these opportunities. For example, they must receive proper training and support both before and during their period working in other countries. This support must extend beyond their work environment to include helping them to integrate their wider social and family lives. Unfortunately, Loosemore and Al Muslmani (1999) found that many managers of construction companies who work overseas are ill prepared for their role and experience significant cultural and communication problems with their subordinates and local management counterparts.

Employing local employees, particularly in management positions, also presents its own challenges and issues with regard to maintaining control and influence over how people work and perform. Many differences will exist between the overseas working climate and that of the home country. Key differences include workplace cultures, employment legislation and methods of working, all of which must be learnt by employees before they seek to exploit these overseas work opportunities. Thus, whilst globalisation may present considerable business opportunities for an ambitious construction company, it also presents considerable challenges from an SHRM perspective.

Future improvement in construction HRM: towards an SHRM approach

It is evident that all five of the areas discussed above are interconnected. *Competitive labour markets* and broader *social and demographic changes* demand *improved working practices* in order that *psychological expectations can be met* and so that *staff are equipped for exploiting the challenges of new global markets*. SHRM has an important part to play in ensuring that construction companies can cope with these challenges in the new millennium. This book has presented evidence that the current reliance on reactive HRM practices in construction is inadequate to meet the challenges of the new millennium. Construction companies must begin to adopt a more strategic outlook in their HRM practices.

SHRM is about taking a longer-term perspective on people-related issues. It demands that an organisation and employee representatives recognise the reciprocal relationship between its future business requirements and the needs of the people who will allow it to achieve these goals. In other words, managers need to consider people's needs when making decisions, and people need to consider managers' needs when responding to those decisions.

The focus of SHRM is on the macro-organisational concerns of structure, culture, effectiveness and performance, matching resources to future requirements and the management of change (Armstrong 1996: 157). According to Armstrong, achieving a strategic vision for HRM demands that the organisation develop the following characteristics:

- strong visionary and charismatic leadership from the top;
- well-articulated missions and values;

- a clearly expressed and implemented business strategy;
- a positive focus on well-understood critical success factors;
- a closely related range of products or services to customers;
- a cohesive top management team;
- an HR director who plays an active role in discussing corporate/business issues as well as making a business-oriented contribution to HR matters.

This provides a useful checklist for construction companies attempting to think more strategically about their people-management practices. A business must, therefore, have an in-depth knowledge of its own market position and what it wants to achieve, an implementation strategy in terms of how it will meet business objectives and a policy in place to cope with the management of change. An approach for evaluating the performance of HRM strategies is discussed in detail below.

Measuring the performance of HRM

Evaluating the performance of SHRM in terms of its contribution to project objectives is crucial for an organisation to determine how healthy it is and to focus attention on crucial success indicators such as satisfaction, motivation, morale and loyalty, which are often neglected in an organisation. It is also essential in helping an organisation continuously improve its practices in line with its other business processes.

The problem in evaluating SHRM performance is that the linkages between good practice and business performance are becoming harder to prove, particularly as organisations increasingly outsource, deregulate and devolve responsibility and authority to SHRM activities (Sparrow and Marchington 1998: 311). However, perhaps the best way to measure SHRM performance is through an audit designed to measure the costs and benefits of the total SHRM programme and to compare this with benchmarks in other industries or organisations. Although SHRM performance information in other sectors is not collected and disseminated in most countries, the purpose of an HRM audit is:

- to help managers evaluate the performance of their people-management activities against that of other organisations;
- to help SHRM make a significant contribution to the organisation's objectives;
- to create value so that a firm is socially responsible, ethical and competitive;
- to provide feedback from employees and operating managers on HRM effectiveness;
- to improve the HRM function by providing a means of deciding when to drop activities and when to add them.

A variety of methods can be used to conduct HRM audits, including interviews, questionnaires and observations. Since the HRM department serves a range of stakeholders it is important to collect data from a wide range of perspectives and to use a combination of these techniques. Obviously, the main criteria on which any assessment must be based will be the main functions of the SHRM function as defined in previous chapters of this book. In essence, these are:

- to provide advice and counsel to management on identifying and solving the problems of individual employees;
- to manage the SHRM cycle (which involves finding good people, utilising them to their full potential, guiding them towards the accomplishment of organisational objectives, integrating their efforts into the organisation, training and developing them, promoting/demoting them and retaining/terminating them;
- to communicate to management the philosophy, legal implications and strategies of employee relations;
- to ensure consistent and equitable treatment of all employees;
- to ensure cooperative industrial relations with unions;
- to ensure good public relations;
- to administer grievance procedure according to policy (identify and analyse problems, review deviations and exceptions, resolve problems);
- to provide advice and counsel to management on staffing policy and related problems;
- to interpret and explain the law relating to human relations and to inform management of the legal implications of policy decisions;
- to ensure compliance with federal and state legislation;
- to monitor and control health and safety performance;
- to communicate policy on sexual harassment and other general equal opportunity philosophy and objectives;
- to involve line management in SHRM decisions;
- to ensure that the SHRM department is open and available to all employees to deal with problems or explain company policies;
- to nurture employees' trust and confidence in an organisation;
- to support all functional departments in their SHRM decisions;
- to ensure that employees are happy, motivated, satisfied and have positive attitudes towards work.

Measuring attainment of the above criteria or some combination of them should indicate the efficiency or effectiveness of the SHRM function. However, to make the process worthwhile it is necessary to measure achievements against specific goals, such as:

- reduce labour costs by 3 per cent this year;
- reduce absenteeism by 2 per cent this year;

- increase the satisfaction index by 5 per cent compared with the results of last year's attitude survey.

These goals should be set relative to past trends, the current achievements of other relevant organisations and the aspirations of an organisation's managers.

Methods of data collection

Once the performance criteria are established and agreed upon, the next decision is to determine which approach to data collection is to be used. There are a number of alternatives, which vary in formality and which can be used in combination or in isolation.

The most frequently used formal evaluation methods are those that examine and analyse an organisation's employment statistics. This approach can be helpful because the statistics gathered can be compared with past performance or with some other external yardstick. Of course, quite often organisations do not collect such information, and it is important to appreciate that quantitative factors alone can only highlight problems, but cannot explain them. However, once problems have been identified more in-depth methods, such as attitude surveys, interviews or focus groups, can be used to explore the reasons for good or bad performance. In this sense, the best approach to performance evaluation is a combination of quantitative and qualitative techniques.

Obviously, one vital factor in any assessment exercise is maintaining confidentiality. To ensure the reliability and validity of the data as far as possible it has become typical to maintain anonymity by questioning groups of employees together and by using an outside consultant. A typical survey is handled in the following way. The project manager or HR department contacts a consultant to administer the survey. The consultant first works with the project manager or HR department to develop the data-gathering approach and design the questionnaires or interview schedules. All stakeholders who may affect or be affected by the process should be consulted in this process to avoid any suspicions and potential problems and to ensure the cooperation of the workforce. Finally, the consultant gathers the data and analyses it independently, producing a confidential report for managers, which highlights both positive and negative areas of SHRM performance and makes alternative recommendations for actions. Typically, no one sees the data or questionnaires completed by the employees and the results are normally presented as aggregate data. This is sometimes important is ensuring anonymity, particularly in organisations which do not have a very diverse workforce, such as construction firms. For example, a female project manager's views or experiences may be easily identifiable if analysis is undertaken by job title and gender. It is important that the ethical implications of analysing data by demographic groups are considered and that any potential risks of data misuse and confidentiality are managed.

The analysis of the collected data is usually carried out in one of a variety of ways. Clearly, the most important issue to explore is how the results compare with organisational objectives and priorities. For example, examining how employees' perceptions compare with the mission and objectives of a firm will reveal whether alignment or gaps exist between the objectives of managers and employees. Other analyses may include comparing present responses with past responses to see if the trends are positive or negative. Alternatively, responses from different sub-units can be compared to see if some are more favourable than others. Responses can also be used to compare results with those of other firms to see what similarities and differences exist. An example of such 'benchmarking' of SHRM practices is currently underway in Australia, where five construction organisations have recently agreed to compare practices and identify best practice with regard to employees' work–life balance. Data will be compared to facilitate the transfer of information between the public and private sectors.

Action

The end result of the SHRM performance-measurement process should be management actions of one type or another. For example, policies might need to be revised, enforced or created and the results communicated to employees. Alternatively, the design of a job may be changed or strategies implemented to empower employees. Indeed, the performance-measurement process itself tends to generate expectations on the part of employees that something will change, and managers need to be sure that some feedback and action follow. As both managers and researchers have found, failure to take action can result in future resistance to participation in surveys and to a feeling that any future efforts to measure performance are ritualistic and not useful.

The best way to communicate results is to prepare a formal report that includes data, comparative statistics and narratives. The report should include information that is important to those who use it to make changes or improvements. For example, reports with recommendations for reducing absenteeism, changing perceptions of a new reward system or reducing the number of grievances about a work condition can be useful for developing methods for improvement. In general, people prefer reports that are easy to read and interpret, reliable and that can result in corrective or proactive steps.

Conclusions

Sustained organisational success can only be achieved through people. In this book a number of techniques and examples have been provided to illustrate how successful firms can think strategically about human resources to accomplish their strategic and operational goals. Unfortunately, few firms in the construction industry use SHRM as a source of competitive advantage. Instead, most rely on financial or technological acumen and scientific management techniques, which place ever-greater expectations on their employees' goodwill and stamina, and take for granted their commitment and skills in being able to rise to any challenge presented. However, considerable evidence is accumulating that this approach is unable to produce the productivity improvements the industry's clients are demanding. As more research becomes available, it is becoming clearer that organisational success is critically influenced by SHRM practices, although in the majority of construction firms there is much more work to be done before SHRM is viewed as anything other than a fringe function.

In answering the question posed at the beginning of this chapter regarding the future for SHRM in construction, we can draw a number of important conclusions. Most importantly, there must be a greater emphasis in the industry on SHRM as a key enabler of organisational objectives. Education clearly has an important role to play in changing attitudes, but companies and industry bodies also have an immense responsibility to create more socially responsible practices and culture. Without a sustained effort to improve the industry's treatment of people it will maintain its negative public image and remain unattractive to the highest performers, holding back the future development and growth of many organisations. For example, efforts must be directed towards the diversification of the construction workforce, because the employment of a workforce which reflects the working population in terms of its age profile, gender balance, ethnic mix and the representation of disabled people will ensure that the industry is reflective of all within society. In addition, it could help to mitigate skills shortages, to improve working practices, and could even help to improve the image of the sector so that higher achievers are attracted in the future. The construction industry must also seek to identify better ways of managing the SHRM function in order that it retains the best employees and ensures a safe, healthy and motivational workplace. To achieve this, companies must gain a better appreciation of the needs of employees, particularly in terms of meeting psychological contract expectations. In the modern employment climate it is naive, grossly simplistic and short-sighted to think that people will tolerate for long, dictatorial, uncaring and insensitive employment practices and a general lack of acknowledgement of their needs both within *and* outside the workplace. Finally, the industry must refine effective ways of coping with the onset of globalisation. Although the real impact has yet to be felt within construction to the extent that it has been in other sectors, it is only a matter of time before a truly global construction market emerges. Organisations that have well-developed SHRM practices will be best positioned to exploit the huge opportunities for growth and development that this will offer.

Discussion and review questions

1 Discuss the advantages of adopting a more strategic approach to the HRM function in terms of ensuring that organisations can take advantage of new global construction markets.
2 Evaluate the extent to which labour market trends will influence the HRM strategies of large construction employers over the next 10 years.
3 Identify three areas of HRM performance critical to your organisation. Identify strategies for evaluating performance in these areas, using both statistical data and discussion with employees.

Bibliography

ABS (Australian Bureau of Statistics) (1994) *Focus on Families: Work and Family Responsibilities*, Cat. No. 4422.0, Australian Government Printing Service, Canberra.

ABS (Australian Bureau of Statistics) (1998a) *Business Register Data*, Commonwealth Government of Australia, Canberra.

ABS (Australian Bureau of Statistics) (1998b) *Labour Force*, Cat. No. 6203.0, May, Australian Government Printing Service, Canberra.

ABS (Australian Bureau of Statistics) (1999a) *Australia Social Trends*, Cat. No. 4102, Australian Government Printing Service, Canberra.

ABS (Australian Bureau of Statistics) (1999b) *Labour Force Status and other Characteristics of Migrants*, Report No. 6250.0, Commonwealth of Australia, Canberra.

ABS (Australian Bureau of Statistics) (2000) *Labour Force Characteristics of Aboriginal and Torres Strait Islander Australians*, Occasional Paper No. 6287.0, Commonwealth of Australia, Canberra.

ABS (Australian Bureau of Statistics) (2001) *Labour Force*, Report No. 6203.0, November, Commonwealth of Australia, Canberra.

Adams, G. A., King, L. A. and King, D. W. (1996) 'Relationships of job and family involvement, family social support and work–family conflict with job and life satisfaction', *Journal of Applied Psychology* 81: 411–20.

Adler, P. (1993) 'Time and motion regained', *Harvard Business Review*, January–February: 97–108.

AFCC (1988) *Strategies for the Reduction of Claims and Disputes in the Construction Industry: A Research Report*, Australian Federation of Construction Contractors, Sydney, Australia.

Agapiou, A., Price, A. D. F. and McCaffer, R. (1995a) 'Planning future construction skill requirements: understanding labour resource issues', *Construction Management and Economics* 13: 149–61.

Agapiou, A, Price, A. D. F. and McCaffer, R. (1995b) 'Forecasting the supply of construction skills in the UK', *Construction Management and Economics* 13: 353–64.

Aldous, J., Osmond, M. W. and Hicks, M. W. (1979) 'Men's work and men's families', in W. R. Burr, R. Hill, F. I. Nye and I. L. Reiss (eds) *Contemporary Theories about the Family*, Free Press, New York.

Alexander, J. A. (1991) 'Professionalism and marketing of civil engineering profession', *ASCE Journal of Professional Issues in Engineering Education and Practice* 117(1), January: 10–20.

Allen, D. S. (1990) 'Less stress, less litigation', *Personnel* 10(1), January: 32–5.

Altmeyer, R. (1988) *The Enemies of Freedom: Understanding Right-wing Authoritarianism*, Jossey-Bass, San Francisco, CA.

Anderson, J. (1998) 'Construction safety: changes needed now to the CDM Regs', *The Safety and Health Practitioner*, May: 26–8.

Ansari, K. H. and Jackson, J. (1996) *Managing Cultural Diversity at Work*, Kogan Page, London.

Ansoff, I. H. (1979) *Strategic Management*, Macmillan, London.

Ansoff, I. H. (1984) *Implanting Strategic Management*, Prentice-Hall International, Englewood Cliffs, NJ.

Anthony, W. P., Perrewé, P. L. and Kacmer, K. M. (1996) *Strategic Human Resource Management*, Dryden Press, Florida.

APESMA (2000) *APESMA Working Hours and Employment Security Survey Report*, PowerPoint Presentation (personal communication).

Armstrong, M. (1991) *A Handbook of Personnel Management Practice*, 4th edn, Kogan Page, London.

Armstrong, M. (1996) *A Handbook of Personnel Management Practice*, 6th edn, Kogan Page, London.

Armstrong, M. (1998) *Managing People: A Practical Guide for Line Managers*, Kogan Page, London.

Arthur, M., Hall, D. and Lawrence, H. (1989) *Handbook of Career Theory*, Cambridge University Press, Cambridge.

Atkinson, J. (1984a) 'Emerging UK work patterns in flexible manning: the way ahead', *IMS Report No. 88*, Institute of Manpower Studies, Brighton.

Atkinson, J. (1984b) 'Manpower strategies for flexible organisations', *Personnel Management*, August.

Atkinson, J. (1985) *Emerging UK Work Patterns*, Institute of Manpower Studies, Brighton.

Australian Financial Review (2000) 'Hard facts on building sites', 5 May: 15–16.

Australian Financial Review (2002a) 'IR laws will harm WA, firms warn', 1 March: 10.

Australian Financial Review (2002b) 'ACTU hails rise in members', 1 March: 11.

Australian Financial Review (2002c) 'Building industry ruled by fear', 5 March: 6.

Australian Financial Review (2002d) 'Scant comfort for unions in Rio shift', 7 March: 58.

Babbage, C. (1832/1971) *On the Economy of Machinery and Manufactures*, Augustus M. Kelley, New York.

Bach, S. (2000) 'From performance appraisal to performance management', in S. Bach and K. Sisson (eds) *Personnel Management: A Comprehensive Guide to Theory and Practice*, Blackwell, Oxford.

Bacon, N. (2001) 'Employee relations', in T. Redman and A. Wilkinson (eds) *Contemporary Human Resource Management*, Pearson, Essex.

Bagilhole, B. (1997) *Equal Opportunities and Social Policy: Issues of Gender, 'Race' and Disability*, London: Longman.

Bagilhole, B. M., Dainty, A. R. J. and Neale, R. H. (1995) 'Innovative personnel practices for improving women's careers in construction companies: methodology and discussion of preliminary findings', *Proceedings of the 11th Annual ARCOM Conference*, University of York.

Baldry, D. (1997) 'The image of construction and its influence upon clients, participants and consumers', *Proceedings of the 13th Annual ARCOM Conference*, Kings College Cambridge, September, vol. 1.

Ball, M. (1988) *Rebuilding Construction*, Routledge, London.

Bangle, G. H. (1964) *The Placing and Management of Building Contracts for Building and Civil Engineering Works*, HMSO, London.

Barbara, S. (1997) 'Team roles and team performance: is there really a link?', *Journal of Occupational and Organisational Psychology* 70(3): 241–58.

Barlow, J., Cohen, M., Jashapara, A. and Simpson, Y. (1997) *Towards Positive Partnering: Revealing the Realities in the Construction Industry*, Policy Press, Bristol.

Barnes, M. (1989) *The Role of Contracts in Management in Construction Contract Policy: Improved Procedures and Practice*, Centre for Construction Law and Management, Kings College, London.

Barnes, M. (1991) 'Risk sharing in contracts, in civil engineering project procedure in the EC', *Proceedings of the Conference Organised by the Institution of Civil Engineers*, Heathrow, London, 24–5 January.

Barnett, R. (1994) 'Home-to-work spillover revisited: a study of full time employed women in dual-earner couples', *Journal of Marriage and the Family* 56: 647–56.

Barnett, R., Marshall, N. L. and Pleck, J. (1992) 'Men's multiple roles and their relationship to men's psychological distress', *Journal of Marriage and the Family* 54: 358–67.

Barrett, R. (1998) *Small Firm Industrial Relations: Evidence from the Australian Workplace Industrial Relations Survey*, Working Paper No. 57, National Key Centre in Industrial Relations, Monash University, Melbourne.

Batstone, E., Boraston, I. and Frenkel, S. (1977) *Shop Stewards in Action*, Blackwell, Oxford.

Beardwell, I. and Holden, L. (eds) (1997) *Human Resource Management: A Contemporary Perspective*, Pitman, London.

Beer, M., Spector, B., Lawrence, P. R., Mills, Q. D. and Walton, R. E. (1984) *Managing Human Assets*, Free Press, New York.

Belbin, R. (1996) *Management Teams: Why They Succeed or Fail*, Butterworth-Heinemann, Oxford.

Belout, A. (1998) 'Effects of human resource management on project effectiveness and success: toward a new conceptual framework', *International Journal of Project Management* 16(1): 21–6.

Berkeley, D., Humphreys, P. C. and Thomas, R. D. (1991) 'Project risk action management', *Construction Management and Economics* 9(1): 3–17.

Bernard, C. (1938/68) *The Functions of the Executive*, Harvard University Press, Cambridge, MA.

Bernardin, H. J. and Russell, J. E. (1998) *Human Resource Management*, 2nd edn, McGraw-Hill, New York.

Bernardin, J. and Cascio, W. (1987) 'Performance appraisal and the law', in R. S. Schuler, S. A. Youngblood and V. Huber (eds) *Readings in Personnel and Human Resource Management*, West, St Paul, MN.

Betts, M. and Lansley, P. (1993) 'Construction management and economics: a review of the first ten years', *Construction Management and Economics* 11(2): 221–45.

Bignell V., Peters G. and Pym C. (1977) *Catastrophic Failures*, Open University Press, Milton Keynes.

Blanchflower, D. and Freeman, R. B. (1992) 'Unionism in the United States and other advanced OECD countries', in M. Bognanno and M. M. Kleiner (eds) *Labor Market Institutions and the Future of Unions*, Blackwell, Oxford.

Bolman, L. G. and Deal, T. E. (1995) 'The organisation as theater', in H. Tsoukas (ed.) *New Thinking in Organisational Behaviour*, Butterworth-Heinemann, Oxford.

Booth, C. (1996) 'Breaking down barriers', in C. Booth, J. Darke and S. Yeandle (eds) *Changing Places: Women's Lives in the City*, Paul Chapman, London.

Bourke, J. (2000) 'Corporate women, children, careers and workplace culture: the integration of flexible work practices into the legal and finance professions', *Studies in Organisational Analysis and Innovation Monograph Number 15*, Industrial Relations Research Centre, University of New South Wales, Sydney.

Bowley, M. (1966) *The British Building Industry: Four Studies in Response and Resistance to Change*, Cambridge University Press, Cambridge.

Bresnen, M. J., Wray, K., Bryman, A., Beardsworth, A. D., Ford, J. R. and Keil, E. T. (1985) 'The flexibility of recruitment in the construction industry: formalization or recasualization?', *Sociology* 19(1): 108–24.

Brett, J. M. and Stroh, L. K. (1994) 'Turnover of female managers', in M. J. Davidson and R. J. Burke (eds) *Women in Management: Current Research Issues*, Paul Chapman, London.

Briscoe, G. (1990) 'Skills shortages in the construction sector', *International Journal of Manpower* 11, part 2/3: 23–8.

Bronzini, M. S., Mason, J. M. and Tarris, J. P. (1995) 'Choosing a civil engineering career: some market research findings', *ASCE Journal of Professional Issues in Engineering Education & Practice* 121(3), July: 170–6.

Brown, A. D. (1995) *Organisational Culture*, Pitman, London.

Bruce, W. and Reed, C. (1994) 'Preparing supervisors for the future workforce: the dual-income couple and the work–family dichotomy', *Public Administration Review* 54: 36–43.

Bunker, B. B., Zubeck, J. M., Vanderslice, V. J. and Rice, R. W. (1992) 'Quality of life in dual-career families: commuting versus single-residence couples', *Journal of Marriage and the Family* 54: 399–407.

Burba, F. J., Petrosko, J. M. and Boyle, M. A. (2001) 'Appropriate and inappropriate instructional behaviors for international training', *Human Resource Development Quarterly* 12: 267–83.

Burke, R. (1997) 'Are families damaging to careers?', *Women in Management Review* 12: 320–4.

Burke, R. J. and Greenglass, E. (1995) 'A longitudinal study of psychological burnout in teachers', *Human Relations* 48(2): 187–202.

Burns, T. and Stalker, G. (1961) *The Management of Innovation*, Tavistock, London.

Butruille, S. G. (1990) 'Corporate caretaking', *Training and Development Journal* 44: 48–55.

Cargill, J. (1996) 'Rising sum', *Building*, 15 November: 42–7.

Cass, B. (1993) *The Work and Family Debate in Australia*, paper presented at

AFR/BCA Conference on Work and Family: The Corporate Challenge, 1 December, Melbourne.

Caudron, C. and Rozek, M. (1990) 'The wellness payoff', *Personnel Journal* 69: 7, 54–6.

Cavill, N. (1999) 'Where have all the QSs gone?', *Building*, 26 March: 24–5.

Cavill, N. (2000) 'Purging the industry of racism', *Building Magazine*, 14 May: 20–2.

CCTA (1995) *Management of Programme Risk: The Government Centre for Information Systems (CCTA)*, HMSO, London.

Chandler, A. (1966) *Strategy and Structure*, Antler Books, New York.

Chapman, R. J. (1999) 'The likelihood and impact of changes of key project personnel on the design process', *Construction Management and Economics* 17: 99–106.

Cheetham, G. and Chivers, G. (2001) 'How professionals learn in practice: an investigation of informal learning amongst people working in professions', *Journal of European Industrial Training* 25/5: 248–92.

Chevin, D. (1995) 'Sex appeal', *Building*, 13 January: 18–21.

CIB (Construction Industry Board) (1996) *Tomorrow's Team: Women and Men in Construction*, Report of the Construction Industry Board Working Group 8 (Equal Opportunities), Thomas Telford, London.

CIDB (1998) *Construction Economics Report: Third Quarter 1998*, Construction Industry Development Board, Singapore.

CITB (Construction Industry Training Board) (2002) *CITB Skills Foresight Report February 2002*, King's Lynn.

CITB/Royal Holloway (1999) *The Under-representation of Black and Asian People in Construction*, CITB/Centre for Ethnic Minority Studies, Royal Holloway University of London, King's Lynn.

Claydon, T. and Doyle, M. (1996) 'Trusting me, trusting you? The ethics of employee empowerment', *Personnel Review* 25: 13–25.

Clegg, H. (1976) *The System of Industrial Relations in Great Britain*, Blackwell, Oxford.

COA (Commonwealth of Australia) (1998) Building for Growth, Department of Industry, Science and Resources, Commonwealth of Australia, Canberra.

Coleman, T. (1965) *The Railway Navvies: A History of the Men Who Made the Railways*, Hutchinson, London.

Commission of the European Communities (1993) *Safety and Health in the Construction Sector*, Office for Official Publications of the European Communities, Luxemburg.

Cook, S. (1994) 'The cultural implications of empowerment', *Empowerment in Organizations* 2(1): 9–13.

Cooper, C. and White, B. (1995) 'Organisational behaviour', in S. Tyson, *Strategic Prospects for HRM*, IPD, London.

Cooper, D. and Robertson, I. T. (1995) *The Psychology of Personnel Selection*, Routledge, London.

Cordes, C. L. and Dougherty, T. W. (1993) 'A review and an integration of research on job burnout', *Academy of Management Review* 18: 621–56.

Cornelius, N. (2001) *Human Resource Management: A Managerial Perspective*, Thomson, London.

Cotton, P. (1995) *Psychological Health in the Workplace: Understanding and*

Managing Occupational Stress, Australian Psychological Society Ltd, Melbourne.

Court, G. and Moralee, J. (1995) *Balancing the Building Team: Gender Issues in the Building Professions*, Institute for Employment Studies/CIOB, University of Sussex.

Coussey, M. and Jackson, H. (1991) *Making Equal Opportunities Work*, Pitman Publishing, London.

Craig, R. (1996) 'Manslaughter as a result of workplace fatality', *Proceedings of the 12th Annual ARCOM Conference*, Sheffield Hallam University, UK.

Crainer, S. and Obeng, E. (1995) 'Re-engineering', in S. Crainer (ed.) *The Financial Times Handbook of Management*, Pitman Publishing, London.

Crichton, C. (1966) *Interdependence and Uncertainty: A Study of the Building Industry*, Tavistock, London.

Croucher, R. and Druker, J. (2001) 'Decision-taking on human resource issues', *Employee Relations* 23(1): 55–74.

Cully, M. and Woodland, S. (1996) *Trade Union Membership and Recognition: An Analysis of Data from the 1995 Labour Force Survey. Labour Market Trends*, May, HMSO, Norwich.

Culp, G. and Smith, A. (1992) *Managing People for Project Success*, Van Nostrand Reinhold, New York.

Cunningham, I. and Hyman, J. (1999) 'The poverty of empowerment? A critical case study', *Personnel Review* 28(3): 192–207.

Dainty, A. R. J. (1998) 'A grounded theory of the determinants of women's under-achievement in large construction companies', unpublished PhD thesis, Loughborough University.

Dainty, A. R. J., Bagilhole, B. M. and Neale, R. H. (1998) 'Innovative personnel practices for improving womens' careers in construction companies: methodology and discussion of preliminary findings', *Proceedings of the 11th ARCOM Conference*, University of York.

Dainty, A. R. J., Bagilhole, B. M. and Neale, R. H. (2000a) 'A grounded theory of women's career under-achievement in large UK construction companies', *Construction Management and Economics* 18: 239–50.

Dainty, A. R. J., Bagilhole, B. M. and Neale, R. H. (2000b) 'The compatibility of construction companies' human resource development policies with employee career expectations', *Engineering, Construction and Architectural Management* 7(2): 169–78.

Davey, C., Davidson, M., Gale, A., Hopley, A. and RhysJones, S. (1998) *Building Equality in Construction: Good Practice Guidelines for Building Contractors and Housing Associations*, UMIST/DETR, Manchester.

De Bono, E. (1993) *Teach Your Child How to Think*, Penguin, Harmondsworth.

Debrah, Y. A. and Ofori, G. (1997) 'Flexibility, labour subcontracting and HRM in the construction industry in Singapore: can the system be refined?', *The International Journal of Human Resource Management* 8(5): 690–709.

Deming, W. E. (1994) 'Leadership for quality', *Executive Excellence* 10(6), June: 3–5.

Denham, N., Ackers, P. and Travers, C. (1997) 'Doing yourself out of a job? How middle managers cope with empowerment', *Employee Relations* 19: 147–59.

Denning, P. J. (1992) 'Educating a new engineer', *Communications of the Association for Computing Machinery* 35: 82–98.

Devanna, M. A., Fombrun, C. J. and Tichy, N. M. (1984) 'A framework for strategic human resource management', in C. J. Fombrun, N. M. Tichy and M.A. Devanna (eds) *Strategic Human Resources Management*, John Wiley, New York.

Devine, P. G. (1989) 'Stereotypes and prejudice: their automatic and controlled components', *Journal of Personality and Social Psychology* 56(1): 5–18.

Dolan, S. L. (1995) 'Individual, organizational and social determinants of managerial burnout: theoretical and empirical update', in R. Crandall and Pamela L. Perrewé (eds) *Occupational Stress: A Handbook*, Taylor & Francis, Philadelphia, PA.

Donovan, P., Hannigan, K. and Crowe, D. (2001) 'The learning transfer system approach to estimating the benefits of training: empirical evidence', *Journal of European Industrial Training* 25: 221–8.

Doyle, M. (1997) 'Management development', in I. Beardwell and L. Holden (eds) *Human Resource Management: A Contemporary Perspective*, Pitman, London.

DPWS (1996) *The Construction Industry in NSW: Opportunities and Challenges*, NSW Department of Public Works and Services, Sydney, Australia.

Drake, B. (1984) *Women in Trade Unions*, Virago, London.

Druker, J. and White, G. (1996a) *Managing People in Construction*, IPD, London.

Druker, J. and White, G. (1996b) 'Constructing a new reward strategy: reward management in the British construction industry', *Employee Relations* 19(2): 128–46.

Druker, J., White, G., Hegewisch, A. and Mayne, L. (1996) 'Between hard and soft HRM: human resource management in the construction industry', *Construction Management and Economics* 14: 405–16.

Dufficy, M. (1998) 'The empowerment audit – measured improvement', *Industrial and Commercial Training* 30: 142–6.

Dufficy, M. (2001) 'Training for success in a new industrial world', *Industrial and Commercial Training* 33: 48–53.

Dunlop, J. T. (1958) *Industrial Relations Systems*, Southern Illinois University Press, Carbondale, IL.

Eagly, A. H. and Chaiken, S. (1993) *The Psychology of Attitudes*, Harcourt Brace, Sydney.

Eddy, E., Stone, D. and Stone-Romero, E. (1999) 'The effects of information management policies on reactions to human resource information systems: an integration of privacy and procedural justice perspectives', *Personnel Psychology* 52(2): 335–59.

Egan, J. (1998) *Rethinking Construction: The Report of the Construction Task Force*, Department of the Environment, Transport and the Regions, London.

Eisenberger, R., Fasolo, P. and Davis-LaMastro, V. (1990) 'Perceived organizational support and employee diligence, commitment and innovation', *Journal of Applied Psychology* 75: 51–9.

Eisenberger, R., Huntington, R., Hutchison, S. and Sowa, D. (1986) 'Perceived organizational support', *Journal of Applied Psychology* 71: 500–7.

Eldukair, Z. A. and Ayyub, B. M. (1991) 'Analysis of recent U.S. structural and construction failures', *Journal of Performance of Constructed Facilities* 5(1): 57–73, American Society of Civil Engineers.

Ely, R. J. and Thomas, D. A. (2001) 'Cultural diversity at work: the effects of diversity perspectives on work group processes and outcomes', *Administrative Science Quarterly* 46(2): 229–73.

Emmott, M. and Hutchinson, S. (1998) 'Employee flexibility: threat or promise', in P. Sparrow and M. Marchington (eds) *Human Resource Management: The New Agenda*, Pitman, London.

EOC (Equal Opportunities Commission) (1990) 'Women in the construction industry', *Equal Opportunities Review*, March/April: 16–22.

EOC/CRE/DRC (Equal Opportunities Commission/Commission for Racial Equality/Diability Rights Commission) (1999) *Equal Opportunities is Your Business Too*, EOC/CRE/DRC, Manchester.

Eylon, D. and Bamberger, P. (2000) 'Empowerment cognitions and empowerment acts: recognising the importance of gender', *Group & Organization Management* 25: 354–72.

Fayol, H. (1949) *General and Administrative Management*, Pitman, London.

Feldman D. C. (1980) 'A socialisation process that helps new recruits succeed', *Personnel* 57, March–April: 11–23.

Field, S. and Jorg N. (1991) 'Corporate liability and manslaughter: should we be going Dutch?', *Criminal Law Review* 156: 156–71.

Finch, P. (1994) 'Construction women hit glass ceiling', *Architects' Journal*, November: 11.

Fisher, G. (1988) *Mindsets: The Role of Culture and Perception in International Relations*, Intercultural Press, Yarmouth, Maine.

Follet, M. P. (1924) *Creative Experience*, Longmans, London.

Fombrun, C. J., Tichy, N. M. and Devanna, M. A. (1984) *Strategic Human Resources Management*, John Wiley, New York.

Ford, D. N., Voyer, J. and Gould Wilkinson, J. M. (2000) 'Building learning organizations in engineering cultures: case study', *Journal of Management in Engineering*, July/August: 72–83.

Ford, J. (1997) 'Better jobs, bad bosses and biscuit tins: the hopes and bugbears of construction professionals', *Building*, 23 May: 30–3.

Foster-Fishman, P. G. and Keys, C. B. (1995) 'The inverted pyramid: how a well-meaning attempt to initiate employee empowerment ran afoul of the culture of public bureaucracy', *Academy of Management Journal Best Papers Proceedings 1995*: 364–72.

Fox, A. (1966) 'Industrial society and industrial relations', *Royal Commission on Trade Unions and Employers' Associations Research Paper No. 3*, HMSO, London.

Foy, N. (1994) *Empowering People at Work*, Gower, Hampshire.

Francis, V. and Lingard, H. (2001) 'Work–life employment practices – meeting the needs of construction professionals in the 21st century', *Proceedings of the Australasian University Building Educators' Association, Constructing and Managing the Built Environment: Education and Research for the Future, 26th Annual Conference*, Adelaide, Australia.

Freeman, R. B. and Kleiner, M. M., (1990) 'Employer behavior in the face of union organizing drives', *Industrial and Labor Relations Review* 43: 351–65.

Freeman, R. B. and Medoff, J. L. (1984) *What do Unions Do?*, Basic Books, New York.

Freudenberger, H. J. (1974) 'Staff burnout', *Journal of Social Issues* 30: 159–65.

Furze, D. and Gale, C. (1996) *Interpreting Management – Exploring Change and Complexity*, International Thomson Business Press, London.

Gale, A. W. (1992) 'The construction industry's male culture must feminise if conflict is to be reduced: the role of education as a gate-keeper to a male construction industry', in P. Fenn and R. Gameson (eds) *Construction Conflict Management and Resolution*, E. & F. N. Spon, London.

Gale, A. W. (1994) 'Women in construction: an investigation into some aspects of image and knowledge as determinants of the under representation of women in construction management in the British construction industry', unpublished PhD thesis, Bath University.

Gantt, H. (1919) *Organizing for Work*, Harcourt Brace Jovanovich, New York.

Garner, J. B. and McRandal, S. (1995) 'Women in the construction industry: a tale of two countries', in CIB W89, *Construction and Building Education and Research Beyond 2000*, Orlando, FL, 5–7 April: 331–8.

Gerber, R. (2001) 'The concept of common sense in workplace learning and experience', *Education and Training* 43: 72–81.

Gilbreth, F. (1911) *Motion Study*, Van Nostrand Rheinhold, New York.

Gleick, J. (1987) *Chaos*, Abacus, London.

Goldstein I. L. (1993) *Training in Organizations*, 3rd edn, Brookes/Cole, Pacific Grove, CA.

Gomez, C. and Rosen, B. (2001) 'The leader–member exchange as a link between managerial trust and employee empowerment', *Group & Organization Management* 26: 53–69.

Gordon, M. E. and De Nisi, A. S. (1995) 'A re-examination of the relationship between union membership and job satisfaction', *Industrial Relations Review* 48: 222–36.

Gorman, E. H., (1999) 'Bringing home the bacon: marital allocation of income-earning responsibility, job shifts and men's wages', *Journal of Marriage and the Family* 61: 110–22.

Greed, C. (1997) 'Cultural change in construction', *Proceedings of the 13th Annual ARCOM Conference*, vol. 1, September, Kings College Cambridge.

Green, S. (1998) 'The technocratic totalitarianism of construction process improvement: a critical perspective', *Engineering, Construction and Architectural Management* 5(4): 376–86.

Green, S. D. (1999) 'The missing arguments of lean construction', *Construction Management and Economics* 17: 133–7.

Green, S. D. (2001) 'Towards a critical research agenda in construction management', *Proceedings of the CIB World Building Congress*, Wellington, New Zealand, CD-ROM Paper, 23 November.

Greenhaus, J. H. and Callanan, G. A. (1994) *Career Management*, 2nd edn, Dryden Press, Orlando, FL.

Greiner, L. E. (1998) 'Evolution and revolution as organizations grow', *Harvard Business Review*, May–June, 76(3): 55–63.

Gretton, I. (1993) 'Striving to succeed in a changing environment', *Professional Manger*, July: 15–17.

Griffith, A. and Headley, J. D. (1995) 'Developing an effective approach to the procurement and management of small building works within large client organizations', *Construction Management and Economics* 13: 279–89.

Groenweg, J. (1994) *Controlling the Controllable*, DSWO Press, Leiden University, Leiden.

Guest, D. (1987) 'Human resource management and industrial relations', *Journal of Management Studies* 24(5): 503–21.

Haas, L. and Hwang, P. (1995) 'Company culture and men's usage of family leave benefits in Sweden', *Family Relations* 44: 28–36.

Hall, G., Rosenthal, J. and Wade, J. (1993) 'How to make reengineering really work', *Harvard Business Review*, November–December: 119–31.

Hall, L. and Torrington, D. (1986) ' "Why not use the computer?" The use and lack of use of computers in personnel', *Personnel Review* 15(1): 3–7.

Hall, L. and Torrington, D. (1998) *The Human Resource Function: The Dynamics of Change and Development*, Financial Times, London.

Hamberger, J. (1995) 'Individual contracts: beyond enterprise bargaining?', *Working Paper No 39, Australian Centre for Industrial Relations Research and Training*, University of Sydney.

Hamilton, A. (1997) *Management by Projects, Achieving Success in a Changing World*, Thomas Telford, London.

Hammer, M. (1990) 'Re-engineering work: don't automate, obliterate', *Harvard Business Review* 68: 104–12.

Hancock, M. R., Yap, C. K. and Root, D. S. (1996) 'Human resource development in large construction companies', in D. A. Langford and A. Retik (eds), *The Organization and Management of Construction* (CIB W65), vol. 1, University of Strathclyde.

Handy, C. (1984/94) *The Future of Work: A Guide to a Changing Society*, Blackwell, London.

Handy, C. (1989) *The Age of Unreason*, Business Books, London.

Hannaford, I. (1997) *Race: The History of an Idea in the West*, Johns Hopkins University Press, Baltimore, MD.

Hansard Society (1990) *Report of the Hansard Society on Women at the Top*, Hansard Society, London.

Hanson, M. (1995) 'Building a better future', *London Evening Standard*, 2 May.

Harley, B. (1995) *Labour Flexibility and Workplace Industrial Relations: The Australian Evidence*, Monograph No. 12, Australian Centre for Industrial Relations Research and Teaching, University of Sydney.

Harris Research Centre (1988) *Factors Affecting Recruitment for the Construction Industry: The View of Young People and Their Parents*, CITB, Kings Lynn.

Hearn-Mackinnon, B. (1997) 'The Weipa Dispute: ramifications for the spread of Australian workplace agreements', in A. Frazer, R. McCallum and P. Ronfeldt (eds) *Individual Contracts and Workplace Relations*, Australian Centre for Industrial Relations Research and Training, University of Sydney.

Heller, R. (1993) 'TQM – not a panacea but a pilgrimage', *Management Today*, January: 37–40.

Hendry, C. and Pettigrew, A. M. (1990) 'Human resource management: an agenda for the 1990s', *International Journal of Human Resource Management* 1(1): 17–43.

Herriot, P. (1998) 'The role of the HR function in building a new proposition for staff', in P. Sparrow and M. Marchington (eds) *Human Resource Management: The New Agenda*, Pitman, London.

Herriot, P. and Pemberton, C. (1997) 'Facilitating new deals', *Human Resource Management Journal* 7(1): 45–56.

Herzberg, F. (1959) *The Motivation to Work*, Wiley, New York.

Herzberg, F., Mausner B. and Snyderman, B. (1959/93) *The Motivation to Work*, Transaction Publishers, New Brunswick, NJ.

Hillebrandt, P. M. and Cannon, J. (1990) *The Modern Construction Firm*, Macmillan, Basingstoke.

Hilmer, F. G. and Donaldson, L. (1996) *Management Redeemed – Debunking the Fads that Undermine Corporate Performance*, Free Press, London.

Hindle, R. D. and Muller, M. H. (1996) 'The role of education as an agent of change: a two fold effect', *Journal of Construction Procurement* 3(1): 56–66.

Hirsch, B. T. and Addison, J. T. (1986) *The Economic Analysis of Unions: New Approaches and Evidence*, Allen & Unwin, Boston, MA.

Hofstede, G. (1980) *Cultural Consequences: International Differences in Work Related Values*, Sage Publications, Beverly Hills, CA.

Holden, L. (1997) 'Employee involvement', in I. Beardwell and L. Holden (eds) *Human Resource Management: A Contemporary Perspective*, Pitman, London.

Hollingsworth, D., McConnochie, K. and Pettman, J. (1988) *Race and Racism in Australia*, Social Sciences Press, Wentworth Falls, NSW.

Holt, G. D., Love, P. E. D. and Nesan, L. J. (2000) 'Employ empowerment in construction: an implementation model for process improvement', *Team Performance Management: An International Journal* 6(3/4): 47–51.

Holton, E. F. III (1996) 'The flawed four-level evaluation model', *Human Resource Development Quarterly* 7: 5–25.

Holton, E. F. III, Bates, R. A. and Ruona, W. E. A. (2000) 'Development of a generalised learning transfer system inventory', *Human Resource Development Quarterly* 11(4): 333–60.

Honold, L. (1997) 'A review of the literature on employee empowerment', *Empowerment in Organisations* 5(4): 202–12.

Hooper, N. and Skeffington, R. (1999) 'Trends: grey shift', *Business Review Weekly* 21(34): 69–81.

Hopkins, A. (1995) *Making Safety Work*, Allen & Unwin, St Leonards.

Hornstein, H. A. (1986) *Managerial Courage*, John Wiley & Sons, New York.

HSE (Health and Safety Executive) (1993) *The Costs of Accidents at Work*, HMSO, London.

HSE (Health and Safety Executive) (1997) *The Costs of Accidents at Work*, HMSO, London.

HSE (Health and Safety Executive) (2000) *Successful Health and Safety Management*, HMSO, London.

Huang, Z., Olomolaiye, P. O. and Ambrose, B. (1996) 'Construction company manpower planning', in A. Thorpe (ed.) *Proceedings of the 12th Annual ARCOM Conference*, September, vol. 1, Sheffield Hallam University.

Huczynski, A. A. (1993) 'Explaining the succession of management fads', *The International Journal of Human Resource Management* 4(2): 443–63.

Huczynski, A. A. and Buchanan, D. (2001) *Organizational Behaviour: An Introductory Text*, 4th edn, Pearson, Essex.

Hughes, R. E. (2001) 'Deciding to leave but staying: teacher burnout, precursors and turnover', *International Journal of Human Resource Management* 12: 288–98.

Hunter, J. A., Stringer, M. and Watson, R. P. (1991) 'Intergroup violence and inter-group attribution', *British Journal of Social Psychology* 30(1): 261–6.

Industry Commission (1995) *Work, Health and Safety*, Report No. 47, Common-wealth of Australia, Canberra.

IPD (1993) *Code on Employee Involvement and Participation in the United Kingdom*, IPD, London.

Jarman, A. and Kouzmin, A. (1990) 'Decision pathways from crisis – a contin-gency theory simulation heuristic for the Challenger space disaster (1983–1988)', in A. Block (ed.) *Contemporary Crisis – Law, Crime and Social Policy*, Kluwer Academic Press, Netherlands.

Jehn, K. A., Northcraft, G. B. and Neale, M. A. (1999) 'Why differences make a difference: a field study of diversity, conflict and performance in workgroups', *Administrative Science Quarterly* 44: 741–63.

Jensen, R. S. (1982) 'Pilot judgement: training and evaluation', *Human Factors* 34: 61–73.

Kandola, R. and Fullerton, J. (1998) *Diversity in Action, Managing the Mosaic*, 2nd edn, Institute of Personnel and Development, London.

Kanter, E. (1983) *The Change Masters*, Allen & Unwin, London.

Kanter, R. M. (1977) *Women and Men of the Corporation*, Basic Books, New York.

Katzenbach, J. R. and Smith, D. K. (1993) *The Wisdom of Teams: Creating the High-Performance Organization*, Harvard Business School, Boston, MA.

Keep, E. and Rainbird, H. (2000) 'Towards the learning organization?', in S. Bach and K. Sisson (eds) *Personnel Management: A Comprehensive Guide to Theory and Practice*, Blackwell, Oxford.

Kirk-Walker, S. (1994) *Women in Architecture 1992: A Synopsis of the Main Find-ings of the Report*, Institute of Advanced Architectural Studies, York, September.

Kirk-Walker, S. and Isaiah, R. (1996) 'Undergraduate student survey: a report of the survey of first year students in construction industry degree courses', *Entrants to Academic Year 1995–1996*, Institute of Advanced Architectural Studies/CITB, York.

Kleiner, M. M., (2001) 'Intensity of management resistance: understanding the decline of unionization in the private sector', *Journal of Labour Research* 22: 519–40.

Knutt, E. (1997a) 'Careers: recruitment drivers', *Building*, March: 4–6.

Knutt, E. (1997b) 'Tax returns', *Building* 11, 21 March: 44–8.

Knutt, E. (1997c) 'Thirtysomethings hit thirtysomething', *Building*, 25 April, 1997: 44–50.

Koberg, C. S., Boss, R. W., Senjem, J. C. and Goodman, E. A. (1999) 'Antecedents and outcomes of empowerment', *Group & Organization Management* 24: 71–91.

Kolb, D. A. (1984) *Experiential Learning: Experience as the Source of Learning and Development*, Prentice-Hall, Englewood Cliffs, NJ.

Koskela, L. (1997) 'Lean production in construction', in L. Alarcón (ed.) *Lean Construction*, A. A. Balkema, Rotterdam.

Kossek, E. E. and Lobel, S. A. (eds) (1996) *Managing Diversity*, Blackwell, Oxford.

Kululanga, G. K., McCaffer, R., Price, A. D. F. and Edum-Fotwe, F. (1999)

'Learning mechanisms employed by construction contractors', *Journal of Construction Management and Engineering*, July/August: 215–23.

Landes, L. (1994) 'The myth and misdirection of employee empowerment', *Training* 31(3): 116–17.

Langford, D., Hancock, M. R., Fellows, R. and Gale, A. W. (1995) *Human Resources Management in Construction*, Longman, Essex.

Lansley, P (1996) 'Aspirations, commitment and careers in construction management', in D. A. Langford and A. Retik (eds) *The Organization and Management of Construction* (CIB W65), University of Strathclyde, vol.2.

Larson, J. H., Wilson, S. M. and Beley, R. (1994) 'The impact of job insecurity on marital and family relationships', *Family Relations* 43: 138–43.

Latham, M. (1994) *Constructing the Team*, HMSO, London.

Laufer, A., Tucker, R. L., Shapira, A. and Shenhar, A. J. (1994) 'The multiplicity concept in construction planning', *Construction Management and Economics* 11: 53–65.

Lawrence, P. R. and Lorsch, J. W. (1967) *Organisation and Environment: Managing Differentiation and Integration*, Harvard University Press, Cambridge, MA.

Lee, Hyo-Soo. (2001) 'Paternalistic human resource practices: their emergence and characteristics', *Journal of Economic Issues* 35: 841–69.

Legge, K. (1989) 'Human resource management: a critical analysis', in J. Storey (ed.) *New Perspectives on Human Resource Management*, Routledge, London.

Lewis, H. G. (1986) *Union Relative Wage Effects*, University of Chicago Press, Chicago, IL.

Lewis, S. (2001) 'Restructuring workplace cultures: the ultimate work–family challenge?', *Women in Management Review* 16: 21–9.

Likert, R. (1961) *New Patterns of Management*, McGraw-Hill, New York.

Lim, E. C. and Alum, J. (1995) 'Construction productivity: issues encountered by contractors in Singapore', *International Journal of Project Management* 13(1): 51–8.

Lindell, M. K. (1994) *Occupational Medicine: State of the Art Reviews* 9(2): 211–40, Hanley & Belfus Inc., Philadelphia, PA.

Lingard, H. (in press) 'The effect of first aid training on Australian construction workers' occupational health and safety motivation and risk control behaviour', accepted for publication in *Safety Research*.

Lingard, H. (2002) 'The effect of first aid training on Australian construction workers' occupational health and safety knowledge and motivation to avoid work-related injury or illness', *Construction Management and Economics* 20: 263–73.

Lingard, H. and Rowlinson, S. (1997) 'Behavior-based safety management in Hong Kong's construction industry', *Journal of Safety Research* 28: 243–56.

Lingard, H. and Rowlinson, S. (1998) 'Behaviour-based safety management in Hong Kong's construction industry: the results of a field study', *Construction Management and Economics* 16: 481–8.

Lingard, H. and Sublet, A. (in press) 'The impact of job and organisational demands on marital or relationship satisfaction and conflict among Australian civil engineers', *Construction Management and Economics*.

Lingard H., Gilbert, G. and Graham P., (2001) 'Improving construction workers' solid waste reduction and recycling behaviour using goal setting and feedback', *Construction Management and Economics* 19: 809–17.

Lo, T. Y. (1988) 'Training programme for supervisors: an element in quality assurance of the construction industry', *Journal of Manpower Development* 17: 576–82.

Lobel, S. A., Googins, B. K. and Bankert, E. (1999) 'The future of work and family; critical trends for policy, practice and research', *Human Resource Management* 38: 243–54.

Lomas, L. (1997) 'The decline of liberal education and the emergence of a new model of education and training', *Education and Training* 39: 111–15.

Loosemore, M. (1999a) 'Power, responsibility and construction conflict', *Construction Management and Economics* 17(6): 699–711.

Loosemore, M. (1999b) 'The problem with business fads', in K. Karim and M. Morosszeky (eds), *Construction Process Re-engineering: Proceedings of the International Conference on Construction Process Re-engineering*, Sydney, Australia, 12–13 July.

Loosemore, M. (2000) *Construction Crisis Management*, American Society of Civil Engineers, New York.

Loosemore, M. and Al Muslmani, H. S. (1999) 'Construction project management in the Persian Gulf – inter-cultural communication', *International Journal of Project Management* 17(2): 95–101.

Loosemore, M. and Chau, D. W. (2002) 'Racial discrimination towards Asian workers in the Australian construction industry', *Construction Management and Economics* 20(1): 91–102.

Loosemore, M. and Hughes, K. (1998) 'Emergency systems in construction contracts', *Engineering, Construction and Architectural Management* 5(2): 189–99.

Loosemore, M. and Tan, Chin Chin (2000) 'Occupational bias in construction management research', *Construction Management and Economics* 18(7): 757–66.

Loosemore, M., Nguyen, B. T. and Dennis, N. (2000) 'Encouraging conflict in the construction industry', *Construction Management and Economics* 18(4): 447–57.

Lorenz, C. (1993) 'Uphill struggle to become horizontal', *Financial Times*, 5 November.

Loughlin, C. and Barling, J. (2001) 'Young workers' work values, attitudes and behaviours', *Journal of Occupational and Organizational Psychology* 74: 543–58.

McAlpine, Sir A. (1998) 'Vroom for improvement', *Building*, 14 August: 22–3.

McColgan, A. (1994) 'The Law Commission consultation document on involuntary manslaughter – heralding corporate liability', *Criminal Law Review* 41: 547–57.

McGeorge, D. and Palmer, A. (1997) *Construction Management – New Directions*, Blackwell Science, London.

McGregor, D. (1960) *The Human Side of Enterprise*, Penguin, Harmondsworth.

MacGregor, R. (1999) *Work and Family Policies: a Win–Win Formula for Business and Society*, Minnesota Center for Corporate Responsibility, Minnesota.

McLeod, M. (2001) 'Surfers' paradigm', *People Managemen* 7(5): 44–5.

McVittie, D., Banikin, H. and Brocklebank, W. (1997) 'The effect of firm size on injury frequency in construction', *Safety Science* 27: 19–23.

Makin, P., Cooper, C. and Cox, C. (1996) *Organisations and the Psychological Contract: Managing People at Work*, British Psychological Society, Leicester.

Marchington, M. (1995) 'Involvement and participation', in J. Storey (ed.) *Human Resource Management: A Critical Text*, Routledge, London.

Marchington, M. and Grugulis, I. (2000) ' "Best practice" human resource management: perfect opportunity or dangerous illusion?', *International Journal of Human Resource Management* 11(4): 905–25.

Marchington, M. and Wilkinson, A. (2000) *Core Personnel and Development*, CIPD, London.

Marginson, P. and Sisson, K. (1988) 'The management of employees', in K. Sisson, P. Marginson, P. K. Edwards, R. Martin and J. Purcell (eds), *Beyond the Workplace: Managing Industrial Relations in the Multi-Establishment Enterprise.* Blackwell, Oxford.

Marks, A. (2000) 'Lifelong learning and the "breadwinner ideology": addressing the problems of lack of participation by adult, working class males in higher education on Merseyside', *Educational Studies* 26(3): 303–19.

Maskell-Pretz, M. (1997) 'Women in engineering: toward a barrier-free work environment', *ASCE Journal of Management in Engineering* 13(1), January/February: 32–7.

Maslach, C. and Leiter, M. (1997) *The Truth about Burnout: How Organizations Cause Personal Stress and What to Do About It*, Jossey-Bass, San Francisco, CA.

Maslach, C., Jackson, S. E. and Leiter, M. P. (1996) *Maslach Burnout Inventory Manual*, 3rd edn, Consulting Psychologists Press, Palo Alto, CA.

Maslach, C., Schaufeli, W. B. and Leiter, M. P. (2001) 'Job burnout', *Annual Review of Psychology* 52: 397–422.

Maslow, A. H. (1943) 'A theory of human motivation', *Psychological Review* 50: 370–96.

Mason, P. (2001) 'It's time to change', *Building*, 27 April: 45.

Mauno, S. and Kinnunen, U. (1999) 'The effects of job stressors on marital satisfaction in Finnish dual earner couples', *Journal of Organizational Behavior* 20: 879–95.

Mayo, E. (1933/60) *The Human Problems of an Industrial Civilisation*, Viking Press, New York.

Mills, A. E. (1967) *The Dynamics of Management Control Systems*, London Business Publications, London.

Mintzberg, H. (1976) 'Planning on the left side and managing on the right side', *Harvard Business Review* 54(2): 49–58.

Mintzberg, H. (1979) *The Structuring of Organisations*, Prentice-Hall, Englewood Cliffs, NJ.

Mintzberg, H. (1983) *Structures in Fives: Designing Effective Organisations*, Prentice-Hall, Englewood Cliffs, NJ.

MLSA (1998) *The Annual Statistics Report of Labourers*, Ministry of Labour and Social Affairs, Riyadh, Saudi Arabia.

MMS (1999) *Re-inventing Construction – Construction 21*, Construction 21 Steering Committee, Ministry for Manpower Singapore, Singapore.

Mohamed, S. (1997) 'BPR critic – CPR advocate', in K. Karim and M. Morosszeky (eds), *Construction Process Re-engineering: Proceedings of the International Conference on Construction Process Re-engineering*, Sydney, Australia, 12–13 July.

Moore, D. R. and Dainty, A. R. J. (1999) 'Integrated project teams' performance in managing unexpected change events', *Team Performance Management* 5(7): 212–22.

Moore, D. R. and Dainty, A. R. J. (2001) 'Intra-team boundaries as an inhibitor to performance improvement in the UK construction industry', *Construction Management and Economics* 16(6): 559–62.

Moorehead, A., Steele, M., Alexander, M., Stephen, K. and Duffin, L. (1997) *Changes at Work: The 1995 Australian Workplace Industrial Relations Survey*, Longman, Australia.

Morgan, C. S. (1992) 'College students' perceptions of barriers to women in science and engineering', *Youth and Society* 24(2), December: 228–36.

Mphake, J. (1989) 'Management development in construction', *Management Education and Development* 18(3): 223–43.

Mullins, L. J. (1999) *Management and Organisational Behaviour*, 5th edn, Pearson Education, Essex.

Mullins, L. J. and Peacock, A. (1991) 'Managing through people: regulating the employment relationship', *Administrator*, December: 32–3.

Mumford, A. (1993) *Management Development: Strategies for Action*, IPD, London.

Munro, W. D. (1996) 'The implementation of the Construction (Design and Management) Regulations 1994 on UK construction sites', in L. M. Aves Dias and R. J. Coble (eds) *Implementation of Safety and Health on Construction Sites*, A. A. Balkema, Rotterdam.

Murdoch, J. and Hughes, W. (1992) *Construction Contracts – Law and Management*, E. & F. N. Spon, London.

Musgrave, E. C. (1994) 'The organisation of the building trades of Eastern Brittany 1600–1790: some observations', *Construction History* 10: 1–13.

Naoum, S. G. (2001) *People and Organizational Management in Construction*, Thomas Telford, London.

Napoli, J. (1994) *Work and Family Responsibilities: Adjusting the Balance*, CHH Australia, North Ryde.

NEDO National Economic Development Office (1988) *Faster Building for Commerce*, HMSO, London.

New Builder (1995) 'Top posts still elude women', *New Builder*, 24 March.

Niehoff, B. P., Moorman, R. H., Blakely, G. and Fuller, J. (2001) 'The influence of empowerment and job enrichment on employee loyalty in a downsizing environment', *Group & Organisation Management* 26: 93–113.

Noe, R. A., Hollenbeck, J. R., Gerhart, B. and Wright, P. M. (2000) *Human Resource Management*, 3rd edn, McGraw-Hill, New York.

NOHSC (1999) *OHS Performance Measurement in the Construction Industry: Development of Positive Indicators*, National Occupational Health and Safety Commission, Commonwealth of Australia, Canberra.

Nykodym, N., Simonetti, J. L., Nielsen, W. R. and Welling, B. (1994) 'Employee empowerment', *Empowerment in Organizations* 2(3): 45–55.

O'Conner, D. D. (1990) 'Trouble in the American workplace: the team player concept strikes out', *ARMA Record Management Quarterly* 24(2): 12–15.

O'Connor, L. and Harvey, N. (2001) 'Apprenticeship training in Ireland: from time-served to standards based; potential and limitations for the construction industry', *Journal of European Industrial Training* 25/6: 332–42.

O'Donoghue, A. (2001) 'Motivational training hits new heights', *Industrial and Commercial Training* 7: 255–9.

O'Rourke, J. (1998) 'Union's plea: racism not all right on site', *Herald Sun*, Sydney, 11 October: 19.

Oliver, J. (1993) 'Shocking to the core', *Management Today*, August: 18–22.

Oram, M. (1998) 'Re-engineering's fragile promise: HRM prospects for delivery', in P. Sparrow and M. Marchington (eds) *Human Resource Management: The New Agenda*, Pitman, London.

Panati, C. (1991) *Panati's Parade of Fads, Follies and Manias – The Origins of Our Most Cherished Obsessions*, Harper Perennial, New York.

Parasuraman, S. (1982) 'Predicting turnover intentions and turnover behavior: a multivariate analysis', *Journal of Vocational Behavior* 21: 111–21.

Pascale, R. T. (1991) *Managing on the Edge*, Penguin, Harmondsworth.

Pastor, J. (1997) 'Empowerment: what it is and what it is not', *Empowerment in Organizations* 4(2): 5–7.

Patton, T. (1977) *Pay*, NY Free Press, New York.

Perrow, C. (1970) *Organisational Analysis: A Sociological Approach*, Wadsworth, Belmont, CA.

Perry-Smith, J. E. and Blum, T. C. (2000) 'Work–family human resource bundles and perceived organizational performance', *Academy of Management Journal* 43: 1,107–17.

Personick, Martin E. and Taylor-Shirley, Katherine (1989) 'Profiles in safety and health: Occupational hazards of meatpacking', *Monthly Labor Review* 5(2): 3–12.

Peters, T. (1989) *Thriving on Chaos – Handbook of Management Revolution*, Pan Books, London.

Peters, T. J. and Waterman, R. H. (1982) *In Search of Excellence: Lessons from America's Best Run Companies*, Harper & Row, New York.

Phinney, J. S. (1996) 'When we talk about American ethnic groups, what do we mean?', *American Psychologist* 51(5): 918–27.

Popenoe, D. (1993) 'American family decline, 1960–1990: a review and appraisal', *Journal of Marriage and the Family* 55: 527–55.

Porter, M. E. (1990) *The Competitive Advantage of Nations*, Macmillan, Basingstoke.

Powell, C. (1983) 'Maid for the job: some women in building', *Building Technology and Management*, December: 6.

Preece, C. N., Moodley, K. and Cavina, C. (1999) 'The role of the planning supervisor under new health and safety legislation in the United Kingdom', in A. Singh, J. Hinze and R. J. Coble (eds) *Implementation of Safety and Health on Construction Sites*, Rotterdam: A. A. Balkema.

Presser, H. B. (2000) 'Nonstandard work schedules and marital instability', *Journal of Marriage and the Family* 62: 93–110.

Proctor, S. and Ackroyd, S. (2001) 'Flexibility', in T. Redman and A. Wilkinson (eds) *Contemporary Human Resource Management*, Pearson, Essex.

Proverbs, D., Holt, G. and Olomolaiye, P. (1999) 'European construction contractors: a productivity appraisal of institute concrete operations', *Construction Management and Economics* 17: 221–30.

Purcell, J. and Sisson, K. (1983) 'Strategies and practices in the management of industrial relations', in G. S. Bain (ed.) *Industrial Relations in Britain*, Blackwell, Oxford.

Quick, J. C., Nelson, D. L. and Quick, J. D. (1990) *Stress and Challenge at the Top: The Paradox of the Successful Executive*, Wiley, Chichester.

Quinlan, M. and Bohle, P. (1991) *Managing Occupational Health and Safety in Australia*, Macmillan, South Melbourne.

Rainbird, H. (1991) 'The self-employed: small entrepreneurs or disguised wage labourers?', in A. Pollert (ed.) *Farewell to Flexibility*, Blackwell, Oxford.

Rainbird, H. (1998) 'Skilling the unskilled: access to work-based learning and the life-long agenda', in T. Lange (ed.) *Skilling the Unskilled: Achievements and Under-Achievements in Education and Training*, Robert Gordon University, Aberdeen.

Rasmussen, J. (1990) 'Human error and the problem of causality in analysis of accidents', in D. E. Broadbent, A. Baddeley and J. T. Reason (eds) *Human Factors in Hazardous Situations*, Oxford Science, Oxford.

Rasmussen, J., Pejtersen, A. M. and Goodstein, L. P. (1994) *Cognitive Systems Engineering*, John Wiley, New York

Ray, R., Hornibrook, J., Skitmore, M. and Zarkada-Fraser, A. (1999) 'Ethics in tendering: a survey of Australian opinion and practice', *Construction Management and Economics* 17: 139–53.

RDA (1975) *Racial Discrimination Act*, Canberra, Australia.

Redman, T. and Wilkinson, A. (2001) *Contemporary Human Resource Management*, Pearson, Essex.

Renzetti, C. M. and Curran, D. J. (1992) *Women, Men and Society*, 2nd edn, Allyn & Bacon, Boston, MA.

Reshef, Y. (2001) 'The logic of labour quiescence', *Journal of Labour Research* 22: 635–52.

Respect for People (2000) HMSO, London.

Rethinking Construction (2000) *A Commitment to People: 'Our Biggest Asset'*, Report of the Movement for Innovation's Working Group on Respect for People, HMSO, London.

Richardson, B. (1996) 'Modern management's role in the demise of sustainable society', *Journal of Contingencies and Crisis Management* 4(1), March: 20–31.

Ridley, J. and Channing, J. (1999) *Risk Management*, Butterworth Heinemann, Oxford.

Robinson, R. (1997) 'Loosening the reins without losing control', *Empowerment in Organisations* 5(2): 76–81.

Rogers, E. M. and Kincaid, D. L. (1981) *Communication Networks: Towards a New Paradigm for Research*, Free Press, London.

Rosenthal, U. and Kouzmin, A. (1993) 'Globalization: an addenda for contingencies and crisis management, an editorial statement', *Journal of Contingencies and Crisis Management* 1(1): 1–11.

Rousseau, D. M. (1995) *Psychological Contracts in Organisations: Understanding the Written and Unwritten Agreements*, Sage, London.

Rowe, C. (1995) 'Introducing 360 degree feedback, the benefits and pitfalls', *Executive Development* 7(1): 14–20.

Runeson, G. (1997) 'The role of theory in construction management research: comment', *Construction Management and Economics* 15(3): 299–302.

Russell, G. and Bowman, L. (2000) *Work and Family: Current Thinking, Research and Practice*, Department of Family and Community Services, Commonwealth of Australia.

Russell, J., Pfatteicher S. K. A. and Meier, J. R. (1997) 'What can you do to improve engineering education?', *ASCE Journal of Management in Engineering* 13(6): 37–42.

Ryan, K. D. and Oestreich, D. K. (1998) *Driving Fear Out of the Work-Place*, 2nd edn, Jossey Bass, San Francisco, CA.

Sagan, S. D. (1993) *The Limits of Safety: Organizations, Accidents and Nuclear Weapons*, Princeton University Press, Princeton, NJ.

Sambrook, S. (2001) 'Factors influencing learners' perceptions of the quality of computer based learning materials', *Journal of European Industrial Training* 25: 157–67.

Sawin, G. (1995) 'How stereotypes influence opinions about individuals', in C. P. Harvey and M. J. Allard *Understanding Diversity*, HarperCollins, New York.

Schein, E. H. (1992) 'Career anchors and job/role planning: the links between career planning and career development', in D. H. Montross and C. J. Shinkman (eds) *Career Development: Theory and Practice*, Charles C. Thomas, Springfield, IL.

Schwochau, S. (1987) 'Union effects on job attitudes', *Industrial Relations Review* 40: 219–20.

Scott, N., Ponniah, D. and Saud, B. (1997) 'A window on management training within the construction industry', *Industrial and Commercial Training* 29: 148–52.

Selznick, P. (1957) *Leadership in Administration*, Row Peterson, Evanston, IL.

Senge, P. M. (1990) *The Fifth Discipline*, Doubleday, New York.

Sheen, B. (1995) '*The Herald of Free Enterprise* – corporate manslaughter?', *Medico Legal Journal* 64/2: 55–69.

Sheldrake, J. (1996) *Management Theory – From Taylorism to Japanization*, Thomson Business Press, London.

Sheridon, J. H. (1991) 'A scarcity of engineers?', *Industry Week* 240(9), 6 May: 11–14.

Shillito, G. (1992) 'Women Into Science and Engineering (WISE) vehicle programme', in T. V. Duggan (ed.) *3rd World Conference on Engineering Education*, vol. 2, University of Portsmouth.

Shrivastava, P. (1992) *Bhopal: Anatomy of a Crisis*, Paul Chapman Publishing, London.

Siegall, M. and Gardner, S. (2000) 'Contextual factors of psychological empowerment', *Personnel Review* 29: 703–22.

Silberman, M. (2001) 'Developing interpersonal intelligence in the workplace', *Industrial and Commercial Training* 7: 266–9.

Silverman, D. (1970) *The Theory of Organisations*, Heinemann, London.

Sisson, K. (1994) 'Personnel management: paradigms, practice and prospects', in K. Sisson (ed.) *Personnel Management*, Blackwell, Oxford.

Sisson, K. and Storey, J. (2000) *The Realities of Human Resource Management: Managing the Employment Relationship*, Open University Press, Birmingham.

Skinner, B. F. (1938) *The Behavior of Organisms*, Prentice-Hall, Englewood Cliffs, NJ.

Smith, A. (1776/1986) *The Wealth of Nations*, Penguin, Harmondsworth.

Smith, B. N., Honsby, J. S. and Shirmeyer, R. (1996) 'Current trends in performance appraisal – an examination of managerial practice', *SAM Advanced Management Journal* 11(3): 10–15.

Smith, L. J. (2000) 'The need for pro-active methodology to reduce health and safety inefficiencies within a construction organization', in A. Akintove (ed.) *Proceedings of the 16th ARCOM Conference*, vol. 2, Glasgow, Scotland.

Smithers, G. L. and Walker, D. H. T. (2000) 'The effect of the workplace on motivation and demotivation of construction professionals', *Construction Management and Economics* 18: 833–41.

Sommerville, J. (1996) 'An analysis of recruitment sources and employee turnover in Scottish construction organizations', *Construction Management and Economics* 14(2): 147–54.

Sommerville, J., Kennedy, P. and Orr, L. (1993) 'Women in the UK construction industry', *Construction Management and Economics* 11: 285–91.

Sparrow, P. R. and Hiltrop, J. M. (1994) *European Human Resource Management in Transition*, Prentice-Hall, London.

Sparrow, P. and Marchington, M. (1998) *Human Resource Management: The New Agenda*, Pitman, London.

Spencer, L. M. and Spencer, S. M. (1993) *Competence at Work*, Wiley, New York.

Spreitzer, G. M. (1995) 'Psychological empowerment in the workplace: dimensions, measurement and validation', *Academy of Management Journal* 38: 1,442–65.

Spreitzer, G. M. (1996) 'Social structural characteristics of psychological empowerment', *Academy of Management Journal* 39: 483–504.

Springel, W. and Myers, C. (eds) (1953) *The writings of the Gilbreths*, Richard Irwin, Holmwood, IL.

Srivastava, A. and Fryer, B. (1991) 'Widening access: women in construction', *Proceedings of the 7th Annual ARCOM Conference*, Bath, September.

Stacey, R. (1992) *Managing Chaos*, Kogan Page, London.

Stoeckel, A. and Quirke, D. (1990) *Services: setting the agenda*, Report No. 2, Centre for International Economics, DITAC.

Stopford, J. M. and Wells, L. T. (1972) *Managing the Multi National Enterprise: Organisation of the Firm and Ownership of Subsidiaries*, Basic Books, New York.

Storey, J. (1993) 'The take-up of human resource management by mainstream companies: key lessons from research', *International Journal of Human Resource Management* 4(3), September: 529–53.

Stuckenbruck, L. C. (1986) 'Project management framework', *Project Management Journal*, August: 25–9.

Suraji, A. and Duff, R. (2000) 'Construction management actions: a stimulant of construction accident formation', in A. Akintoye (ed.) *Proceedings of the 16th ARCOM Conference*, vol. 2, Glasgow, Scotland.

Swenson, D. X. (1997) 'Requisite conditions for team empowerment', *Empowerment in Organisations* 5(1); 16–25.

Sykes, G., Simpson, M. and Shipley, E. (1997) 'Training and empowerment improve performance: a case study', *Integrated Manufacturing Systems* 8(2): 90–102.

Tailby, S. and Winchester, D. (2000) 'Management and trade unions: towards social partnership', in S. Bach and K. Sisson (eds) *Personnel Management: A Comprehensive Guide to Theory and Practice*, Blackwell, Oxford.

Taylor, F. W. (1911) *Principles of Scientific Management*, Harper & Row, New York.

Taylor, S. (1998) *Employee Resourcing*, CIPD, London.

Teicher, J. and Svensen, S. (1997) *The Nature and Consequences of Labour Market Deregulation in Australasia*, Working Paper No. 51, National Key Centre in Industrial Relations, Monash University, Melbourne.

Teicher, J. and Svensen, S., (1998) *Legislative changes in Australasian Industrial Relations: 1984–1997*, Working Paper No. 56, National Key Centre in Industrial Relations, Monash University, Melbourne.

Thomas, K. W. and Velthouse, B. A. (1990) 'Cognitive elements of empowerment: an "interpretive" model of intrinsic task motivation', *Academy of Management Review* 15: 666–81.

Thompson, D. (1990) 'Wellness programs work for small employers too', *Personnel*, March: 23–33.

Thompson, N. (1997) *Anti-discriminatory Practice*, Macmillan, London.

Thompson, R. and Mabey, C. (1994) *Developing Human Resources*, Butterworth-Heinmann, Oxford.

Thorlakson, A. J. H. and Murray, R. P. (1996) 'An empirical study of empowerment in the workplace', *Group & Organization Management* 21: 67–83.

Toffler, A. (1970) *Future Shock – A Study of Mass Bewilderment in the Face of Accelerating Change*, Bodley Head, London.

Toohey, R. P. and Whittaker, J. (1993) 'Engineering women: a view from the workplace', *ASCE Journal of Management in Engineering* 9(1), January: 27–37.

Torrington, D. and Hall, L. (1995) *Personnel Management: HRM in Action*, 3rd edn, Prentice-Hall, Hemel Hempstead.

Tracey, J. B., Hinkin, T. R., Tannenbaum, S. and Mathieu, J. E. (2001) 'The influence of individual characteristics and the work environment on varying levels of training outcomes', *Human Resource Development Quarterly* 12: 5–23.

Tsoukas, H. (1995) *New Thinking in Organisational Behaviour*, Butterworth-Heinemann, Oxford.

Uff, J. (1995) 'Contract documents and the division of risk', in J. Uff and A. M. Odams (eds) *Risk Management and Procurement In Construction*, Centre for Construction Law and Management, London.

United States Department of Labor, Bureau of Labor Statistics (1999) *National Census Of Fatal Occupational Injuries, 1998*, Washington, DC (http://stats.bls.gov/special.requests/ocwc/oshwc/cfoi/cfnr0005.pdf).

Vinson, M. N. (1996) 'The pros and cons of 360 degree feedback: making it work', *Training and Development* 6(1): 11–12.

Vojtecky, M. A. and Schmitz, M. F. (1986) 'Program evaluation and health and safety training', *Journal of Safety Research* 17: 57–63.

Von Bertalanffy, L. (1950) 'The theory of open systems in physics and biology', *Science* 1(6): 23–9.

Wall, C. (1997) 'Editorial note', *Construction Labour Research News Special Edition: Women in Construction* 3: 1–2.

Wallace, J. E. (1999) 'Work-to-nonwork conflict among married male and female lawyers', *Journal of Organizational Behavior* 20: 797–816.

Walton, R. and McKersie, R. (1965) *A Behavioral Theory of Labor Negotiations*, McGraw-Hill, New York.

Waxler, R. and Higginson, T. (1993) 'Discovering methods to reduce workplace stress', *Industrial Engineering* 25(6): 19–21.

Weber, M. (1947) *The Theory of Economic and Social Organisations*, translated by A. M. Henderson and T. Parsons, Oxford University Press, Fair Lawn, NJ.

Wells, J. (1990) *Female Participation in the Construction Industry*, ILO Sectoral Activities Programme Working Paper, Industrial Activities Branch, International Labour Office, Geneva, March.

Westman, M., Etzion, D. and Danon, E. (in press) 'Job insecurity and crossover of burnout in married couples', *Journal of Organizational Behavior*.

White L. and Keith B. (1990) 'The effect of shift work on the quality and stability of marital relations', *Journal of Marriage and the Family* 52: 453–62.

Wildavsky, A. (1988) *Searching for Safety*, Transaction Books, New Brunswick, NJ.

Wilkinson, A. (1998) 'Empowerment: theory and practice', *Personnel Review* 27: 40–56.

Wilkinson, A. (2001) 'Empowerment', in T. Redman and A. Wilkinson (eds) *Contemporary Human Resource Management*, Pearson, Harlow.

Wilkinson, S. J. (1992) *Construction and the Recruitment of Female Labour* (rev. edn) Oxford Polytechnic.

Williams, S. W. (2001) 'The effectiveness of subject matter experts as technical trainers', *Human Resource Development Quarterly* 12: 91–7.

Winch, G. (1994) 'The search for flexibility: the case of the construction industry', *Work, Employment and Society* 8(4): 593–606.

Wood, J. T. (1997) 'Gendered media: The influence of media on views of gender', in Davis, M. H. (ed.) *Social Psychology Annual Editions 97/98*, 162–71.

Wysocki, L. (1990) *Implementation of Self-Managed Teams Within a Non-Union Manufacturing Facility*, paper presented to the International Conference on Self-managed Work Teams, Denton, TX, September.

Yamnill, S. and McLean, G. N. (2001) 'Theories supporting transfer of training', *Human Resource Development Quarterly* 12: 195–208.

Yates, J. K. (1992) 'Women and minorities in construction in the 1990s', *Cost Engineering* 34(6), June: 9–12.

Yoder, J. D. (1991) 'Rethinking tokenism: looking beyond numbers', *Gender and Society* 5: 178–92.

Young, B. A. (1988) 'Career development in construction management', PhD thesis, UMIST, Manchester.

Young, K. (2001) 'The effective deployment of e-learning', *Industrial and Commercial Training* 33: 5–11.

Zarkada-Fraser, A. and Skitmore, M. (2000) 'Decisions with moral content', *Construction Management and Economics* 18: 101–11.

Index